DOGS: DOMESTICATION AND THE DEVELOPMENT OF A SOCIAL BOND

This book traces the evolution of the dog, from its origins about 15,000 years ago up to recent times. The timing of dog domestication receives attention, with comparisons between different genetics-based models and archaeological evidence. Allometric patterns between dogs and their ancestors, wolves, shed light on the nature of the morphological changes that dogs underwent. Dog burials highlight a unifying theme of the whole book: the development of a distinctive social bond between dogs and people. The book also explores why dogs and people relate so well to each other. Though the book is cosmopolitan in overall scope, greatest emphasis is on the New World, with an entire chapter devoted to dogs of the arctic regions, mostly in the New World. Discussion of several distinctive modern roles of dogs underscores the social bond between dogs and people.

Darcy F. Morey received his Ph.D. in anthropology, with a specialization in archaeology, in 1990 from the University of Tennessee in Knoxville. Subsequently, he spent a year as a guest researcher at the University of Copenhagen Zoological Museum in Denmark. He was there for the express purpose of studying dog remains from archaeological sites in arctic Greenland. In addition to participating in archaeological fieldwork there in 1990, he has worked in Norway, France, and Denmark, as well as numerous places in the United States. He has published actively on a variety of topics, with his work on dogs being especially prominent. On that general topic, he has published as sole or senior author many articles and book reviews in journals such as *Arctic*, *Journal of Archaeological Science*, *Quarterly Review of Biology*, *Archaeozoologia*, *Current Anthropology*, and *Journal of Alabama Archaeology*. Dr. Morey has also published on the topic of dogs in popular science outlets, including the *American Scientist* and *La Recherche*. He joined the faculty at the University of Kansas in Lawrence in 1998. There, in addition to his ongoing research activities, he was selected by students as the most notable teacher of undergraduates in his department (Anthropology) in 2000. In addition, in 2002 he was elected to the Alpha Pi chapter of Phi Beta Delta, The Honor Society for International Scholars. He resigned from the University of Kansas in 2006 and began working at the University of Tennessee in Martin. He is presently a Research Associate with the Forensic Science Institute at Radford University in Radford, Virginia.

Photograph of dog burial # 1 from the prehistoric Indian Knoll
Site in Kentucky (some 5,000 years old), taken more than fifty
years ago. This picture was previously published in the *Journal
of Archaeological Science* (Morey 2006: 163, figure 3). The original
photograph was enhanced and made available by George M.
Crothers, William S. Webb Museum of Anthropology, University of Kentucky, Lexington, Kentucky. More information on the
broader phenomenon of dog burials is contained in Chapter 7.

DOGS

Domestication and the Development of a Social Bond

DARCY F. MOREY

Radford University

CAMBRIDGE
UNIVERSITY PRESS

CAMBRIDGE
UNIVERSITY PRESS

Shaftesbury Road, Cambridge CB2 8EA, United Kingdom

One Liberty Plaza, 20th Floor, New York, NY 10006, USA

477 Williamstown Road, Port Melbourne, VIC 3207, Australia

314–321, 3rd Floor, Plot 3, Splendor Forum, Jasola District Centre, New Delhi – 110025, India

103 Penang Road, #05–06/07, Visioncrest Commercial, Singapore 238467

Cambridge University Press is part of Cambridge University Press & Assessment, a department of the University of Cambridge.

We share the University's mission to contribute to society through the pursuit of education, learning and research at the highest international levels of excellence.

www.cambridge.org
Information on this title: www.cambridge.org/9780521757430

First published 2010

A catalogue record for this publication is available from the British Library

Library of Congress Cataloging-in-Publication data
Morey, Darcy, 1956–
Dogs: domestication and the development of a social bond / Darcy Morey.
     p.   cm.
Includes bibliographical references and index.
ISBN 978-0-521-76006-5 (hardback) – ISBN 978-0-521-75743-0 (pbk.)
1. Dogs – History.   2. Dogs – Evolution.   3. Dogs – Behavior.   4. Human-animal relationships – History.   I. Title.
SF422.5.M67   2010
636.709 – dc22   2009023645

ISBN   978-0-521-76006-5   Hardback
ISBN   978-0-521-75743-0   Paperback

This book is gratefully dedicated to the Emergency Medical Technicians associated with the Fire Department in Washburn, North Dakota, with special thanks to Clayton Verke and Mary Devlin. But for the crucial, competent, and caring roles they played on one fateful night, July 13, 2000, this book simply could not exist. On that night, though, they did their jobs exceptionally well, above and beyond the call of customary professional duty, making it possible for this book to exist. And so, now it does.

Darcy F. Morey

CONTENTS

Contents

LIST OF FIGURES AND TABLES

FIGURES

TABLES

List of Figures and Tables

FOREWORD

It was headline news for the BBC not long ago: "If you want to live a healthier life, get a dog." Those who own dogs would not have been surprised to read this, and those who study the relationships between dogs and their owners have known this particular bottom line for quite some time. As dog scientist Deborah Wells has observed, dogs can prevent us from becoming ill, can help us recover from being ill, and can even alert us that we may be about to become ill. Dog owners who suffer heart attacks are nearly nine times more likely to survive the following year than those who do not own dogs (cats do not help at all here). Therapy dogs decrease the stress levels, and increase the social interactions, of people lucky enough to be visited by them. The list goes on and on.

It is not just dog owners who benefit from interactions with their canine companions. Quite obviously, the dogs themselves benefit. In fact, the mutual benefits are so great that the phrase "dog owner" is not really an appropriate one because dogs own us as much as we own them. As Darcy Morey points out in the book you are about to read, the process of dog domestication was one in which members of different wolf societies adapted themselves to living in the environments that people created. At the same time as this was happening, members of different human societies adapted themselves to being in close presence of the offspring of wolves. The result was something we call dog domestication, but the situation is far richer than that since in a very real way dogs and people were domesticating each other for very mutual benefits.

We have no way of knowing what the human benefits were when this process was first happening. But as Morey points out, there are many,

many possibilities: dogs to guard, dogs to help hunt, dogs to transport goods, dogs as food, perhaps dogs as symbols of relationships and beliefs that we cannot even imagine. There are so many possibilities, and so many possible ancestral wolf groups, that there is every reason to believe that the process happened multiple times. What we can be sure of, though, is that the ancestors of dogs were wolves, and that wolves had begun the process of converting themselves into dogs, and of people into dog consorts, by at least 14,000 years ago. The results of that process are all around us today, from Chihuahuas and Pomeranians to Saint Bernards and Great Danes. That all dogs were ultimately sired by wolves is made clear by multiple studies of the genetics of the group of mammals to which both dogs and wolves belong.

The U.S. Humane Society tells us that there are some 75 million dogs in the United States today and that about 40 percent of U.S. households have at least one of them. I do not know of any estimates of the number of free-ranging dogs in the United States, but numbers provided by Margaret Slater and her colleagues show that roughly 10 percent of the dogs in Italy are free-ranging. If the same proportions apply to the United States, that means that 90 percent of the dogs in this country have their very own human households. That is what domestication has done for these descendants of wolves, and I have already mentioned a few of the things they can do for us.

I have no idea how many people in those households know anything about the deeper history of their canine coresidents, but I would be surprised if many of them have thought about the ultimate origins of dogs and of the ways in which the complex interrelationships we have with them have developed through time. For those who are interested, this book will be of enormous value.

In his acknowledgments, Morey credits me with suggesting that he should write a book on dogs. It is true that I made this suggestion, but my motivation was selfish. Darcy Morey is one of the world's great experts on the history of dogs. When I want to check my facts on the earliest known dogs of any given part of the world, it is to his work that I turn. When I want to consult a critical evaluation of new claims for an ancient domesticated dog, it is to his work that I turn. When I want my students to read a balanced evaluation of dog prehistory, it is to his work that I turn.

What makes his work so valuable is not simply that he knows so much about dog skeletons and appropriate ways to analyze them. Others know such things as well. What makes it so valuable is that he adds to these technical skills knowledge about dog sociology and biology that is broad and deep, including an understanding of the relationships between dogs and people on a global scale.

In short, I suggested to Darcy that he write this book because I wanted to read it. Now that he has done so, and I have read it, I am thankful for the rare stroke of lucidity that led me to the suggestion. Before I read it, I thought I knew a lot about dogs. Now, I know a lot more and so will you. So, sit down with this book in hand, your dog next to you, and enjoy the rich pages that follow. This book is so interesting that you may not want a break from it. But if you need one, take your dog for a walk. The exercise will do you good, you will enjoy the social interactions you have along the way, and you will live longer. And remember that just as our ancestors domesticated the dog, so did your dog's ancestors domesticate us.

Donald K. Grayson
Department of Anthropology
University of Washington, Seattle

PREFACE AND ACKNOWLEDGMENTS

This work is about what I have come to think of as the journey of the dog. That is, a major goal is to clarify just when and how the dog came into being, and what steps it took along the way to arrive at its modern destination. In a curious way, though, this is also a story about my own journey through the world of dog-related research for more than two decades. To a certain extent, the progression of topics covered here roughly parallels the course of developments in my dog-related research work.

In an ultimate sense, work on this book began more than twenty years ago, though I was not aware of it then. At that time I published my first paper on dogs (Morey 1986), as a graduate student at the University of Tennessee in Knoxville, a study concerning matters of taxonomic resolution from archaeological bones in the North American plains. Those as well as other taxonomic issues receive attention in Chapter 3. Subsequently, I was awarded a doctoral degree from the University of Tennessee, with a dissertation devoted to the evolution of the domestic dog as revealed especially from archaeological cranial remains (Morey 1990). That was my first synthetic effort devoted to the dog, and though it has its weaknesses, some of which bear noting, I also draw from it at several junctures during the course of this book. Those two early works presaged what became the regular production of published dog-related research, ranging from local and regional levels, all the way to the genuinely international level. Much of that work is addressed in this book, sometimes extensively. Prior to this book, my most recent study concerned the phenomenon of dog burials (Morey 2006). Chapter 7 in this volume elaborates and expands upon this topic. It is also the point at which the focus expands beyond archaeological considerations, and into the domains of biology and physiological psychology, including neuroscience. The objective in branching out is to address meaning-fully the question of just why people and dogs related to each other so remarkably well, a circumstance leading to the routine burial of dogs when they die.

So many people have contributed to this effort that it is really not possible to do justice to them all. But rather than completely leave out anyone, I have chosen to divide them into two basic groups. The first group consists of those people who have either provided or steered me toward one or two sources that I have consulted and utilized, or they contributed in some other specific way. Those people I merely list alphabetically, in order to conserve space. This tactic should not be taken as a sign that expressing gratitude is merely a formality, for I am genuinely grateful for their help. I trust they understand that, as I thanked them warmly at the time. In any case, those people are Dan Amick, Mark Beech, Cliff and Donna Boyd, Susan Crockford, Chris Curcio, Chris Darwent, Jason Flay, Holger Funk, Elizabeth Garrett, Erika Hill, Jack Hofman, John Hoopes, Libby Huber, Dimitry Ivanoff, Noel Lanci, Karen Lange, Sophia Maines, Barbara Matt, Ann-Janine Morey, Donald and Martha Morey, Jennifer Myer, Ray Pierotti, Anthony Podberscek, Ivana Radovanoviç, Randy Ramer, Carolyn Rebbert, Ger-ald Schroedl, Mary Sorrick, Don Stull, Lyudmila Trut, Renee Walker, Diane Warren, Dixie West, and Elizabeth Wing. Beyond these individ-uals, the interlibrary services at the University of Tennessee in Martin and University of Kansas libraries have been instrumental in obtaining some sources that were not in their holdings. I am genuinely grate-ful to all of these individuals and library services, and now wish to acknowledge a second group of individuals who made especially major contributions. I indicate these individuals alphabetically as well.

First on this list is my Danish colleague Kim Aaris-Sørensen, a coau-thor on a published paper that plays a conspicuous role in Chapter 6. But one of Kim's other notable contributions to this book is contained in Chapter 5. First, Kim sent to me a 1977 edited publication in Danish, with a piece by him. For a relevant passage in that piece, I roughed out a translation into English, and Kim substantially refined it. The trans-lated passage is directly quoted in Chapter 5. Had Kim not assisted, the English translation would have been crude, at best, and inaccurate on a specific point. Kim also plays a notable role in Chapter 6, but in that case his distinctive contributions involve more than providing a source, or assisting with translation. For one thing, he retrieved from storage in Copenhagen a pair of distinctively modified archaeological dog mandibles that I had not seen in many years and rendered his judgment as to their significance, by that helping me avoid an unfortunate interpretive error. As for his other role in Chapter 6, I identify that contribution only at the appropriate juncture in that chapter. Elsewhere in the world, Claus Andreasen is deputy at the Greenland National Museum and Archives, in Nuuk. His relevance is that I was hoping to obtain a recent photograph of a distinctive set of bones representing a simulated dog sled and team from a particular archaeological site in Greenland. A picture of that set had been published earlier, but in an obscure 1933 report. I originally approached Bjarne Grønnow at the National Museum of Denmark, in Copenhagen, about this matter. Grønnow informed me that the entire collection from that particular site was no longer stored in Copenhagen, but had been transferred in recent years back to Greenland. Consequently, he provided the contact information for Andreasen, who located this distinctive set of old bones and arranged for the curator at that museum, Mikkel Myrup, to produce the desired new photograph, arranged much like the original. This excellent photograph appears in Chapter 6, and its special relevance becomes quite clear during the course of that chapter.

From Germany, Norbert Benecke provided some key assistance concerning an archaeological specimen known as the Bonn-Oberkassel dog, an important case that receives attention at different points in this volume. In fact, in summarizing Benecke's role, I can do little better than repeat my own words from the acknowledgments section of my most recently published dog paper: "I am especially grateful to Norbert Benecke, who generously shared important information with me, and in doing so, patiently accommodated my extraordinarily rudimentary capabilities in the German language" (Morey 2006: 171). In this instance he identified an important source in German (Street 2002), and kindly summarized its content in English. The importance of this knowledge becomes clear at different points, beginning with Chapter 2.

Linda Carnes-McNaughton was instrumental in facilitating my capacity to obtain a series of photographic images of prehistoric dog burials from the recently excavated Broad Reach site in North Carolina. One such dog burial appears in Chapter 7. It is one that I chose among several alternatives provided by Heather Millis. George Crothers, director of the William S. Webb Museum of Anthropology at the University of Kentucky, in Lexington, has been especially helpful. First, there appears in Chapter 7 a photographic image of a dog burial from the prehistoric Ward site in Kentucky. An image of this burial originally appeared in a report from long ago (Webb & Haag 1940: 82, figure 9). From the original negative of that photograph, curated at the Webb Museum, George arranged for the production of an enhanced image that appears in Chapter 7. In addition, George provided me with one of the last known copies of the original report, to clarify the context. Beyond that image, the front of this book is graced with an image of dog burial No. 1 from Indian Knoll, also in Kentucky, as previously provided by George for a publication in the *Journal of Archaeological Science* (Morey 2006: 163, figure 3). Like the Ward dog burial, this image was made from one of William Webb's original negatives that he enhanced. At a broader level, George saw to it that the Webb Museum's dog burial holdings were inventoried carefully, revealing some discrepancies between the originally reported numbers and the museum's holdings. I address those circumstances more fully in Chapter 7, and needless to say, I am grateful to be able to report the information as accurately as possible. Doing so has been possible only because of George.

Mark Derr, a professional writer and longtime devotee of dog-related work, has played an instrumental role. Derr has written entire books on dogs for the general reader (e.g., Derr 2004a, 2004b), the first of those initially appearing in the 1990s. Moreover, in his capacity as a writer, he regularly comments on ongoing scientific research, in magazine columns and newspapers. In conjunction with this aspect of his work, he has access to recent primary publications in the scientific literature, and he has provided me with several examples. Beyond Derr, I would be genuinely remiss not to thank Carl Falk. Though Carl provided only one specific piece of literature, he was, first of all, directly behind my very first foray into dog-related research (Morey 1986), the study that prompted the comment near the beginning of this preface, that in an ultimate sense, this book got started more than twenty years ago. Beyond that important role, Carl has shown almost unbelievable support and kindness in the aftermath of a genuinely horrific event in 2000 that nearly cost me my life (Morey et al. 2004; Maines 2006). Without such support and kindness, I doubt that this book would have happened.

I am also grateful to my Danish colleague, Anne Birgitte (Gitte) Gotfredsen, for her important help on more than one front. First, she directly provided copies of several relevant publications for this work. Because two of these are in Danish, quoting passages from them, which I have done, required translation into English. Like the work by Aaris-Sørensen noted earlier, I initially did the translation work, and Gotfredsen herself fine-tuned my work. As well, Gitte provided some organizational information that has been quite important in Chapter 6. Next, it was Don Grayson who initially put the idea in my head to accomplish such a book. Our association stems from late 1980s work in France, and although that work does not figure into this book, other work of Grayson's does, especially in Chapter 5. Tim Griffith also provided some indispensable help, calling attention to some important sources and directly providing one concerning the Ashkelon site, prominently featured in Chapter 7. As part of the presentation on the Ashkelon site, Chapter 7 also includes several original photographic images from there, taken during field work. Though Tim did not take the pictures, he directly facilitated my acquisition of them through his own work at Ashkelon and his association with Brian Hesse and Paula Wapnish, who kindly provided the photos.

Walter Klippel, my doctoral supervisor in the Department of Anthropology at the University of Tennessee, accessed an unpublished M.A. thesis in that department's holdings. From that document, he provided firsthand information that I call attention to in Chapter 7. My sister, Noralane (Laney) Lindor, has been especially helpful in a particular area. Laney is a medical doctor at the Mayo Clinic in Minnesota and has regular access to electronic search engines tailored to biomedical literature. By using different combinations of search terms, she was able to produce lengthy lists of references to potential sources. I identified many, tracked some down, and have used quite a few of them. Though several of these sources crop up at different points, by far the majority of them figure into a substantial section of Chapter 10. Quite simply, I am indebted to her.

Mike Logan, with the Department of Anthropology at the University of Tennessee, steered me to several sources that figure into Chapters 7 and 10. Mike also put me in contact with the right person, Phil Snow, whereby I obtained a photograph of the War Dog Memorial in Knoxville, Tennessee, that appears in Chapter 10. Additionally, he directly facilitated my capacity to acquire the image of a prehistoric anthropomorphized ceramic dog effigy from Mexico that appears in Chapter 5. The actual image resulted from the combined efforts of Randy Ramer, Shane Culpepper, and Jeremy Planteen, at the Gilcrease Museum in Tulsa, Oklahoma. In short, Mike Logan's role has been substantial on more than one front. I also received some valuable help from Georg Nyegaard at the Greenland National Museum and Archives in Nuuk. Georg provided some useful points of information as well as translation help on a newspaper piece in Danish that is featured near the end of Chapter 6. Additionally, he provided some useful information about a dog burial in Denmark that he helped to excavate some years ago. Bill Turner is also responsible for an important contribution. Working in Montgomery, Alabama, Bill was able to make complete photocopies of several 1940s vintage reports by William Webb and David DeJarnette that I almost certainly would not have obtained otherwise. I have drawn from all of these reports, and am truly grateful to him.

Finally, though they are certainly out of alphabetical order, I wish to express genuine gratitude to some other people. First, I thank Chris Darwent for her helpful input on an entire draft of this volume, input that has strengthened the presentation in several places. I am also grateful to the editor of this volume at Cambridge University Press, Chris Curcio. From the initial proposal to the final production, Chris has been a source of routine encouragement and advice, especially during a particularly vexing and unexpected complication that threatened to derail this project before it could be completed. I really don't have words that can adequately convey my gratitude. So I offer a simple thanks and trust he understands that I really mean that. Likewise, his editorial assistant, Glendaliz Camacho, has been genuinely helpful in certain stages of the process. Similarly, Shelby Peak, Ernie Haim, and Sara Black have been genuinely helpful in the later stages of production. Finally, a word to my wife, Beth McClellan, who has stuck with me through the worst of times and provided seemingly endless assistance with the development of this entire project. Thank you, Beth.

And now, it's time to stop talking about people and start talking about dogs. I hope you enjoy this effort.

PREAMBLE TO THE DOG'S JOURNEY
THROUGH TIME

> Like the Dachshund that is a dog and a half long and half a dog high,
> the state of Tennessee has peculiar proportions.
>
> (Kneberg 1952: 190)

For one who lived and worked in the state of Tennessee for many years,
Madeline Kneberg's deliciously phrased perspective holds special resonance for getting started on a book about dogs. Her comment, of course,
concerns a modern breed, but modern breeds are not the focus of this
book. This book is concerned mostly with how and why dogs in general came into being, and why they have the basic characteristics that
they do, irrespective of modern breed standards. That said, it is still
reasonable to ask a simple question: Is the world really in need of yet
one more book about dogs? Over approximately the past half-century
many books that center on dogs have been published, so at face value
the question is legitimate. I, of course, believe that what is offered here
should be desired and welcomed. But to make clear how this offering
differs from previous works, it is useful to consider briefly the nature
of those other contributions. Following that, it is important to spell out
this book's framework, an approach that should help serve to highlight
how this volume is different from others. There is no claim to exhaus-
tiveness here, and the purpose is simply to draw attention to a series
of works, mostly prominent, that I am acquainted with. Most of these
works receive attention, to various degrees, at different point in this
volume.

The essential goal here is that this volume be useful to specialists with
particular areas of interest, and simultaneously accessible to interested
lay readers. In keeping with this goal, I have attempted to write in a
style that is clear and avoids pretentious academic jargon. That means

avoiding cumbersome technical presentations as much as possible, in an effort to express the content in a clear and, hopefully, engaging style. In briefly characterizing some previous volumes below, I largely bypass (with exceptions, including the first one) those written for the public in general. To be sure, several of those efforts receive attention at different points, but the primary goal here is to identify and compare some more specialized efforts with this one.

PREVIOUS VOLUMES ABOUT DOGS

First, covering this ground entails bypassing volumes that deal with the topic of domestication broadly enough that dogs are but one component. In briefly covering several previous dog books, it is useful to take a basic chronological approach, beginning just over half a century ago. In 1954, the well-known Austrian ethologist Konrad Lorenz published a book on dogs (first published in German), intended to be accessible to the public (Lorenz 1954). In that book Lorenz covered, among other things, the behavioral features of dogs, augmented with accounts of his own dogs, emphasizing the fundamental compatibility between dogs and people. In fact, early on he made a statement that is appropriate for different parts of this book: "The whole charm of the dog lies in the depth of the friendship and the strength of the spiritual ties with which he has bound himself to man" (Lorenz: 1954: ix). This is a theme that he developed subsequently (e.g., Lorenz 1975), providing some important insights. Given that this book was published some half a century ago it did not, by definition, benefit from all that has been learned since then. In fact, at that time Lorenz advanced an erroneous case that the golden jackal, *Canis aureus*, was the wild ancestor of most dogs. It is notable that Lorenz himself later recognized this position as erroneous, and retracted it (Lorenz 1975). Lorenz's retraction predated the empirical demonstration of the true ancestry of dogs, covered in the next chapter, and his reasoning at that time is especially notable. That aspect of Lorenz's work is featured later in this volume, especially in Chapter 9.

About a decade after Lorenz's book, John Paul Scott and John Fuller produced a book on dogs intended to be more relevant to professional workers (Scott & Fuller 1965), but accessibly written, as well. This book was based on some thirteen years of research at the Jackson Laboratory, in Bar Harbor, Maine. They were especially focused on the role of heredity in shaping behavior. They dealt with several dog breeds, and working especially with crosses between Basenjis and Cocker Spaniels, they found that heredity is involved in virtually every trait tested. They emphasized, however, that there can be a complex link between genetic endowment and the behaviors that take place. For present purposes their findings regarding the formation of social bonds are especially relevant. This aspect of their work is especially important in Chapter 4 here, where domestication as a process is the focus. A few years after Scott and Fuller's work, Michael Fox (1971) published a book that focused on how brain development was integrated with behavior in the dog. Fox was working from the vantage point of physiological psychology, a theme he developed more in later work. In 1975, he edited the volume on wild canids containing the foreword by Konrad Lorenz, in which Lorenz rescinded his earlier view that the golden jackal was a primary ancestor of the dog. Subsequently, Fox (1978) published a book that was intended to be a rather comprehensive account of the domestication of the dog. Not being fully familiar with areas outside his realm of expertise in modern canids, Fox made an unsupported inference about dog ancestry. At the same time, though, he developed some noteworthy insights, stemming from his background in physiological psychology, concerning the behavioral capacities of dogs, highlighted here in Chapter 8.

In 1985, Stanley J. Olsen (1985) produced a book on dogs for which his vantage point was primarily the archaeological record. And just as Michael Fox's earlier work reflects his lack of primary background in that area, Olsen's work, understandably, reflects his own lack of primary background in Fox's areas of expertise. Olsen was no newcomer to dogs, though, having produced several published papers about them during the years prior to his book, and on domestication in general. In his book Olsen was dealing especially with the origins of the dog, as understood primarily from their archaeologically recovered skeletal remains. Because of this emphasis, Olsen's book is relevant here, especially during the earlier portions of this volume. Although now somewhat outdated, more than twenty years having passed since its publication, Stanley Olsen's book was very important for its time.

Moving to the 1990s, Stanley Coren (1994), a dog trainer and professional psychologist, published an engagingly written book in which he purported to examine the capacities of dogs that can be can be regarded as intelligence. Well into the book, he conceptualizes his central model of canine intelligence as occurring along three basic dimensions: Instinctive, Adaptive, and Working/Obedience intelligence. In earlier sections of the book, however, he addresses a number of issues that don't really seem to contribute to his framework. For example, a chapter on the natural history of dogs doesn't seem particularly germane to his objectives. He is also prone to express his views in anecdotal style, though ironically, it is in one of those sections that he offers some information that is useful here. Specifically, he provides an account of the receptive vocabulary of his own dogs, information that gains credence when one considers the later work of others, done with more methodological rigor.

Also in that decade, 1995 saw the publication of a collection of different pieces on dogs, edited by James Serpell (1995a). This volume was an outcome of a 1991 conference hosted by the Companion Animal Research Group in Cambridge, England. Its express goal was to offer a thoroughly modern synopsis of the behavior and natural history of the dog, from a scientific standpoint. Sixteen different contributions were distributed among three basic sections: (1) domestication and evolution, (2) behavior and behavior problems, and (3) human–dog interactions.[1] The year 1996 saw the publication of three noteworthy books about dogs. One was anthropologist Mary Thurston's (1996) engagingly written account of dogs, ranging from scenarios about their origins all the way to their standing in modern societies. Thurston's greatest strength lies in the more recent realm, and of primary importance to this volume is her treatment of how favored dogs are often dealt with in modern times when they die. How dogs have been treated upon death is a topic that figures prominently into this book, with Chapter 7 devoted to that subject alone. The second 1996 book is by Michael Lemish and deals exclusively with the role of dogs in warfare activities (Lemish 1996). That topic is important in Chapter 10 here, when the focus is on modern dogs. Finally, for the year 1996, Hank Whittemore and Caroline Hebard (1996) produced a thoroughly appealing account of search-and-rescue dogs. Written for a general readership, rather than a specialized audience, the authors aimed (successfully) for this book to be enjoyable, as well as informative regarding just how useful dogs can be in certain kinds of situations.[2]

[1] This is a volume that I have decided familiarity with, having formally reviewed it in print (Morey 1997). Several of the chapters in Serpell's volume receive attention at different points in this book, since aspects of all three of the principal areas are important here, at different junctures.

[2] The book is primarily accounts of Hebard's search-and-rescue dogs in operation, on an international basis. Beyond the fact that this book is simply a good read, its special relevance here lies with the accounts of the psychological impact of certain missions on the dogs themselves. This is an aspect of the book that has occasioned

In 1997, Marion Schwartz produced a more academically oriented book on early dogs in the Americas (Schwartz 1997). Schwartz considers their roles in different places in the New World, aided by relevant ethnohistorical information, including an entire chapter devoted to one of the subjects addressed here in Chapter 5, the use of dogs as food. She also devotes an entire chapter to how dogs were conceived of and then dealt with in death, a topic that has relevance in Chapter 7 of this volume. Schwartz also includes a chapter on the representation of dogs in artistic expressions, a topic that gets brief attention in Chapter 5, here. Because the topic has been dealt with ably before, this volume addresses it only in limited fashion, focusing especially on the ways in which artistic representations of dogs can entail a conspicuously anthropomorphic theme.

In a distinctly nonarchaeological vein, Linda Case (1999) produced a book on the care and management of modern dogs. That it is distinctly nonarchaeological is reflected in occasional inaccurate background comments. An obvious one is her statement that "During the Mesolithic Period, human culture developed the use of weapons for hunting" (Case 1999: 9). In fact, people developed hunting weaponry well before the Mesolithic Period, which began about 10,000 years ago in Europe, and slightly before that in certain other regions. Aside from missteps such as that, Case does provide some genuinely useful information, for example about common health disorders in dogs, though in some cases she seems to be stressing the obvious.[3]

Of considerably more relevance here is an edited volume published the next year (Crockford 2000a). This volume was the product of the first symposium held by the International Council for Archaeozoology (ICAZ) on the history of the domestic dog. The symposium was held in Victoria, British Columbia, and the resulting volume covers diverse

some professional attention, and that factor is important in Chapter 9, here. A more recent account pertains to the roles of search-and-rescue dogs in the aftermath of the fateful destruction of the World Trade Center in New York City on September 11, 2001 (Bauer 2006).

[3] An example of that is when she covers proper feeding strategies: "If a dog gains too much weight (energy surplus) the amount fed should be decreased. Conversely, if weight is lost, an increased amount of food is provided" (Case 1999: 314). She is also concerned with the conformity of any given dog to pure breed "show" standards, but with little attention to the overall well-being of the animals. Her contribution plays no real role here, and bringing this book up mostly reflects a desire to be thorough, given that I formally reviewed it in print (Morey 2000). A new edition was released in 2005, and one hopes that it has been updated with, if nothing else, greater accuracy regarding the archaeological record.

topics within the realm of how dogs are represented through time, as known archaeologically. It is worth calling attention to one particular contribution, by the editor herself (Crockford 2000b). This brief paper, about the role thyroid hormone physiology in dog domestication was in fact an initial effort of hers that she expanded, revised, and elaborated on in subsequent publications (Crockford 2002, 2006).[4] Shortly after that, Raymond and Lorna Coppinger (2001) produced a biologically oriented book on dogs, intended to be accessible to educated readers and relevant to a variety of specialists. With a backlog of previous experience writing about dogs, the Coppingers develop two themes that are especially relevant here. The first theme is that dogs' characteristics have changed as a consequence of domestication, and they are fundamentally different animals from their ancestors. Their second theme is the idea that dogs evolved from wolves to be efficient feeders at human waste dumps. This theme receives some critical attention here, especially in Chapter 4.

Finally, for present purposes, three newer volumes should be identified. One is an edited volume of pieces stemming from a 2002 conference session on dogs focused on archaeozoological matters (Snyder & Moore 2006). Different chapters in it are relevant to parts of this book. Another useful book is a volume of separate contributions, edited by Per Jensen (2007a). The chapters in his book concern mostly modern dogs, and accordingly, I draw from several in my later chapters. Other chapters in his book do have some archaeological relevance, and so I refer to them earlier in this volume. Lastly, for present purposes, is a recent book by Ádám Miklósi (2007a), also concerning mostly modern dogs, but including some archaeological considerations.[5]

---

[4] In general the goal in this introductory chapter is to highlight volumes that are specifically about dogs; however, Crockford's (2006) most recent contribution is a book that, while it goes well beyond dogs, calls on dogs as her primary example. She discusses at some length how and why evolution occurs under domestication, and just what constitutes domestication. Accordingly, Crockford's book receives substantial commentary in Chapter 4, here, which deals with the process of domestication, and it plays a limited role in Chapter 8 as well. She also wrote an earlier volume devoted to archaeological dogs (Crockford 1997), but only to a structural aspect of dogs inhabiting a quite specific region of North America.

[5] Archaeological considerations are of primary concern in a work originally written in German close to a century ago, by Kontondo Hasebe, recently translated into English by Holger Funk (Hasebe 2008). The bulk of the book, Hasebe's work, is about osteological characteristics of some prehistoric Japanese dogs. Funk's lengthy introduction, though, offers a variety of useful insights that concern matters covered here especially in Chapter 7.

As mentioned earlier, this is by no means an exhaustive account of dog books, but a sample of works produced over about the past half century that focus explicitly on dogs, that I have had occasion to become familiar with. To be sure, the foundation of this book lies in knowledge of the archaeological record, and a goal is to be comprehensive in identifying archaeological books about dogs. As far as other works previously described, some address specialized nonarchaeological audiences, though they are sometimes relevant.[6] I also appeal to modern general books at different points. Though examples of each are incorporated, the sheer quantity of material available about dogs prohibits one from meaningfully covering everything.

THIS VOLUME ABOUT DOGS

A central, unifying theme of this book is how and why dogs and people developed a distinctive social bond with each other. But it is appropriate initially to explore some background topics that are important as a frame of reference for what follows, leading up to the book's end. That said, it may be common knowledge that the wolf, *Canis lupus*, is the immediate ancestor of domestic dogs. For that reason, beginning by delving into the topic of immediate ancestry (Chapter 2) may at first seem like old news. In part, however, it is important to establish the basic taxonomic framework, an important first step. Then, as to the issue of immediate ancestry, going over the history of uncertainty on that topic and ultimately its modern resolution is an exercise full of useful lessons. To anticipate one, the genetic near-identity of wolves and dogs stands in sharp contrast to the fundamentally different animals that they are. Beyond basic wolf ancestry, I also consider the issue of which variety or varieties of wolf were involved.[7]

---

[6] An example is the recent comprehensive assessment of the genome of the dog, in a volume edited by Ostrander et al. (2006). As a specialized collection dealing with a topic that is not a central concern here, it does not warrant substantial coverage, though I call on one chapter in it later in this volume. To be sure, as one reviewer of this book pointed out "The dog is poised to be a guide for geneticists" (Lyons 2007: 64). But I am not a geneticist, and this book is not really about dog genetics.

[7] One of the points emphasized is how absolute body size has historically figured into considerations of this issue. For example, small Indian or Arabian wolves have often been invoked as ancestral, partly because most dogs have been considered too small to be derived from larger wolves. The issue of size reduction under domestication is taken up in Chapters 3 and 4, so is set it aside for now. Instead, it is worth making the simple point that the earliest definitively identified dog is not from one of the regions

Chapter 3 deals with the morphology of dogs versus wolves, both how they are similar, and how they are different. Drawing extensively from archaeological research, it highlights the use of quantitative data to explore in some detail the nature of the differences between dog and wolf skulls, and circumstances that can render the distinction difficult to achieve. An additional point of emphasis is the evolutionary significance of the changes that have occurred, specifically what biological mechanisms they reflect. A central overall objective is to make clear how different kinds of archaeological information speak rather directly to the question of when domestication of the dog first occurred. Those kinds of information are primarily in the form of skeletal morphology and the initial occurrence of dog burials, the latter a topic to be dealt with much more fully at later points in the volume. The understanding gained from those sources of information is at direct odds with some inferences based on the genetic studies of modern animals. It is important to explain how that is the case, along with why the archaeological evidence can be regarded as more reliable when considering the issue of domestication timing.

Having elucidated the changes that occurred in dogs, along with considering the timing of dog domestication, Chapter 4 then turns to the very process of domestication. An initial step is to spell out the traditional view of domestication in general, as associated with the scholarly tradition in North American social sciences, including anthropology, the field in which I received my basic training. This chapter highlights several examples of that approach in traditional, and in certain prominent cases, recent literature. Perhaps not surprisingly, that conception of domestication is not favored here. What may be surprising, though, is the explanation of why, which begins with several well-documented examples of domestic relationships that do not involve people at all, a point at direct odds with dominant themes in the social science tradition. Therefore, it is important to consider just what is meant by a domestic relationship.[8] Turning to dogs specifically, the first and foremost task is to spell out what seems to be the most likely general scenario for dog domestication, one that accounts for the changes that are documented

with smaller wolves, but from Europe. That fact is important toward the close of the next chapter.

[8] A common denominator, whether the topic is people or not, is that both organisms derive an evolutionary benefit from the close association. The particular manner in which it got started in any given case varies depending on which domestic association is under consideration. And it may not always be reliably determined.

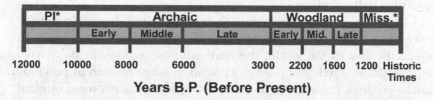

FIGURE 1.1. Schematic representation of major periods in the prehistory of interior eastern and midwestern North America, as commonly conceived (e.g., Fagan 1991). Time boundaries within major periods are arbitrary, reflecting certain trends over time. Major periods themselves are approximations corresponding to fundamental changes in human economic strategies that are important in this volume at several different points. *PI = Paleoindian, Miss. = Mississippian.

archaeologically. This effort is informed by some important biological principles. Also explored in this chapter is the issue of why dog domestication, although apparently the first domestic relationship with animals that people developed, occurred relatively late in the larger sweep of the human past. That is, behaviorally (apparently) modern humans had been on the scene for thousands of years before they became involved in documented domestic relationships, with any organisms. Chapter 4 ends with an effort to encapsulate, in simple and graphic terms, what constitutes the primary basis for the domestic relationship between people and dogs.

Having established my take on how, why, and when dogs and people entered into a domestic relationship, Chapter 5 then considers the various roles that dogs played among past people. In part, this effort is guided by a substantial body of empirical archaeological information, and given my own primary background, much (not all) of it is from North America. Given that factor, it is useful to introduce right now the basic cultural historical framework that structures considerations of a significant portion of the course of North American prehistory (Figure 1.1) and is relevant at several junctures throughout this volume. How certain non–North American schemes correspond to this framework is indicated at appropriate points. Chapter 5 also makes use of relevant ethnohistoric information on different peoples. In doing so, it is important to be clear as to the distinction between secure archaeological information and compelling though not entirely definitive ethnohistorical information. At the same time, some of the roles inferred for dogs of the past are substantially conditioned by what dogs get used for in modern or recent times, and in some cases, logical supposition comes into play in these ideas.

Chapter 6 represents the point at which this effort departs in a major way from topics covered in other books. This chapter is devoted to a specific area of the world, the northernmost regions, or the Arctic. Although the Arctic has been dealt with in some fashion in previous books on dogs, coverage here is extensive, and presents some original information that stems from my own past experiences. The North American Arctic, especially archaeological work in Greenland, is the focal point of this original effort, though other arctic areas are selectively incorporated. The unique aspects of the coverage involve the later periods of prehistory there and stem primarily, though not exclusively, from information gathered by me some two decades ago on archaeological dog bones from many different sites in Greenland.

And that brings this sequence to a chapter devoted to dog burials, another distinctive aspect of this book. Although other books have dealt with this phenomenon, sometimes substantially (e.g., Schwartz 1997), this volume is distinctive in that it is cosmopolitan in scope. It elaborates on a portion of an article previously published (Morey 2006), but includes many more dog burials than I reported then (certainly there are more yet to be identified). In addition to examples from around the world, an entire section of Chapter 7 is focused on a particular region in North America, to highlight how routine dog burials are in some contexts. Coverage of modern dog cemeteries, dealt with briefly in the 2006 article, is delayed to a more substantial treatment in Chapter 10.

Having gotten the dogs dead and buried, so to speak, the next step in the sequence, Chapter 8, is distinctive as well. The focus on dog burials leads rather directly to the question of just why dogs and people relate so closely. Working toward an answer to that question involves explicitly linking specific categories of archaeologically derived information, with information on modern animals. Obviously, other works do that to one degree or another, but this effort focuses especially on the neurological characteristics of dogs as compared to other animals, and especially on the linking of distinctive behavioral traits of dogs with their underlying neurological and neurohormonal characteristics. Notably, this effort also makes use of information on experimentally domesticated foxes in Russia, in modern times, and develops what I believe is a compelling argument for why foxes could only be domesticated experimentally, such apparently never having taken place in prehistory. Another conspicuous topic is the question of what relevant neurological characteristics separate dogs from cats, given the infrequency with which cats were treated ritualistically at death. It is a matter of contrasts, since both dogs and cats are popularly maintained animals in modern times. The difference concerns differential patterns of sociality toward people, and the chapter does offer a resolution to that issue. In all, Chapter 8 helps set the stage for developments in Chapter 9, which focuses on some of the characteristics of dogs that are similar to people, beyond sociality per se. One such characteristic is made especially clear by considering the impact of search-and-rescue activities on trained dogs. To anticipate, it turns out that search-and-rescue dogs want to find living people, and they are adversely affected when the outcome is different. Yet another distinctive characteristic of dogs takes the focus initially back to their wolf ancestors and concerns, perhaps surprisingly, their musical aptitude. That theme is represented with suitable documentation, and also includes an anecdotal account that pertains to one particular dog.

Having dealt more fully with some of the distinctive behavioral capabilities of dogs, the final chapter (Chapter 10) concerns dogs in modern times. This effort covers some ground that has been covered by others, but focuses especially on those roles of dogs that draw attention to their social bond with modern or recent people. As a fitting way to wind up this volume, this chapter offers an account, in some detail, of the use of dogs in a mortuary role, highlighting the burial of dogs in modern or recent times. In keeping with that closing theme, the true close of the book is a final, brief, and purely personal epilogue account of the recent journey of one special dog.

To hark back to the first page, in some ways it seems that this brief opening chapter was about half a dog long. Others will be more like a full dog long, or occasionally maybe even a dog and a half long. Regardless, welcome to the pages of this book on dogs. I hope you enjoy it, learn some useful information in the process, and are led to think about some of the issues raised. If you do all three, I must consider this volume a success.

2

IMMEDIATE ANCESTRY

AS NOTED IN CHAPTER 1, IT IS LIKELY COMMON KNOWLEDGE BY NOW that the gray wolf, *Canis lupus*, is the direct ancestor of the domestic dog. This chapter covers the history of work that makes such a statement possible, including some of the useful lessons learned in reaching that point, but it is initially important to impart some basic background. Accordingly, the present objective is to set the stage by clarifying the phylogenetic position of dogs. Within the taxonomic class Mammalia, they are assigned to the order Carnivora, or primarily flesh-eating carnivores. This order contains a variety of animals, including the cat, hyena, bear, weasel, seal, mongoose, civet, and dog families (Wayne 1993a: 218). The dog family is Canidae, and the Canidae, to put matters simply, "is a morphologically diverse family of dog-like carnivores" (Wayne & O'Brien 1987: 339). Based especially on genetic data, Wayne (1993a: 219, 1993b: 18; Wayne & Ostrander 2007: 557–559) divides the canid species into three basic groups: the wolf-like canids, the South American canids, and the foxlike canids. Figure 2.1 illustrates the basic phylogenetic affinities of different canid species and indicates the approximate timing of the origin of different root stocks leading to a particular species.

As indicated in their Figure 2 (Wayne & Ostrander 2007: 559), the dog falls within the wolf-like canid grouping, with particular affinities to a certain wild wolf-like canid. In fact, Wayne (1993a: 220) has stated that "Dogs are gray wolves [*Canis lupus*], despite their diversity in size and proportion; the wide variation in their adult morphology probably results from simple changes in developmental rate and timing" (see also Wayne 2001). "Developmental rate and timing" are key factors in the next chapter.

The immediate ancestry of the dog was, however, uncertain for a long time. In fact, as recently as about two decades ago, with the

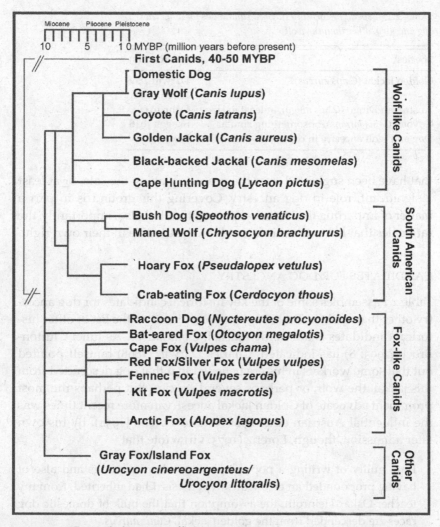

Miocene   Pliocene Pleistocene

10   5   1 0 MYBP (million years before present)

First Canids, 40-50 MYBP

Domestic Dog

Gray Wolf (*Canis lupus*)

Coyote (*Canis latrans*)

Golden Jackal (*Canis aureus*)

Black-backed Jackal (*Canis mesomelas*)

Cape Hunting Dog (*Lycaon pictus*)

Wolf-like Canids

Bush Dog (*Speothos venaticus*)

Maned Wolf (*Chrysocyon brachyurus*)

Hoary Fox (*Pseudalopex vetulus*)

Crab-eating Fox (*Cerdocyon thous*)

South American Canids

Raccoon Dog (*Nyctereutes procyonoides*)

Bat-eared Fox (*Otocyon megalotis*)

Cape Fox (*Vulpes chama*)

Red Fox/Silver Fox (*Vulpes vulpes*)

Fennec Fox (*Vulpes zerda*)

Kit Fox (*Vulpes macrotis*)

Arctic Fox (*Alopex lagopus*)

Fox-like Canids

Gray Fox/Island Fox
(*Urocyon cinereoargenteus/
Urocyon littoralis*)

Other Canids

FIGURE 2.1. Basic phylogenetic affinities among modern canids, with approximate timing of divergences indicated. This framework is adapted and modified from both Morey (1994a: 340) and Wayne & Ostrander (2007: 559).

accomplishment of my own first comprehensive study of the dog (Morey 1990), one could reasonably conclude that although Gray Wolf ancestry was the most convincing possibility, it was not yet established. Only several years later (e.g., Wayne 1993a) did it become possible to say more definitively that the Gray Wolf is the immediate ancestor of the dog. It is, nevertheless, worth covering the other principal candidates

TABLE 2.1. *Several historically invoked candidates for domestic dog ancestry, other than the wolf*

Species	Reference(s)
Golden Jackal (*Canis aureus*)	Darwin 1868
	Lorenz 1954
Australian Dingo (*Canis dingo*), at least in size	Dahr 1942
Coyote (*Canis latrans*), or something similar	Skaggs 1946
Dog and wolf ancestry in common	Fox 1978

that have been suggested over the years as playing the sole, or at least a significant, role in dog ancestry. Covering this ground is in part a matter of imparting the historical perspective, but also, importantly, the rationales that led to these inferences are instructive in their own right.

CANDIDATES FOR DOG ANCESTRY

Table 2.1 assembles some of the key historical candidates for dog ancestry other than wolves. Probably most prominent on the list of other historical candidates is the Golden Jackal, *Canis aureus*. As Juliet Clutton-Brock (1995: 8) has indicated, Charles Darwin (1868) himself pointed out that some workers in his era believed that that dog descended from this jackal, the wolf, or perhaps several species. But perhaps the most prominent advocate of Golden Jackal ancestry in more recent times was the influential Austrian ethologist Konrad Lorenz (1954). By his own later admission, though, Lorenz (1975: vii) wrote that

> I am guilty of writing a popular book on domestic dogs and also of having propounded an erroneous hypothesis: I had inherited, from my teacher Oskar Heinroth, the assumption that the bulk of domestic dog races are descended from the golden jackal, *Canis aureus*.

Lorenz went on, in the same brief essay, to favor the wolf as the most likely ancestor of most breeds. Interestingly, bearing in mind that this modified view of his predated the definitive establishment of the wolf as the ancestral species, his new view was based especially on similarities in the manner in which dogs and wolves howl. That is indeed a notable similarity, and it will receive more attention at a later point in this volume (Chapter 9). It is also worth noting that in that same essay, Lorenz maintained that several of what he had called "lupus dogs"

(Lorenz 1975: viii), for example the chow, the husky, and Greenland dogs, had a different wild ancestor than most or all of the other breeds. The question, of course, is what that other ancestry was, and why it was invoked. His statement refers to different subspecies of *Canis lupus*, covered shortly, but it continues to be instructive to consider some of the other historical nonwolf candidates. There is in fact a noteworthy thread that unites the logic behind invoking the various canids as playing a major role in the ancestry of the dog.

For example, anticipating Lorenz's idea of a different ancestry for larger lupus dogs versus other varieties, a good many years earlier Dahr (1942) set about to arrive at an inference about primary ancestry by investigating the relationship between the length and the breadth of the braincase in dogs and wild canids. Addressing the idea that smaller dogs are descended from jackals, while larger ones are descended from wolves, he inferred from his data that the original size was smaller than true wolves, but larger than jackals. He suggested that "the most plausible size of the ancestral forms coincides with the size of the Australian dingo" (Dahr 1942: 54). He implied, without being explicit, that it could have been ancestral, especially since he raised the possibility that the Dingo may have been originally widespread in Eurasia. But derivation of the Dingo has long been unclear. Several writers have dealt with the Dingo (e.g., Macintosh 1975; Newsome et al. 1980; Corbett 1995), and one of them has explicitly written about "The origin of the dingo: An enigma" (Macintosh 1975). In recent years, however, that enigma may have been solved, at least partially. Specifically, Savolainen et al. (2004) have analyzed mitochondrial DNA sequences of Australian Dingoes, hundreds of modern dogs, some Eurasian Wolves, and some pre-European dogs from Polynesia and inferred that dingoes originated from domestic dogs of East Asia, possibly about 5,000 years ago, after which time they were isolated from other dog populations. Whatever the true origin is, given that a Dingo is usually smaller than a Gray Wolf, it could indeed serve as a general model for the initial size of early dogs, for reasons developed more clearly later.[1]

---

[1] A related issue concerns the enigmatic New Guinea Singing Dog, apparently known to science beginning only in the 1950s. This animal is classified variously as "*Canis dingo, Canis lupus dingo, Canis familiaris dingo,* or *Canis familiaris hallstromi*" (Koler-Matznick et al. 2000: 239). Rather clearly, it is held to be closely related to the Australian Dingo, perhaps itself a true Dingo. That factor of course renders its origin problematical.

Also considering cranial dimensions, Skaggs (1946) advanced a case for the North American Coyote (*Canis latrans*), or a "coyote-like ancestor" (Skaggs 1946: 345), perhaps from Asia, as progenitor of North American dogs. Skaggs recorded measurements on the crania of seventeen dogs from the Indian Knoll site in Kentucky (Webb, 1946), some 5,000 years old. She compared them, by virtue of proportional ratios between the measurements, to type specimens of both modern wolves and Coyotes and found that the Indian Knoll dogs were more similar to the Coyotes. Even ignoring the miniscule sample size of wolves and Coyotes (two "type specimens" of each species, a male and a female), such a comparison could not, in and of itself, account for allometric differences in proportions. Such differences can stem from the different sizes of the dogs and the wolves, in comparison to the more comparable overall size of the dogs and the Coyotes. Gould (1966) explained why different sizes result in different proportions, the phenomenon of allometry. Quite simply, the phenomenon of allometry refers to the study of size and its consequences. That topic played a central role in an earlier comprehensive study of the evolution of dogs (Morey 1990) and in a paper that directly followed from it (Morey 1992). The phenomenon of allometry constitutes a well-established principle; however, for present purposes it is simply worth noting that, historically, absolute size differences have probably often been accorded greater significance than they should, both directly and indirectly.

Finally, Michael Fox (1978: 248) once suggested the possibility that dogs and wolves share a common ancestor prior to domestication, and "the dog was a dog before it was domesticated." In that vein, he also suggested that the dog arose "from a more primitive prototype," one that was possibly dingo-like (Fox 1978: 243). Moreover, this primitive prototype would have evolved from the basic wolf/jackal stem prior to any initial association with humans. Summarizing this work here is mostly for the sake of historical completeness, for this idea was advanced in a conspicuous source. Even at the time that it was advanced, however, it had no empirical support, and now there are definitive data that contradict it (see discussion that follows).[2]

[2] In more recent times, Raisor (2005) has resurrected this untenable suggestion. In proposing what she calls an alternative hypothesis, she accepts wolf ancestry, but writes as follows: "I suggest that dogs were dogs long before man even considered the possibility of exploiting these animals through selective breeding to produce an animal with characteristics tailored to human needs" (Raisor 2005: 86). She does not cite that 1978 work of Michael Fox's and is apparently unfamiliar with it. Much

THE GENETIC NEAR-IDENTITY OF DOGS AND WOLVES

For a good many years prior to the establishment of the point summarized here, archaeological data, in form of skeletal morphology of canid bones, suggested strongly that the wolf was the primary ancestor of the dog. The nature of that evidence, particularly how it relates to an understanding of the timing of the appearance of the domestic dog, is important in the next chapter. There, it will be seen that archaeological considerations directly contradict certain genetic inferences, but the genetic data at issue here, centered on modern animals, pose no problem.

In the early 1990s, Robert K. Wayne synthesized information on the molecular evolution of the doglike carnivores (Wayne 1993a; see also Wayne 1993b). In this review, he summarized work by himself and several colleagues (e.g., Wayne et al. 1992) with a mitochondrial DNA (mtDNA) restriction fragment analysis of seven dog breeds and twenty-six gray wolf populations from different places in the world. His review of this work resulted in the following statement:

> The domestic dog is an extremely close relative of the gray wolf, differing from it by at most 0.2% of mtDNA sequence.
>
> In comparison, the gray wolf differs from its closest wild relative, the coyote, by about 4% of mitochondrial DNA sequence. Therefore, the molecular genetic evidence does not support theories that domestic dogs arose from jackal ancestors. (Wayne 1993a: 220)

In making that last statement, Wayne cited Konrad Lorenz's 1954 book, in which Lorenz indeed argued for jackal ancestry. In that light, it is unfortunate that Wayne did not call attention to Lorenz's brief 1975 essay in which he rescinded that view. Regardless, this body of work, subsequently elaborated along various lines (e.g., Wayne & Vilà 2001), established that the Gray Wolf is indeed the basic ancestor of domestic dogs.

---

the same untenable idea was also advanced by Koler-Matznick (2002), though Koler-Matznick, unlike Raisor, doubts wolf ancestry. Koler-Matznick does not cite Fox's 1978 work either. Interestingly, Raisor cites Koler-Matznik (Raisor 2005: 86) for, among other things, including the 2002 article, a personal communication about this very issue ("Koler-Matznick 2003b"), apparently at Raisor's personal residence in College Station, Texas.

Overall, the importance of Wayne's original body of work is under-scored when one considers its influence in more recent times:

> the domestic dog is an extremely close relative of the grey wolf, differing from it by *at most* 0.2% of the mtDNA sequence.... In comparison the wolf differs from its closest wild relative, the coyote, by about 4% of the mtDNA sequence. (Brewer 2001a:19, original emphasis)

Thus, the wolf was established as the basic ancestor of the dog.[3] In fact, in the same year that Wayne's influential publications came out the Smithsonian Institution, in conjunction with the American Society of Mammalogists, formally reclassified the dog to recognize it as a mere variety of the wolf (Wozencraft 1993: 281). One startling implication of this new standard is that one's toy poodle, or Chihuahua, or what have you, is now a mere variety of the wolf, *Canis lupus*. That is, dogs and wolves can now be regarded as the same species, although there are some workers who still prefer to regard the dog as a separate species, *Canis familiaris*, the Latin designation it had long borne (e.g., Coppinger & Coppinger 2001: 281; Nowak, 2003: 257; Crockford 2006: 100; Björnerfeldt 2007: 21).[4]

One complication of this new standard centers on the fact that regard-ing the dog as a variety of the wolf means it can be referred to as *Canis lupus familiaris*. Unfortunately, it remains unclear whether the use of this trinomial term should carry the implication that it usually does in other

---

[3] Dobney & Larson (2006) have pointed out that even though dogs basically descended from wolves, one should not lose sight of the probability that a variety of different wild canid species surely contributed genetic material to the dog lineage. The reason is that interfertility is the rule rather than the exception within the genus *Canis*, and so other canids surely made some contribution to the dog stock. As they say, "Domestic dogs, therefore, can be viewed not simply as designer wolves, but as a chimeric species possessing DNA from several ancestral sources" (Dobney & Larson 2006: 267). Their point is well taken, but does not negate the fact that Gray Wolves are the basic ancestors of the dog. Dobney & Larson's work receives more attention later in this volume, especially in Chapters 3 and 4.

[4] It is not hard to appreciate the logic of both ways of viewing the dog taxonomically. On the one hand, if genetic near-identity is the main criterion, it is fully justified to regard dogs and wolves as the same species. On the other hand, dogs and wolves are clearly different animals, exhibiting some biologically based behavioral changes that only indirectly concern genetics. That aspect of dogs is covered at some length in Chapter 8, and although no vested interest in either taxonomic viewpoint obtains here, it would be helpful if all researchers could reach a consensus. In the meantime, continuing difficulties in resolving appropriate taxonomic nomenclature, for dogs as well as other domestic species, have been thoroughly covered by Gentry et al. (2004).

contexts. That is, can the dog be regarded as a formal subspecies of the wolf? The wolf itself has a variety of formally designated subspecies, but in those cases the general place and timing of the appearance of that kind of wolf has been worked out. With dogs, it remains unresolved whether the dog has a single or multiple evolutionary origins. If a single origin is the case, the trinomial term just indicated would suffice. On the other hand, if multiple origins are the case, then a different formal subspecific designation should be given to each variety of dog that has arisen. In that case, the general place and timing of appearance associated with each variety, perhaps formally recognized as a breed, would need to be worked out. The difficulty such a task would pose underscores the complexity of the issues involved.

In any case, with the wolf established as the basic ancestor, the next question is, of course, which wolf (or wolves)? As noted, wolves come in a variety of subspecies, most associated with different geographic areas of the world. Moreover, these different varieties of wolves tend to have different modal sizes, and, as mentioned earlier, the role of sheer size in framing historical scenarios about ancestry is sometimes misleading. Certainly sheer size is relevant, but it must be taken into account judiciously. At an early point, for example, Werth (1944) accepted the wolf as the basic ancestral stock for the dog, based mostly on morphological considerations regarding skeletal remains. In doing so, he was led to wonder which of the small local varieties of wolf it was. In some cases, this is probably a legitimate question. In other cases, though, it may not be so meaningful.

WHICH WOLF, OR WOLVES?

Well before definitive genetic data established the wolf as the ancestral species, many, if not most, authorities were inclined to accept that probability. Not surprisingly, a good deal of effort was invested in trying to determine which variety or varieties of wolf were primarily involved. Moreover, this is not a simple matter of deciding among a very few. In one recent compilation, Nowak (2003: 246) indicates at least fifteen different subspecies. In arriving at that compilation, he did not list as separate subspecies some cases regarded as such by other authorities, but regarded by him as synonymous terms for some varieties. The ambiguity is further underscored by the fact that Dewey & Smith (2002) suggested that as of 1997, thirty-two different subspecies could be recognized. The present account is restricted to dealing with several varieties that have over the years been commonly invoked when this issue is considered.

Historically, it was widely held that most dogs are too small to have been derived from the larger, northern varieties of *C. lupus*. This is an example of the prevalence that sheer body size has often been given in these considerations. To be sure, most early dogs were systematically smaller than large, northern wolves. At the same time, however, other domestic animals often exhibit similar changes in comparison to their wild ancestors (see Zeuner 1963; Epstein 1971; Clutton-Brock 1981, 1999: 33–34). For example, Clutton-Brock: (1981: 22) stated that

> The early stages of domestication of any species of mammal are almost always accompanied by a reduction in size of the body. This is so generally true that it is used as the main criterion to distinguish the skeletal remains of domestic from wild animals when these are retrieved by archaeological excavation of early prehistoric sites.
>
> During the later stages of domestication, animals that are either very much larger or very much smaller than the wild progenitor are selected for and breeds developed from them, for example, shire horses and Shetland ponies.

Most people are surely familiar with exceptionally large and small breeds of dogs, as products of the later stages of dog domestication.

There are, in fact, sound theoretical reasons to expect body size reduction under conditions of early domestication (Morey 1992), in keeping with pattern commented on by Clutton-Brock and pointed out by others (e.g., Crabtree 1993). Later considerations highlight those reasons, especially as they apply to the case of dogs. For present purposes, however, there is no inherent reason why, in principle, one or more small southern varieties might not have been involved in dog ancestry, as coverage in the next section indicates.

Principally Invoked Wolf Subspecies

It should be pointed out at the outset that the basic Gray Wolf of the Old World is regarded as *Canis lupus lupus* (e.g., Nowak 2003: 246). It is also known sometimes as the Eurasian Wolf, given that it is known to occur in Europe, eastwards through an undetermined expanse of Russia, central Asia, southern Siberia, China, Mongolia, Korea, and the Himalayan region (Nowak 2003: 246). Most of these Gray Wolves are medium-sized to somewhat, though not dramatically, large in size. Historically, the preconception that most dogs are too small to have been derived from the larger, northern varieties of *Canis lupus* led a good many investigators to suggest small Indian or Arabian wolves, *C. l. pallipes* or *C. l. arabs*, as being prime candidates for progenitor of the dog (e.g., Werth 1944; Zeuner 1963; Lawrence 1967; Epstein 1971; Hemmer 1976; Clutton-Brock 1984). In certain cases, this inference appears justified at face value. An example is a partial canid mandible, about 12,000 years old and tentatively identified as dog, from Palegawra Cave in Iraq, as reported by Turnbull & Reed (1974). Assuming that the morphological criteria that were used to arrive at this determination were valid, it is clear how an inference of *C. l. arabs* as the source stock was reasonable.

Other noteworthy examples are from the same general region, and consist of canid finds from two Natufian (ca. 12,000–10,000 B.P.) sites in northern Israel, Ein Mallaha (Davis & Valla 1978) and Hayonim Terrace (Tchernov & Valla 1997). From Ein Mallaha, there is a diminutive canid carnassial tooth, as well as an adult mandible, dating within the range of 12,000–11,000 B.P. In general, they were identified as domestic dog based on their smaller size compared to recent wolves of the region. Another find there from the same time frame, however, was a dog or wolf puppy skeleton, too small for a secure identification to be made, and in direct association with a burial of an elderly person. Davis & Valla (1978: 608) published a photograph of this burial, showing the human skeleton's hand resting on the puppy skeleton. At Hayonim Terrace, dating between about 11,000 and 10,500 B.P., remains of two dogs were associated with three human burials. It is again clear how smaller southern *C. lupus* stock could reasonably be inferred to have given rise to these animals.

Continuing with smaller variants of *C. lupus*, S. J. Olsen (1974, 1985, Olsen & Olsen 1977), argued, based on morphological criteria, that the relatively small Chinese Wolf, *C. l. chanco*, gave rise to most Asian and North American dogs. Moreover, in this model dogs arrived in the New World with people, by way of the Bering Strait, where a land bridge existed during the late Pleistocene Period. This is an instance in which shifting taxonomic standards affect terminology, as Nowak (2003: 246) now lists *C. l. chanco* as synonymous with *C. l. lupus*. Whichever taxonomic standard is preferred, Leonard et al. (2002), based on mitochondrial DNA (mtDNA) sequences from archaeological dog remains from both Latin America and Alaska, suggested that New World dogs arose from multiple Old World lineages that accompanied people across the Bering Strait land bridge. At the same time, Savolainen et al. (2002) suggested, based on mtDNA sequences among over 600 modern dogs, representing all major dog populations, that dogs share a common origin from an East Asian stock. Inferences based on modern dogs are always more problematical and more recent work indicates why. Specifically, Verginelli et al. (2005) extracted DNA samples from five prehistoric Italian canids, ranging in age from about 15,000 to 3,000 years ago, and compared those samples to corresponding samples from several hundred purebred dogs and modern wolves. What they found in their study was as follows:

> Genetic data obtained comparing the ancient sequences with extant dog and wolf sequences and early archaeozoological evidences concur in suggesting that Late Glacial/Early Holocene wolf populations of the West Eurasian steppes (that stretched over South-Eastern Europe and West Asia) contributed to the origins of the dog.... Genetic data also suggest multiple independent Asian and European domestication events. (Verginelli et al. 2005: 2549)

Clearly, the integration of genetic data with archaeological data played an important role in arriving at those inferences, a general topic that merits subsequent attention (see also Chapter 3). But for now, it is important to cover more closely the archaeological evidence that speaks against the idea of only an East Asian origin for dogs, beyond the examples noted above (e.g., Ein Mallaha) when considering size differences.

Other Non–East Asian Lineage Probabilities

In introducing this topic, the primary source of evidence will be skeletal remains of early dogs, or putative dogs, from regions other than East Asia. Granting that some early non East Asian dogs could logically be derived from East Asian stock, the logical next step is to identify the earliest known dog and determine where it comes from. Before delving into specific information bearing on this topic, it is important to note that, as of this writing, the earliest securely established dog is not from East Asia. Even if this case (see discussion that follows) stands the test of time, though, this does not eliminate a substantial role for East Asian dogs. With little question, for example, East Asian populations gave immediate rise to New World dogs, as covered earlier. A dog with a securely established age of just over 10,500 years old is from

extreme eastern Asia, specifically at the Ushki-1 Site, on the Kamchatka
Peninsula of Siberia (Dikov 1996: 245; Vasil'evskiy 1998: 291; Goebel &
Slobodkin 1999: 131–133). The presence of this animal, of such an age,
speaks further to the probability that New World dogs were immediately derived from an East Asian source. Therefore, if the first dog
originated someplace other than East Asia, it would be older than the
Ushki-1 dog.

In broaching the topic of what constitutes the earliest securely documented domestic dog, from East Asia or elsewhere, it is important to
bear in mind that, whatever date is associated with it, canid domestication began, by definition, somewhat earlier than that. With that caveat
in mind, there are some notable non East Asian cases, including those
in Israel. And it should be pointed out that the case advanced for the
smaller southern varieties has never been taken to exclude larger varieties from having played a role (e.g., Olsen & Olsen 1977; Clutton-Brock
1984: 199; 1999: 57–58). In particular, some mainland European contexts are especially worth emphasizing. In doing that, coverage is, as
indicated, restricted to cases that are known, or reasonably inferred, to
exceed about 10,500 years in age. Some of the tentative or secure cases
that slightly predate that time, as known in recent decades, have been
covered by both Morey (1990: 19–22) and S. J. Olsen (1985: Chapter 6).
One example is the Ein Mallaha case from Israel, covered previously.

With the standard just noted as a guide, as early as the 1970s Musil
(1974, 1984) reported on canid remains from Kniegrotte in Germany, that
date in the range of 13,000–12,000 B.P. He suggested that the remains
represent the domestic dog at an early stage in the domestication process. Much more recently (Musil 2000), he elaborated on what had been
found, still tentatively suggesting the presence of domestic dog, and
also briefly covering finds from two additional central European sites.
The *Canis* remains from Kniegrotte included an incomplete maxilla from
the topmost site horizon and three mandible fragments from the middle
horizon. He provided measurements on these remains but illustrated
only the partial maxilla. In making his case that early dogs were represented, Musil (2000: 27) enumerated the traits that led him to this inference. They included generally reduced size, a shorter snout region and
associated facial parts of the skull, and crowded teeth, some of which
were positioned obliquely in the tooth row. Those are in fact traits that
have long been known to be associated with early dogs (e.g., Clutton-
Brock 1970: 305–306; Morey & Wiant 1992; Clutton-Brock & Jewell 1993:
23–24). Size reduction and associated morphological changes occur for

reasons explored in the next chapter, along with a more comprehensive specification of how they manifest themselves.[5] In any case, Musil has presented a noteworthy tentative case for dogs in Central Europe that are 12,000–13,000 years old. The next case, however, need not be regarded as tentative. Moreover, Benecke (1987), responsible for the work in question, measured the Kniegrotte specimens himself as part of his study and cast doubt on their status as genuine dogs, though their proximity to zoo wolves in his DFA studies (see discussion in the next section) did lead him to suggest that they could represent an initial stage in the domestication of the wolf.

THE CASE OF THE BONN-OBERKASSEL DOG

The remains in question were originally reported by Nobis (1979, 1986, 1996) in German and come from a site in Germany known as Bonn-Oberkassel, dating to about 14,000 years ago. Beyond Nobis's work, Benecke (1987) provided a treatment in English, much more accessible to a broader range of readers. The site was excavated in the early 1900s, and the analyzable remains that are still available consist only of a single jaw (mandible) fragment. But there were originally more bones, and in fact this single surviving specimen was directly associated with a human double grave, containing remains of a ca. 50-year-old man and a 20- to 25-year-old woman. Moreover, the dog, now represented by only that single jaw fragment, apparently was a complete skeleton as originally encountered. The loss of bones was largely due to excavation methods and partial destruction of the grave, given that it was discovered in a quarry in 1914. The peculiar history of the Bonn-Oberkassel find has been covered by Street (2002).

But what is of most concern for present purposes is the basis for identifying that single specimen as, taxonomically speaking, a dog.[6] Rather

---

[5] Proportionally large, crowded teeth in early dogs apparently reflect a "lack of tight developmental integration between dental growth and overall growth" (Morey 1992: 198). In keeping with that pattern, large modern dogs often have conspicuously small teeth, while small modern dogs often have dental anomalies stemming from teeth that are crowded into undersized jaws (e.g., Smythe 1970: 43–45; McKeown 1975). Quite simply, it seems that tooth size changes have lagged behind skull and body size changes in rapidly evolving domestic dogs.

[6] Here it is relevant to acknowledge an oversight in my own previous synthetic treatment of dog evolution (Morey 1990). Specifically, Morey cited Benecke's study only in superficial fashion and drew mostly from Nobis's (1986) work in German, in spite of rudimentary skills in German. Nobis, in his work, based his inference on the shortness

than rely on comparisons of simple proportional dimensions, Benecke subjected the Bonn-Oberkassel specimen to multivariate discriminant function analysis, or DFA (see Tatsuoka 1971; D. G. Morrison 1974). DFA is a complex statistical technique that is used successfully to address questions about the taxonomic identity of certain archaeological or paleontological canid specimens (e.g., Higham et al. 1980; Morey 1986; Morey & Wiant 1992; Nowak 2002). For archaeology, this procedure calls for the establishment of reference groups consisting of individuals of taxa that potentially represent the archaeological specimen. On that specimen, and the specimens of a reference group, linear bone dimensions serve as the basis for analysis. The objective of the procedure is to determine which group is most similar to the archaeological specimen, in terms of the combination of recorded measurements. For his reference groups, Benecke used series of recent wolves from Europe and Greenland, seventeen in all, forty-seven Late Paleolithic wolf specimens, several recent zoo wolves, dogs from eighteen Mesolithic Period sites, and some recent dingoes. This thorough analysis, in addition to casting doubt on the Kniegrotte specimen, as just covered, clearly suggested that the Bonn-Oberkassel specimen represents a dog.[7]

Three aspects of this finding warrant emphasis. One, of course, is the age of the specimen, about 14,000 years old. This finding makes it, as of this writing, the oldest securely established dog specimen known in the world.[8] That does not mean that it is the oldest possible specimen,

of the jaw relative to wolves, a trait to be expected in a dog. The jaw in question, how-ever, was characterized by several missing premolars (Nobis 1986: 370–371). Conse-quently, especially given the great antiquity of the specimen (ca. 14,000 years old), caution took the form of suggesting that it "could simply signify an aberrant wild wolf" (Morey 1990: 21). It is unfortunate that I did not consult Benecke's (1987) more comprehensive analysis, reported in English, more carefully, but have for this work and so rectify that situation now.

[7] A photograph of the surviving partial jaw from Bonn-Oberkassel, alongside a slightly larger complete jaw from a modern European wolf, can be found in Morey (1996: 77).

[8] On this point, Raisor has made the following astonishing misattribution: "Morey and Wiant (1992: 227) conclude that the Koster dogs represent the oldest archaeological discovery of domesticated canids in the world" (Raisor 2005: 58). First, on the cited page number, 227, there is a large photograph of canid bones (Figure 2), and brief discussion that has nothing to do with the timing of canid domestication. Where the authors did deal with that issue, they stated as follows: "At present, specimens from Danger Cave, Utah, dating between 9,000–10,000 B.P., are the oldest well-documented remains of domestic dog from North America (Grayson 1988: 23)" (Morey & Wiant 1992: 225). So, they made no such claim even for North America, let alone the world. What Koster does have is the oldest dog *burials* known from North America. The Koster canids receive more attention in Chapter 7, and Raisor's work is covered more in Chapter 4.

and such a case receives coverage in closing this chapter. It is beneficial, however, to call attention to a second ramification of the antiquity of this dog. One should bear in mind, for later purposes, that the people responsible for the Bonn-Oberkassel site were purely hunter-gatherers. Agricultural economies lay in the future. At the same time, one should also bear in mind that it was morphological criteria that were the essential basis for determining it to be, taxonomically speaking, a dog. Finally, it is relevant to emphasize that this earliest securely established dog can reliably be assigned to *Canis lupus lupus*, the basic Gray Wolf, and not one of the smaller subspecies that have often been considered.

OTHER EARLY POSSIBILITIES

Sablin & Khlopachev (2002) have made a case for some early domestic dog remains from a site on the central Russian plain. The site in question, Eliseevichi 1, is situated along the Sudost River, in the Dnieper River Basin. More recently, as part of a team, Sablin has indicated that the remains there are about 13,900 years old, so approximately contemporaneous with Bonn-Oberkassel (Germonpré et al. 2009: 474). What this site yielded was two nearly complete large canid crania, apparently isolated. The skulls were found many years ago in an area of the site that contained a variety of artistically rendered objects. Naturally, skulls of reported dogs from this time frame are bound to merit close scrutiny.

Both skulls are described as those of adults. Based on comparisons to reference samples, the authors find that the crania of these Eliseevichi I canids are almost as large as those of northern wolves and modern Great Dane dogs. At the same time, they have proportionally wide palates and short snouts, traits consistent with dogs. Data plotted in their Figure 2 allow for a comparison of the Eliseevichi I crania with those of modern wolves, Siberian Huskies, and a pair of Great Danes. Specifically, the authors constructed ratios of greatest palatal breadth to skull length, in the form of condylobasal length (see Figure 3.1, this volume). This comparison revealed that the Eliseevichi I specimens were most similar to the Great Danes. But given their large size it should be borne in mind that, as Miklósi (2007a: 103) has pointed out, "Alternatively, they might have been local wolves living in captivity, in close contact with humans the descendants of an even larger wolf subspecies, or hybrids of some sort." Moreover, as Wang & Tedford (2008: 157) have pointed out, the ratio of the third premolar crown length to that of the fourth premolar on one of the skulls, the only one for which those measurements were available, is consistent with that of a Gray Wolf, not a dog. The status of these canids as really being dogs is therefore questionable.

One of the two skulls does exhibit a noteworthy alteration. Specifically, the authors noted cases from elsewhere in which it had been reported that the occipital bones of dog skulls had been removed, apparently to facilitate removal of the brain (e.g., Bökönyi 1974). In that light, they reported, with an illustration, that on one of the Eliseevichi skulls a hole had been made in the side of the skull, presumably for removal of the brain, in order to consume it. So, Sablin & Khlopachev (2002) advance a morphological case for the Eliseevichi I specimens as being dogs, though legitimate doubts remain about that taxonomic status (e.g., Miklósi 2007a: 103; Wang & Tedford 2008: 157).

Most recently, Mietje Germonpré and colleagues, including Sablin, have advanced a provocative case for three older Eurasian dogs, based primarily on analysis of a series of large skulls (Germonpré et al. 2009). One, from Goyet Cave, Belgium, is nearly 32,000 years old, and was originally found in the 1800s. The other two are from Mezhirich and Mezin, the Ukraine, and are younger, some 14,000 to 15,000 years old. Significantly, while the latter two sites are well-known human settlements, from the information presented (Germonpré et al. 2009: 474), the skull from Goyet is not even securely linked to the deposits that reflect human activity there. At any rate, the authors used DFA to make their case, but in a way that frankly renders their results unconvincing. First, one might recall from Benecke's use of DFA, just covered, that he had a relatively robust series for reference groups in his study. In this study, however, the only reference group of archaeological dogs consisted of a mere five early specimens, including the two from Eliseevichi 1 about which reservations remain. To those two, they added three other relatively early (ca. 9000–10,000 B.P.) dogs that apparently are taxonomically secure, but smaller in size. To control for the size differences, they worked only with indexed data, the ratio of one measurement to another, an approach that is sometimes justified (see Chapter 3).

Surprisingly, they had no other archaeological dogs in their reference groups. Other dogs in their reference groups were all recent/modern animals, and they divided those into several categories, such as "recent other dogs with wolf-like snout" (Germonpré et al. 2009: 479). One such group consisting of breeds thought to have ancient origins, primarily Chows and Huskies, was labeled as "recent archaic dogs" (Germonpré et al. 2009: 479). Recalling the minute and questionable reference group of archaeological dogs, they showed that the three early

Eurasian specimens were closest to that reference group in multivariate space, and not to any of the other groups. One thing that is noticeable on the discriminant score plot (their Figure 3) is that those three are closer to the recent wolf reference group than to the different recent dog groups. These authors would have made a far more credible case if they had taken the five specimens in their early dog reference group and treated each as an unknown in a DFA with a legitimate reference sample of established prehistoric dogs (not modern breeds). Then, recognizing that the object of a DFA is to take predefined groups and separate them maximally in multivariate space, one could see where those individuals fell. Regarding those points, the next chapter includes a comprehensive consideration of DFA in efforts to achieve taxonomic resolution in studies of canid skeletal morphology.

For Germonpré et al.'s (2009) case to be convincing, of course, the animals would have to have been classified as dogs. Three of the five surely would, but it is not clear at all that the Eliseevichi specimens would, and it seems likely that the three canids that came out closest to their reference group of five would have been assigned to wolves. And tellingly, the cases that they suggest represent dogs show no appreciable size reduction relative to wolves, casting further doubt on their conclusions. Moreover, in their summary discussion, the authors say their results "are consistent with the hypothesis that changes in dog morphology compared to wolf morphology appeared rather abruptly" (Germonpré et al. 2009: 481). But given the allometric principles covered above the same should be true of size itself, since size is a meaningful component of morphology.[9] So a frankly puzzling implication of the case made by Germonpré and colleagues is that for many thousands of years after ca. 32,000 B.P. dogs showed no appreciable size reduction but then, remarkably, sometime after about 15,000 years ago, they rapidly did. One wonders why. To be sure, at one level the authors are certainly correct in my view: that is, ca 14,000 to 15,000 years ago is when dogs did start to show appreciable size reduction and associated morphological changes, but only because that's when they became dogs. Perhaps these

---

[9] To be sure, the Goyet skull does exhibit some proportions that, at least superficially, are consistent with some of the traits expected of dogs. Those proportions involve the relations between certain cranial length and width dimensions (see Chapter 3). One should bear in mind that it is but one wolf-sized skull, and not even from deposits that can clearly be linked to episodes of human activity at the cave. Considering that factor, and given the other analytical shortcomings covered here, that skull provides a shaky basis, at best, for advancing a case for domestication of the dog that early.

authors are correct in their conclusions, but in the absence of a sound demonstration of the accuracy of their taxonomic inferences, I hold this as a markedly unconvincing case that dog domestication occurred that early and will continue with the later time framework for that process that is still most justified.

In the next chapter I focus on a more comprehensive consideration of morphological changes under domestication, their reason(s) for occurring, and their documented timing. Following that task, I then address a particularly compelling line of contextual evidence that bears directly on an understanding of the timing and fundamental nature of this particular domestic relationship.

3

EVIDENCE OF DOG DOMESTICATION AND ITS TIMING: MORPHOLOGICAL AND CONTEXTUAL INDICATIONS

HAVING ELUCIDATED THE BASIC ANCESTRY OF THE DOG, MY INITIAL purpose in this chapter is to clarify the morphological changes involved in the evolution of the wolf to the dog and how they manifest themselves. In conjunction with this effort, it is important to consider how these changes bear directly on an understanding of the timing of dog domestication. This chapter also explores how challenging it can be to distinguish between dogs and wolves in certain situations. Last, the coverage includes a particular line of contextual evidence, namely dog burials, that bears directly on these issues. As with the coverage of morphological changes, this more conceptual component appeals to factors introduced in the previous chapter. It is empirical in the sense that there is genuine evidence, but the kind of evidence speaks directly to the fundamental nature of this particular domestic relationship.

As noted in Chapter 2, morphological changes are a routine phenomenon among different animals undergoing domestication (e.g., Clutton-Brock 1981, 1999). In the present case, as highlighted, dogs initially became smaller than wolves (Morey 1990, 1992, Dayan 1994; Clutton-Brock 1995, 1999), as did many early domesticates in comparison to their ancestral species (e.g., Epstein 1955; Clutton-Brock 1981; Bökönyi 1983; Tchernov & Horwitz 1991; Crabtree 1993). They also underwent some allometrically associated morphological changes (see discussion that follows). The reasons that dogs underwent size reduction and associated changes in morphology are complicated, but come down to evolutionary changes in developmental rate and timing early in their individual development, or ontogeny. Those changes relate to selection especially for precocious sexual maturation, perhaps along with reduced body size itself, partly in response to the altered diet that they experienced in association with people (Morey 1992; 1994a). Those thoughts are developed more fully in the next chapter, but for the time being it is appropriate to impart a clearer idea of the nature of those changes.

ALLOMETRIC PATTERNS AND MORPHOLOGICAL DISTINCTIONS

Patterns involving size and morphology are best assessed in quantitative terms. Accordingly, it is necessary to begin here by illustrating some of the measurements that have proved to be useful over the years in dealing with these patterns. Figure 3.1 shows schematic illustrations of two different views of a canid skull, highlighting the principal measurements. All measurements reported here were taken with calipers in metric units, to the nearest 0.1 millimeter for the shorter dental dimensions and to the nearest whole millimeter for the others. Unless otherwise specified, I recorded the measurements. This is a potentially important point because, as S. J. Olsen (1985: 93) pointed out, interobserver error is a real possibility when two or more people are responsible for recording measurements. This is a lesson learned the hard way in the past, prompting an acknowledgment of some modest interobserver error in print (Morey 1992: 187). In this chapter the acronyms indicated on Figure 3.1 receive routine use.

To set the stage for this account, and to anticipate part of Chapter 4, dogs were colonizers of a new ecological niche, a domestic association with people. Under conditions of colonization, where animals undergo rapid population growth, early sexual maturation is at a distinct premium (Cole 1954; Lewontin 1965; Meats 1971; Giesel 1976; Gould 1977; Stearns 1977). As spelled out by Gould (1977) in a pioneering work, the accelerated onset of sexual maturity is known as *progenesis*, and it leads directly to adult size reduction (Gould 1977: 293). Wild wolves attain sexual maturity at about two years of age (A. Murie 1944; Novikov 1962; Pulliainen 1975), whereas modern dogs may breed at six months to a year (Scott & Fuller 1965; Fox 1978; Clutton-Brock 1984). Progenesis results in juvenilized morphology in the evolving organism compared to its ancestor, a phenomenon known as *paedomorphosis* (Gould 1977: 255). Morphologically, dogs are indeed just wolves, altered by simple changes in developmental rates and timing (Wayne 1993a). The most prevalent change can be termed *progenetic heterochrony*, the word "heterochrony" referring specifically to evolutionary changes in

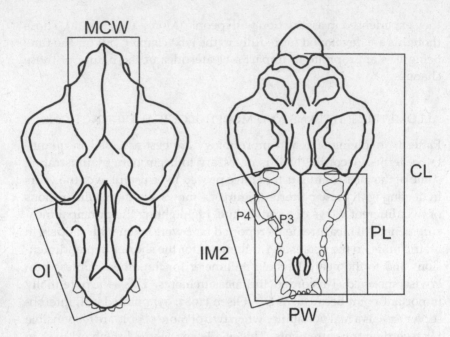

MCW = Maximum Cranial Width
OI = Orbit to Anterior Alveolus of First Incisor
IM2 = Alveolus of First Incisor to Posterior Alveolus
        of Second Molar
PW = Palatal Width
PL = Palatal Length
CL = Condylobasal Length
P4 = Upper Fourth Premolar Length
P3 = Upper Third Premolar Length

FIGURE 3.1. A suite of basic canid cranial measurements that I have employed regularly, beginning in the 1980s. Diagram adapted from Morey (1992:187, Figure 1). Identical or similar measurements are in Von den Driesch (1976) and Haag (1948).

developmental rates and timing (Gould 1977: 2).[1] Thus, unlike other canids dogs are, osteologically speaking, smaller, paedomorphic

---

[1] Not surprisingly, Gould's influential work has prompted a good deal of critical scrutiny. For example, Zelditch et al. (2000) were concerned that the prevalence often attributed to heterochrony as an evolutionary mechanism distracted attention from the possibility that heterotropy, change in the spatial patterning of development, "may be equally common" (Zelditch et al. 2000: 1363). More recently, Vinicius & Lahr (2003) expressed concern that heterochrony, as generally conceived, needed a primary means for establishing homology between two phenomena. In a specific context, they also suggested that a more convincing account of human encephalization than standard

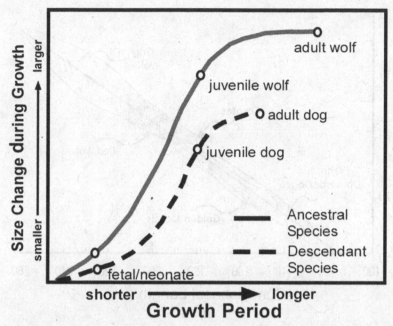

FIGURE 3.2. A schematic depiction of how heterochrony is apparently involved in the changes in adult sizes and proportions that characterize the change from wolf to dog. Adapted from Morey (1994a: 343, Figure 8).

(morphologically juvenilized) wolves. This principle is depicted schematically in Figure 3.2, and in this model "the descendant species grows more slowly very early in life, and finishes growing sooner than the ancestral species" (Morey 1994a: 343). Certainly different kinds of data help to clarify the depicted pattern.

Static Data

Some years ago, Cock (1966: 135–137) outlined a distinction between three kinds of data for allometric studies, one of which is called static. Static data consist of observations on individuals at a single stage of

heterochrony is entailed by sequential hypermorphosis: "delay in offset time of each growth phase while preserving their original allometric coefficients and shape slopes" (Vinicius & Lahr 2003: 2464). These points are well taken in general, and I am ill-equipped to evaluate the human case, but in the present case, homology (ancestor-descendant relationship) between dog and wolf is established independently, and particular changes in developmental timing are demonstrated, or reliably inferred on morphometric grounds. In other words, the case of dogs versus wolves would seem to be one exception to the legitimate concerns raised by these authors.

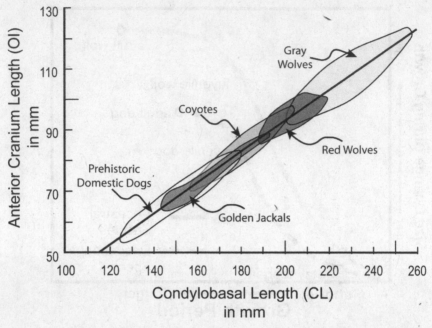

FIGURE 3.3. Range of plotted scores (contours) on two cranial length variables for several groups of modern adult wild canids, and some prehistoric dogs. The regression line that describes this relationship and passes through the groups is depicted as well. Based on logarithmically (base 10) transformed data, associated calculations include a slope of 1.096 and a correlation coefficient (R) of 0.992.

development, usually adults. In beginning to clarify the use of static data, it is useful as a relevant frame of reference to indicate the ways that dogs are allometrically similar to other canids, as gauged by cranial morphology. As an example, Figure 3.3 illustrates the relationship between two cranial measurements, OI and CL, as recorded on several groups of adult canids, including dogs. Table 3.1 summarizes the composition of the samples that make up the adult wild canid series, and over 200 adult wild canid specimens are represented.[2] Appendix A at the end of this volume includes a compilation of all the raw data, metric

[2] Much of the metric data on modern canids reported here are the product of working in 1987, courtesy of a short term visitor's grant, at the Smithsonian Institution, National Museum of Natural History, Washington, D.C. In the same general time frame, I collected data on some of the wolves at the James Ford Bell Museum of Natural History in Minneapolis, Minnesota; the University of Illinois Natural History Museum, Urbana, Illinois; and the Natural History Museum at the University of Kansas, Lawrence, Kansas. I collected data on the Coyotes at the Illinois State Museum, in Springfield,

TABLE 3.1. *Summary of adult wild canid specimens used in quantitative analyses reported in this chapter fn 2*

Species/Subspecies	Geographic Location	No. of Specimens
Canis lupus (Gray Wolf)		**(100)**
C. l. lycaon[4]	Minnesota, Michigan, or Ontario	57
C. l. baylei	Arizona, New Mexico, or northern Mexico	43
Canis rufus (Red Wolf)	Texas	29
Canis latrans (Coyote)	Midwestern United States	62
Canis aureus (Golden Jackal)		**(29)**
C. a. maroccanus	Morocco or Mauritania	6
C. a. algirensis	Morocco	2
C. a. spp.	Morocco	5
C. a. lupaster	Egypt or Libya	6
C. a. indicus	Nepal or India	4
C. a. lanka	Sri Lanka	1
C. a. anthus	Senegal	5
Total		**220**

These data are adapted from Morey (1990: 83, Table 3).

and otherwise, used to generate the original results presented in this chapter. That statement excludes the summary treatment of Morey's (1986) taxonomic study below, based on an earlier data set.[3] Table A.3 contains data on each individual adult wild canid specimen (specimens 1–220). The Gray Wolf, Red Wolf, and Coyote specimens are North American and the jackals are from the Old World. Among the Gray Wolves, two North American subspecies served as a reasonable model

Illinois. All regression statistics based on these data, and data on dogs, were originally computed with the SAS (SAS Institute 1985) General Linear Models Procedure
software on a mainframe computer at the University of Tennessee, Knoxville. Subsequently, some statistics reported here were recomputed based on a slightly expanded
data base. The mainframe computer at the University of Tennessee exclusively generated results of multivariate discriminant function analysis (DFA) summarized here
(Morey 1986), using SPSS (Method = Wilks; Nie et al. 1975), and BMDP7M with
JACKKNIFE (Dixon 1981) for the generation of classification results.

[3] Lengthy Table A.3 is, by necessity, preceded by two briefer tables. Table A.1 summarizes the criteria for assigning a given specimen to an individual age category. That
step is important given that some analytical steps call only on adults, while others
utilize juveniles. Table A.2 is partly of internal importance, as it provides the key for
the numerical codes comprising the different taxa. In doing that, it simultaneously
divulges the different taxa that form the larger data base, inventoried at length in
Table A.3.

for allometric patterns among wolves in general.[4] Table 3.2 summarizes
the composition of the archaeological dog sample, and the raw data
appear in Table A.4 (specimens 268–332).[5] This sample consists of sixty-
five specimens, of which, eighteen are from northern Europe (mostly
Denmark), and the remainder are from North America. With only one
exception, the North American dogs fall within the time range of 8000–
3000 B.P. (the exception is slightly earlier), and the same is true of the
European series. Much of the raw data are in Morey (1990), but with
several measurement estimates on fragmented specimens eliminated
due to concerns about their accurate replication (revised data also in
Morey 1992: 188–189, Table 3). When perusing Table 3.2, it is useful to
bear in mind the following statement: "much of our knowledge about
dogs of the past has been gained from their skeletal remains, as recovered from burial contexts" (Morey 2006: 159). All of the North American
dog specimens summarized in that table are from burials, and the same
is known to be true for at least some of the European specimens, probably most of them. In fact, most of those North American sites play a
conspicuous role in Chapter 7, on dog burials.[6]

---

[4] As noted by Mech & Boitani (2004: 124), Wilson et al. (2003) have suggested, based on
mtDNA from two historically killed wolves in the northeastern United States, thought
to be *Canis lupus lycaon*, that these wolves were not Gray Wolves. They favor a North
American evolution of the Eastern Timber Wolf, originally *C. lycaon*, and the Red Wolf,
C. rufus, independently of the Eurasian Gray Wolf. But in the same year, Nowak (2003:
246), himself an authority on the Red Wolf and wolf taxonomy in general, listed *C.
l. lycaon* as a recognized subspecies of *C. lupus*, the Gray Wolf. That is the standard I
have always used, and I continue to use it here. But specific taxonomic uncertainty in
this case has no real bearing on the basic allometric patterns at issue here.

[5] Data on most of the North American dogs were recorded at the Museum of Anthropology, University of Kentucky, Lexington, Kentucky, in the late 1980s. But some of
the Tennessee specimens are from the Frank H. McClung Museum at the University
of Tennessee, Knoxville, Tennessee, while the two Illinois specimens are from the Illinois State Museum, Springfield, Illinois. Meanwhile, in the same general time frame,
I examined all of the European specimens at the Zoological Museum, University of
Copenhagen, Denmark.

[6] Other North American contexts highlight this reality as well. For example, Diane
Warren's (2000, 2004) studies of paleopathogies evident on prehistoric dog bones in
the southeastern United States (see Chapter 5) drew almost exclusively from skeletons derived from dog burials: "With rare exceptions, the dogs from the Kentucky,
Alabama, and Tennessee sites were intentional burials of what were originally complete, articulated animals" (Warren 2004: 11). Similarly, Handley's (2000) investigation
of different dog types known from archaeological sites in several states of the northeastern United States draws attention to this reality as well. Specifically, among the
more than 60 dogs studied, "All of the dogs represent burials recovered within the
last 112 years of archaeological excavation" (Handley 2000: 206).

TABLE 3.2. *Archaeological sites yielding domestic dog crania used in analysis*[5]

Site[a]	Date (B.P.)	No. of Specimens	Reference(s)
North America			
1. Indian Knoll	8000–3000	14	Webb 1946
2. Carlston Annis	8000–3000	5	Webb 1950a
3. Ward	8000–3000	4	Webb & Haag 1940
4. Chiggerville	8000–3000	1	Webb & Haag 1939
5. Read	8000–3000	1	Webb 1950b
6. Perry	8000–3000	7	Webb & DeJarnette 1942, 1948a
7. Flint River	8000–3000	2	Webb & DeJarnette 1948b
8. Whitesburg Bridge	8000–3000	1	Webb & DeJarnette 1948c
9. Little Bear Creek	8000–3000	2	Webb & DeJarnette 1948d
10. Mulberry Creek	8000–3000	3	Webb & DeJarnette 1942
11. Bailey	8000–3000	1	Bentz 1988
12. Cherry	8000–3000	2	Magennis 1977
13. Eva	8000–3000	2	Lewis & Lewis 1961
14. Modoc	ca. 7000	1	Parmalee 1959
15. Koster	ca. 8500	1	Morey & Wiant 1992
Europe			
Senckenberg	10,000–9000?	1	Mertens 1936; Degerbøl 1961; Benecke 1987
Vedbaek	7300–6500	1	Degerbøl 1946; Aaris-Sørensen 1977
Saltpetermosen	7150–6000	2	Unpublished
Ertebølle	5800–5000	1	Madsen et al. 1900; Andersen & Johansen 1986
Ringkloster	5700–5000?	2	Andersen 1975
Bundsø	4700–4200	7	Degerbøl 1939
Spodsbjerg	ca. 4300	1	Aaris-Sørensen 1985; Nyegaard 1985
Lidsø	4400–4200	3	Hatting 1978
Total		**65**	

[a] For the North American specimens, site numbers 1–5 are in the state of Kentucky, 6–10 are in Alabama, 11–13 are in Tennessee, and 14–15 are in Illinois. For the European specimens, Senckenberg is in northern Germany, and the remainder are in Denmark.

Using CL (the horizontal axis) as a general size standard, the plot on Figure 3.3 indicates that in terms of a particular anterior cranial length dimension (OI), the different canids are allometrically scaled up (or down) versions of each other. As a consequence, the smaller animals do generally have proportionally somewhat shorter faces. While a skull length dimension (CL) cannot be considered an absolute substitute for overall body size, "there is no question that animals with longer skulls (e.g. wolves) have larger bodies than animals with shorter skulls (e.g. dogs or jackals)" (Morey 1992: 187). Additionally, because the data represented in Figure 3.3 are all adults, those data are static.

The earlier studies (Morey 1990: Chapter 5; Morey 1992) included two additional anterior cranial length dimensions beyond OI, specifically PL and IM2 (see Figure 3.1). Despite minor variations, across the adult groups the bivariate relationships tended toward isometry: "In biology, constancy in shape with change in size is termed *isometric growth* (or *isometry*)" (LaBarbera 1989: 98, original emphases). Accordingly, when scaled against CL, these three anterior cranial length dimensions deviated only slightly, though systematically, from approximate proportionality across the size range represented. In all three cases (OI, PL, IM2), the slopes ranged between 0.90 and 1.10 (Morey 1992: Table 5), indicating at most only weak negative or positive allometry, though given a considerable size range, the outcome can be appreciable. Similarly, in a study of cranial allometry among modern canids, Wayne (1986a: 247) found that "All dog breeds are exact allometric dwarfs with respect to measures of skull length." Regarding the data represented in Figure 3.3, it should be noted that the North American Red Wolf, *Canis rufus*, has long posed a taxonomic complication. Nowak (1979: 85–90), based especially on quantitative data in the form of cranial measurements, indicated that the Red Wolf tends to be smaller than the Gray Wolf, a pattern captured by Figure 3.3. At the same time, Nowak (1979: 87) noted as follows: "In nearly all measurements and other features in which *C. rufus* differs from *C. lupus*, the former approached *C. latrans* [the Coyote]." So a principal issue, according to Nowak, was the relationship between the Red Wolf and the Coyote. More recently, based on cranial data again, using DFA, Nowak (2002) has found that the Red Wolf can be regarded as a species distinct from both the Gray Wolf and the coyote.

Red wolf complications aside, it is useful to illustrate the basic differences between a morphologically generalized dog skull and that of a Gray Wolf. Accordingly, Figure 3.4 provides lateral views of a modern Gray Wolf skull, and that of an archaeological dog skull from a historic period (ca. 1650–1750) Native American site in the state of South Dakota. Both specimens are adults, and immediately one can see that the dog is smaller. As is also apparent, the dog has a proportionally shorter snout region. That trait, along with its directly associated steeply rising

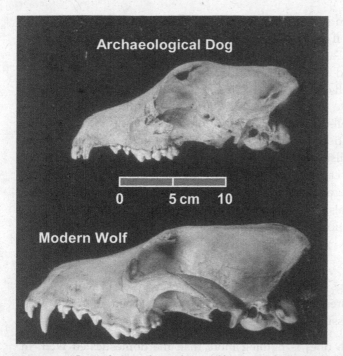

FIGURE 3.4. Lateral view of a modern wolf cranium, lower; University of Tennessee – Knoxville, Department of Anthropology, Zooarchaeology specimen 4590. Corresponding view of an archaeological dog cranium from the Larson Site in South Dakota, dating from about 1650–1750 A.D. A different photograph of these skulls previously appeared in the *Journal of Archaeological Science* (Morey 1986: 135, Figure 6).

forehead region, has long been recognized as the "most doglike" of all characters (Lawrence 1967: 47). For the present (see Chapter 8 for further insights into that trait), it is sufficient to note that it is a component of juvenilized, or paedomorphic, morphology.

Dogs vs. Wolves: Not Always a Straightforward Distinction

At this juncture it is well worth highlighting the difficulties than can arise in efforts to distinguish dogs from wolves. The bones of large, morphologically generalized dogs are so similar to those of wolves that sometimes it is difficult to distinguish them, despite the availability of several useful osteological guides (e.g., Sisson & Grossman 1953: 184–206; Evans 1993). Quantitative approaches, including statistical techniques, are sometimes used to make the distinction, and even this objective can be difficult to attain (e.g., Degerbøl 1927, 1961; S. J. Olsen 1985; Bozell 1988).[7] Morphological similarity throughout the taxonomic Family Canidae is pervasive, and about a century ago Sidney Reynolds was led to recognize "the extreme difficulty in distinguishing between the skull of a wolf and that of a dog" (S. Reynolds 1909: 2).

In the northern plains of North America a real difficulty stems from the practice, well-documented in ethnohistoric literature, of deliberate human selection, including culling of new litters, for large, strong dogs (e.g., G. Wilson 1924: 199). The purpose behind this practice was to promote the capacity of these dogs to be used as draft animals. Coupled with occasional hybridization between wolves and dogs, the result was animals that in many respects could be mistaken for wolves. Historic descriptions of plains dogs of that era draw attention to this similarity:

> The Indian dogs which I saw here so very closely resemble wild wolves, that I feel assured that if I was to meet with one of them in the woods, I should most assuredly kill it as such. (Audubon 1960: 520)

> It [the Arikara dog] is nothing more than the domesticated wolf. In wandering through the prairies, I have often mistaken wolves for Indian dogs. The larger kind has long curly hair, and resembles the shepherd dog. (Brackenridge 1904: 115)

> In shape they [the dogs] differ very little from the wolf, and are equally large and strong. . . . Their voice is not a proper barking, but a howl, like that of the wolf, and they partly descend from wolves, which approach the Indian huts, even in the daytime, and mix with the dogs. (Maximilian 1906: 310)

---

[7] Interestingly, Horsburgh (2008) has recently presented a DNA-based procedure for resolving the taxonomic identity of morphologically ambiguous canid remains from relatively late prehistoric times in South Africa. In that context, the nondog taxon in question is the Black-backed Jackal (*Canis mesolmelas*). After comparing established reference sequences with each other, and then to archaeological canid remains from four sites, she concluded that "These sequences clearly demonstrate that domesticated dogs and black-backed jackals are easily distinguished which is consistent with these species having diverged approximated [sic] 2 million years ago . . . and that all of archaeological canid specimens are derived from black-backed jackals" (Horsburgh 2008: 1477). This is clearly a promising approach, though the genetic near-identity of dogs and wolves leads one to wonder if they, too, can be so easily distinguished, especially since they diverged much later. But I certainly welcome any effort to accomplish such a goal.

From these accounts it is clear that many dogs from the northern North American plains were much like wolves in terms of size and general appearance. The immediate context for dealing with this issue was an effort to identify thirty-three archaeological canid crania from late prehistoric to historic period sites in the Missouri River Valley (Morey 1986), in the states of North and South Dakota. In this case, given that size alone was not an adequate criterion, discriminant function analysis, or DFA, provided a better approach.

This study included three reference groups of dogs: modern Eskimo dogs from Greenland and Alaska, archaeological dogs from St. Lawrence Island, and relatively small Archaic Period dogs from archaeological sites in Kentucky, Tennessee, and Alabama. There were sixty-four specimens total, with all but five specimens' measurements taken from a published source (Haag 1948). This is the situation in which it surfaced only later (Morey 1992: 186) that there was some modest interobserver error between Morey and Haag. Additionally, I measured skulls of two reference groups (subspecies) of North American wolves, thirty-nine in all, and a final reference group consisting of skulls of nineteen Coyotes from scattered locations in the western United States.[8]

In DFA, the statistical results are often adequately portrayed on a two-dimensional plot of the discriminant function scores. The maximum number of functions is one less than the number of groups analyzed, so in this case there were five functions. Classification results are generated purely mathematically and a two-dimensional graphic portrayal facilitates subjective assessment of the results. Although there were five functions, the first two accounted for 90 percent of the variance in the data set, so a two-dimensional plot, by scores on the first two functions, is reasonably adequate. Figure 3.5 represents a version of Figure 3 in the original presentation (Morey 1986: 132). From this illustration it can be appreciated that the archaeological skulls clustered in the vicinity of the large northern dogs, and the wolves. Each individual case was classified by its proximity to the multivariate mean, or centroid, of each reference group. Many cases were clear-cut as far as taxonomic classification, though some involved subjective assessment when associated

---

[8] I made every reasonable effort to follow Haag's (1948) guidelines as precisely as possible. Subsequent familiarity with Von den Driesch's (1976), work was beneficial, but perhaps it can be appreciated that this was an early study, in fact my first ever with dogs. In this study, to maximize consistency with Haag, and as a first effort, thirteen measurements seemed appropriate, including the ones on Figure 3.1, and five others that were not part of later studies (e.g., Morey 1992; Morey & Wiant 1992).

classification probabilities were not robust. But all individual crania were classified as either dogs or wolves, with one exception, closest to a Coyote (Morey 1986: 133).

In that article I then dealt with specimens that classified ambiguously, recognizing, for example, that there was no reason to expect that large, wolf-like dogs from the northern plains would closely resemble the arctic dogs (from Greenland and Alaska) that comprised the most similar dog reference group. Accordingly, the focus then shifted to "The Dog–Wolf Hybridization Problem" (Morey 1986: 133). Given lack of a clear indication as to what a genuine hybrid should look like in a DFA, a then-recent effort to deal with that issue, by Walker & Frison (1982), merited scrutiny. Early in their presentation, those authors made clear the purpose of their article: "it discusses circumstances resulting in a need for a continual re-domestication or taming of the North American aboriginal dogs from early in their North American record throughout the Proto-historic period" (Walker & Frison 1982: 126). They worked with fourteen archaeological canid crania from the state of Wyoming, representing a span of several thousand years, with the majority from late prehistoric sites. Some in this small series had previously been identified as wolves. The pivotal point in their study was a DFA based on several groups of dogs, with data taken from Haag (1948), a sample of twelve modern wolves, and the fourteen archaeological crania. Perhaps the most noteworthy aspect of their procedure is that they pooled the archaeological crania into a separate group for purposes of analysis. In doing so they revealed their desired result, because such a step assumes that the sample forms a taxonomically homogeneous group.

Figure 3.6 represents an effort to generate a version of their Figure 10b (Walker & Frison 1982: 150). Unfortunately, the original was so small in publication that this effort excluded one of their eight groups, though it was not a consequential one for present purposes.[9] It is quite

---

[9] There is no point in claiming to have replicated specific boundaries absolutely correctly. The fact is that in the busier parts of their diagram, the drawing was a confusing cluster of intersecting polygons (group contours) and overlapping geometric symbols, the latter representing individual data points. However, Figure 3.6 does capture the essential relationships. Specifically, the small group of twelve wolves occupies a small polygon on the right side of the graph. Just to the left is the almost-as-small group of fourteen Wyoming canids, for which taxonomic identity was sought. To the left of them are the different groups of reference dogs, the confusing cluster of intersecting polygons. The Wyoming canid group falls between this cluster and the wolf group. This intermediate placement was the primary basis for designating them as hybrids. In fact, recalling that classification results are generated mathematically, two of the

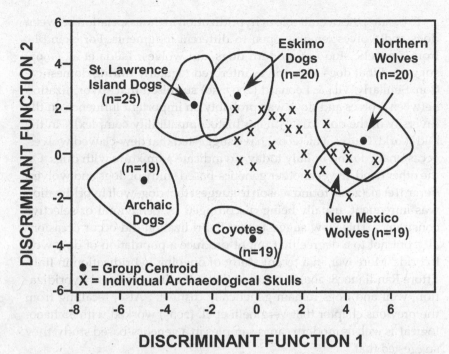

DISCRIMINANT FUNCTION 1

FIGURE 3.5. Plotted ranges of discriminant scores on the first two functions from analysis of thirteen cranial measurements from six reference samples of canids (the contours), ... and twenty-one archaeological canid crania from the North American northern plains, for which all measurements were available. Others were analyzed separately based on the different subsets of measurement available. Adapted from Morey (1986: 132, Figure 3); see text for additional clarification.

noteworthy that two specimens classified as dogs, especially considering the pooling of their small series into a separate group, and treating it as such. Overall, it seems that Walker and Frison had two or three dogs, and several wolves. But they interpreted the entire series as dog–wolf hybrids because, first, they assumed in advance that these represented a homogeneous group. Then they subjectively interpreted the intermediate position of this group on plots as being consistent with that assumption. With little question, it is possible that a given specimen in their study (or mine) could represent a true dog–wolf hybrid, and since there was no known reference group of hybrids in either study, that possibility can't be eliminated. But what seems unrealistic is the inference that there was an entire population of dog–wolf hybrids.

Wyoming canids were classified as dogs, while most of the others fall nearest the wolf group, as does the group centroid.

To be sure, the overall role of hybridization in the association between dogs and wolves remains open to different judgments. For example, from a genetic study of modern dogs and wolves, Tsuda et al. (1997) suggested that dogs and wolves interbred regularly during domestication. Similarly, Vilà & Leonard (2007: 50) surmised that "hybridization between wolves and dogs was probably an important influence on the diversity of the dog MHC [major histocompatibility complex]." In the Old World, Ciucci et al. (2003) have suggested that dew-clawed wolves, occasionally found in Italy today, do indicate admixture with dogs. On the other hand, in yet another genetics-based study of dogs and wolves, Verardi et al. (2006) found reason to suggest that dog–wolf hybridization was infrequent, usually being discouraged by behavioral or selective constraints. This view suggests that hybridization did occur occasionally, but not to a degree that might produce a population of dog–wolf hybrids. Moreover, in a recent study of ongoing hybridization in Italy, Ettore Randi (2008: 290) concluded that "despite occasional hybridization, wolf and dogs remain genetically distinct." Also, recalling from the previous chapter that Verginelli et al. (2005) worked with archaeological as well as modern canid samples in a genetics-based study, they suggested that

> In fact, genetic separation between dogs and wolves is likely to have occurred only after the Neolithic agropastoral revolution (≈ 8,000 YA [years ago]) that resulted in incompatibility between wolves and humans because of the presence of livestock. (Verginelli et al. 2005: 2549)

This suggestion clearly concerns hybridization, those authors believing it to have taken place periodically until about 8,000 years ago, with the inception of human agriculture and livestock husbandry. This position may well have some merit, though the suggested time frame seems applicable only to the Old World. Accurate or not, the relevant human economies emerged only somewhat later in the New World, and animal husbandry was absent in North America. How such a time frame for genetic separation in the New World might apply is not clear. In the New World, it is likely that North American wolves periodically hybridized in the past with local dogs in the High Arctic regions, where agriculture was not a factor (e.g., Clutton-Brock & Kitchener 2000). In fact, dogs and wolves may have hybridized occasionally in numerous settings, though perhaps the wolf populations were affected more than the local dogs (Crockford 2000c: 307–308). So hybridization surely did

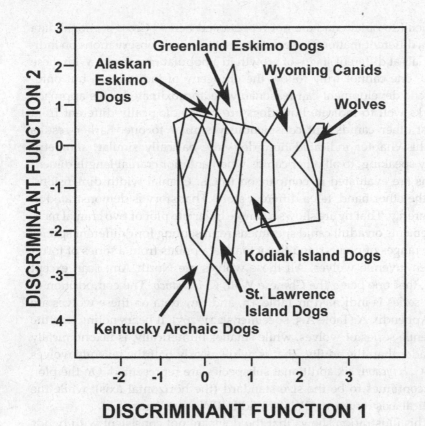

DISCRIMINANT FUNCTION 1

FIGURE 3.6. Approximate ranges of plotted discriminant scores on the first two functions (the polygons) for seven of eight canid groups analyzed by Walker & Frison (1982). Adapted from Walker & Frison (1982: 150, Figure 10b); see text for additional clarification.

sometimes take place, though the implications are apparently not major. But hybridization issues aside, the other kinds of data identified by Cock (1966: 135–137), discussed in more detail next, are relevant to the issues under consideration here.

Longitudinal and Cross-sectional Data

Longitudinal data consist of multiple observations on single individuals at different stages during their individual development, or ontogeny. In principle, they are the best kind of data, but they are also the most difficult to obtain and are not represented here.[10] Cross-sectional data are a different matter. These data consist of single observations on individuals at different stages of growth in a population sample. With these data, one cannot truly follow the ontogeny of individuals, but ontogenetic development can reasonably be approximated. This approach works well to indicate how dogs are morphologically different from most other canids, but conspicuously similar to one. Earlier results in this chapter indicated how dogs are basically similar, allometrically speaking, to all other canids when anterior cranial length dimensions are evaluated in comparison to CL. Cranial width dimensions, on the other hand, tell a different story. That story is demonstrated in Figure 3.7. That figure shows a simple bivariate plot of two cranial measurements on adult canid specimens representing four different species (the ranges of points), and the actual data points from a series of forty-seven juvenile wolves. All those wolves are North American except one, that one being the Chinese Wolf, *C. l. chanco*. The composition of this series is indicated in Table 3.3, and raw data on these wolves are in Appendix A (Table A.3, specimens 221–267). It bears noting that the juvenile series of wolves, while smaller numerically, is taxonomically broader than the adults. That is, while nearly half the juvenile wolves are *C. l. lycaon*, six additional subspecies are represented. On the plot, CL continues to be the size standard (the horizontal axis), while the vertical axis represents Palatal Width (PW – see Figure 3.1).

This illustration shows that the dogs are not consistent with trends shown by the other adult canids, but have broader palates at generally comparable overall sizes. The same pattern characterizes comparisons involving an unillustrated cranial breadth dimension, MCW. The statistical regression results for these comparisons are in Table 3.4. In Figure 3.7, it is noteworthy that the ontogenetic regression line for the series of juvenile wolves passes through only the range of variation shown by the dogs, falling above the other groups. Given that pattern,

---

[10] In comparative studies of the development of limb bone conformation in modern canids, Wayne (1986b, 1986c) was able to gather longitudinal data on growing dogs of several different breeds by means of radiographs taken at periodic intervals as they grew. In those studies, Wayne made a well-founded case that a relatively invariant length of the gestation period in dogs, some 60–63 days, strongly conditions their limb proportions as adults, with differences in limb bone lengths established quite early in their ontogeny, 0–40 days postpartum, or prior to birth. Specific growth rates after that juncture are quite similar regardless of final adult size. Clearly, longitudinal data were of great importance in these investigations, though such an approach is obviously not possible in archaeological studies.

TABLE 3.3. *Summary of subadult (age categories 1, 2, and 3) Gray Wolf specimens used in quantitative analyses reported in this chapter* [2]

Subspecies	Geographic Location	No. of Specimens
Canis lupus lycaon[4]	Minnesota, Michigan, or Ontario	22
C. l. baylei	southern New Mexico, Arizona, or northern Mexico	4
C. l. nubilus	Colorado, southwestern Minnesota	6
C. l. irremotus	Wyoming	4
C. l. youngi	New Mexico	7
C. l. arctos	North American Arctic	3
C. l. chanco	northern Pakistan	1
Total		47

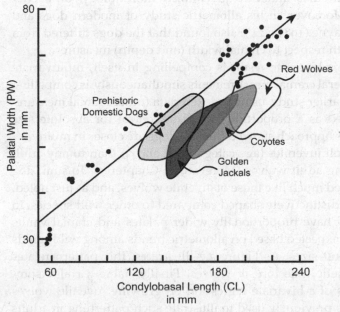

FIGURE 3.7. Bivariate plot of PW by CL for juvenile Gray Wolves (age categories 1–3), with corresponding regression line. Several essentially redundant observations on the juvenile wolves are masked in this plot. Ranges of plotted scores are indicated (contours) for four adult groups as well. Comparing separate group regressions, both the dogs and the adult Gray Wolves (not shown) are, statistically speaking, strongly transposed above the other groups. In addition, the dogs are not significantly different from the adult Gray Wolves in either slope or elevation (P > 0.25 for both tests; see Morey 1992: 191, 194).

TABLE 3.4. *Least squares linear regression statistics from bivariate comparisons between two skull dimensions, CL (independent, or predictor, variable) and different dependent, or predicted, variables, of forty-seven subadult modern Gray Wolves. The calculations are based on values that were logarithmically (base 10) transformed*

Dependent Variable	Slope[a]	Standard Error of Estimate	Y-Intercept	R (correlation) coefficient)
PW	0.603	0.025	0.459	0.980
MCW	0.521	0.021	0.637	0.981
OI	1.173	0.010	−0.729	0.999

[a] The slopes are significantly different from zero ($P < 0.01$).

it is reasonable to infer that dogs show what is called ontogenetic scaling with respect to their ancestral species (see Shea 1985; Shea et al. 1990), a pattern that underlies juvenilized morphology, or paedomorphosis.[11] Moreover, in his allometric study of modern dogs and wild canids, Wayne (1986a: 245) also found that the dogs differed from wild canids with respect to cranial width (and depth) measures.

Although this bivariate result is compelling in itself, multivariate DFA, using several cranial measurements simultaneously, is compelling as well. In an earlier study, using indexed data (other cranial measurements expressed as a proportion of CL) to control for absolute size differences, this approach showed that the dogs are closer in multivariate space to wolf juveniles (age categories 2 and 3) than to any adult group, including adult wolves (Morey 1990: Chapter 7). To sum, dog skulls are shaped much like those of juvenile wolves, and as just noted, adult dogs are distinctively shaped compared to other wild species. In particular, they have proportionally wider palates and cranial vaults than would be expected based on allometric trends among wild canids of different adult sizes. As Figure 3.7 illustrates, this pattern relates to the ontogenetic trajectory of wolves. Finally, Table 3.4 also summarizes results of a bivariate regression on growing juvenile wolves based on OI, as previously used to illustrate such patterning in adults

[11] In comparative studies of the cranial morphology of domestic dogs versus wild canids, both Wayne (1986a), dealing only with modern canids, and Morey (1992), dealing with archaeologically represented dogs and modern wild canids, addressed the issue of ontogenetic scaling in the sense of a pervasive parallel between development and the range of adult morphology in dogs. While that pattern is certainly informative, "the essence of size-related changes is a linkage between ancestral ontogeny and descendent adult morphology" (Morey 1992: 199). Though not dealt with meaningfully some years ago, that issue is taken up in the present chapter.

(Figure 3.3). As anticipated, it scales against CL similarly to how it does in nongrowing adult canids, including dogs, though an elevated slope (1.173) reflects the proportional lengthening of the snout region as the animal grows.[12]

In evaluating the different patterns that are summarized here, it is useful to keep in mind Shea's (1981: 180–181) distinction between what he calls size-required versus size-related changes. Size-required changes simply obey biomechanical laws of size-required shape alteration. As a convenient example, limb bone proportions of animals such as elephants and mice must be different, to accommodate the radically different structural support requirements of the two organisms. Nature abounds with such examples. On the other hand, size-related changes involve "interspecific shape differences that mirror those between young and adults of the larger species" (Shea 1981: 181). In the present case, the distinction really concerns different levels of shared developmental programming. The point is that size-related changes "specify a developmental relationship between ancestral and descendent morphology that may be distinguishable from morphological patterning at a broader taxonomic level" (Morey 1992: 184). That appears to be the situation here, and indeed, one can refer to this pattern as ontogenetic scaling (Morey 1992: 184). The morphological consequence in this case is juvenilized, or paedomorphic, morphology in dogs. Moreover, the significance of juvenilized morphology in dogs is underscored when one considers that during the evolution of canids in general, in the absence of juvenilized morphology "Hypercarnivorous forms tended to acquire a shorter rostrum, paralleling the condition in cats" (Wang & Tedford 2008: 80). But dogs are omnivorous, and the

---

[12] Since growth certainly entails size change, some fundamental similarities between static and ontogenetic allometries are no surprise. At the same time, though, the effects of modern breeding practices can confound basic allometric tendencies. A compelling example is provided by Drake & Klingenberg (2008), who monitored historical patterns of cranial shape change in St. Bernard dogs over the preceding 120 years. The international breed standard specifies, among other things, that "The desired head-shape is imposing, massive and wide, with cheek bones (zygomatic arches) that are strongly developed and high" (Drake & Klingenberg 2008: 72). Those characteristics have been accentuated over time. What the authors point out is that, in modern St. Bernards, allometric shape changes do not account for their morphology. In that light, it is worth emphasizing that these dogs are the product of intensive human selection for animals that conform morphologically to an international breed standard. There is little question that prehistoric dogs were not subject to a comparable regime, and it really is no accident that their morphology, as opposed to that of at least one modern breed, is understandable in terms of basic allometric principles.

mechanism underlying their shorter snout is related to developmental timing rather than directly to diet (see Chapter 4).

Unfortunately, while the juvenilized appearance of dogs has been widely recognized, its evolutionary significance has sometimes not been fully appreciated. For example, under an assumption that domestic animals change mostly as a reflection of human preferences, Clutton-Brock (1984: 205) wrote as follows: "Dogs that remained dependent on their substitute parent, the human owner, and had the large eyes and appealing form of the puppy would be favoured." Consequently, according to her scenario, people systematically favored the perpetuation of these heritable traits, and the result was the evolution of what we know as the dog. This idea is at least plausible, but there is no way at all to determine what traits people personally preferred, and the juvenilized appearance of dogs is efficiently accounted for by established evolutionary processes, as just covered.

MORPHOLOGY, GENETICS, AND DOMESTICATION TIMING

A useful frame of reference for what follows here are some inferences about the timing of dog domestication that are based on genetic patterns in modern animals. Specifically, Vilà et al. (1997) suggested that dog domestication happened at least 100,000 years ago, probably even earlier than that. They made some problematical inferences about the rate that molecular substitutions occurred in some canid lineages, and translated those rates into a "molecular clock" (Mayr 1988: 309). It may be useful to refer back to the latter part of Chapter 2, especially the point that the Bonn-Oberkassel dog, about 14,000 years old, predated agricultural economy. This fact is relevant because Vilà et al. (1997: 1689) suggested that the presence of dogs that could be detected morphologically happened only with a change in selective regime that accompanied the emergence of agricultural economies, which happened some time later than 14,000 years ago. Moreover, the Bonn-Oberkassel dog is the earliest securely identified dog in the world, as of this writing.

Despite the compelling Bonn-Oberkassel case, Vilà and colleagues have continued over the years to advance a case for a greater antiquity of dog domestication than the archaeological record will support (e.g. Vilà et al. 1999; Wayne & Vilà 2001; Wayne et al. 2006: 282). One of the points worth emphasizing here is that dogs are not the only animals for which molecular clock-based estimates of domestication antiquity are turning out to be inaccurate. As a general rule, such studies tend to yield estimates that are far older than the archaeological evidence indicates. For example, a genetics-based estimate on domestic pigs suggested that the ancestral form of this animal diverged some 500,000 years ago (Giuffra et al. 2000), but pig domestication did not happen until about 9,000 years ago. Another conspicuous example is cattle, for which Michael Bruford and colleagues (2003) noted that a molecular clock-based estimate suggested a divergence of the relevant forms several hundred thousand years ago, but "cattle domestication was known to have occurred much later than this, within the last 10,000 years" (Bruford et al. 2003: 904).

In fact, there is a growing awareness among biologists that molecular time-of-divergence estimates tend to substantially overestimate demonstrably recent processes (Ho et al. 2005). Along with that growing awareness are questions about whether molecular clocks can really be used at all, coupled with a need to "mend the unnatural schism that has kept morphological and molecular systematics apart" (Schwartz & Maresca 2006: 369). The two need to be integrated, and the case of dogs can contribute to that goal. And in some quarters, archaeologists are collaborating productively with geneticists. For example, an archaeologist and a zoologist, K. Dobney and G. Larson (2006: 266–267), noted that

> It is tempting to presume that very early dates derived from molecular data suggest that the archaeological dates are significant underestimates and that domestication has been taking place for tens of thousands of years. The more likely explanation is that molecular clocks often do not have the resolution to date very recent events (such as domestication) and that the error bars surrounding the molecular estimates actually envelope the archaeological dates, resulting in no discrepancy between the two. The lack of a statistically significant difference between the two estimates therefore removes the necessity to invent novel and far-fetched implausible justifications for the perceived inconsistency (e.g. Raisor 2005).

Those authors favor the archaeological inferences, and would like to see the two sources of information brought more in line with each other, a development that would be optimal.

In any case, in considering the genetic situation with dogs, it seems important to keep two questions distinct. One question is when the dog and wolf genomes became permanently separated, except for occasional hybridization under idiosyncratic circumstances. The other is when the domestic relationship actually began. Regarding dogs, "It is entirely possible that the genomes became separated for reasons having nothing to do with domestication" (Morey 2006: 166–167). In commenting on that very piece of Morey's, Carles Vilà agreed. Specifically, in the periodical *Science* (Volume 311, 3 February 2006, page 587), he was quoted as saying, "genetic divergence is not the same as domestication." In keeping with this point, if the dog and wolf genomes really did separate as long ago as some of the genetic studies have suggested, or even in that vicinity, "the animals that were destined to become dogs must have made their living for some time essentially in the old-fashioned way, like wolves" (Morey 2006: 166).

Hopefully the genetics-based researchers will increasingly take the archaeological evidence seriously. Too often in the past it seems that archaeologists are simply expected to find what the genetics-based researchers say should be there, and if they don't, they just haven't found it yet, or the genetics-based researchers propose an invalid reason, as did Vilà et al. (1997). The molecular clocks clearly have a theoretical basis, but they need to be calibrated more accurately. It would be genuinely beneficial for genetics-based researchers to incorporate evidence from the archaeological record, the record of what actually happened in the past, into their calibrations. Any reasonable archaeologist should take their suggestions of what should be there seriously, and be willing to acknowledge it if it can be found. The goal should be to bring these two sources of information into better accord.

A useful development as far as accomplishing that objective can be gleaned realistically from more recent genetics-based work. One piece of work suggests a date of domestication for dogs within the span of about 15,000–40,000 years ago (Savolainen et al. 2002). More recently, in commenting on the ambiguity associated with many genetics-based estimates, Savolainen (2007: 34) has suggested that

> Furthermore, in the absence of a clear result from the mtDNA data, the best evidence for the time of the first origin of the domestic dog remains the archaeological record, which indicates an origin approximately 15,000 years ago.

Similarly, Vilà & Leonard (2007: 44) suggest that "integrating both approaches will lead us towards a more accurate view of the domestication process."

But not so long ago, Wayne & Vilà (2003: 225) suggested that "considerable debate still surrounds the date of the domestication of dogs from wolves." That debate, however, seems to characterize the situation "mostly among those who are unduly influenced by molecular genetic data" (Morey 2006: 167). Wayne and colleagues are still puzzled by the lack of congruence between the earlier genetic inferences and the archaeological evidence (Wayne et al. 2006), as are Zeder and colleagues (Zeder et al. 2006: 9–10). In fact, as Wayne and colleagues put the matter: "these molecular estimates imply an ancient origin of domestic dogs from wolves . . . much older than suggested by the archaeological record" (Wayne et al. 2006: 282). They also hold that the documented morphological shift, some 14,000 to 15,000 years ago, resulted from a change to a more sedentary life among people (Wayne et al. 2006: 282–283). But increasingly sedentary life seems relevant to the very existence of dogs, not to a shift within dogs (see Chapter 4).

One factor that should not be lost sight of is that, genetically speaking, dogs and wolves can now be regarded as the same species. Thus the question becomes whether one can even talk meaningfully of a genetic split between them. Quite simply, for some years now, many genetics-based researchers seem to have missed the most important point when it comes to identifying the essence of dog domestication accurately. Genetic patterns are certainly one piece of the puzzle, but they are only one piece. Far more consequential biological changes have occurred during dog domestication that are only indirectly related to genetic patterns. Those changes are covered later in this volume, but it is appropriate now to shift attention to an important line of contextual evidence that bears directly on an understanding of the timing of dog domestication, and the fundamental nature of this particular domestic relationship.

DOG BURIALS AND DOMESTICATION TIMING

In beginning this last section it is relevant to point out that Carles Vilà, in the same 2006 column in *Science* in which he agreed that genetic divergence did not equal domestication, suspected that dogs were tamed well before they started being ceremonially buried. With that point in mind, it is worth emphasizing again that skeletal morphology was the initial primary basis for inferring that the 14,000-year-old canid from a burial at Bonn-Oberkassel was in fact a dog. Secure evidence that would support Vilà's position of great antiquity of dog domestication does not exist. To be sure, the first recognized dog would be represented only some time after the domestic relationship was underway. However, according to Vilà & Leonard (2007: 44), in referring to morphological evidence, how much later is unclear.

Regarding contextual evidence for the domestic relationship, one can refer to Figure 3.8, a generalized map of the world showing the locations of some well-documented dog burials, as recovered from specific sites or known to be associated with particular complexes.[13] This is by no means a comprehensive depiction, as there are many more known than are illustrated there. The main value of Figure 3.8 is to highlight the worldwide distribution of burials, a major point emphasized previously (Morey 2006). But for the present, the issue is the bearing of this phenomenon on an understanding of domestication timing.

As covered, the Bonn-Oberkassel dog is about 14,000 years old, and this fact makes it, as of this writing, the oldest securely documented dog in the world. So by definition, it is also the oldest known dog burial in the world. And those two facts together go hand in hand. Norbert Benecke once put the matter this way: "In Europe, the practice of dog burials started right at the beginning of canid domestication at the transition from the Pleistocene to the Holocene. The earliest evidence comes from Bonn-Oberkassel (c. 12000 cal BC)" (Benecke & Hanik 2002: 20). Given that Bonn-Oberkassel represents the earliest securely documented dog in the world, Benecke and Hanik's statement applies not just to Europe, but the world in general. Put simply, the burial of dogs directly reflects the domestic relationship, while genetic patterns among modern dogs do not.

Another point that should be stressed is that, by definition again, all other known dog burials are less than 14,000 years old. From that fact, it can be appreciated that the archaeological evidence is directly at odds with many recent genetics-based inferences. As noted, a welcome exception is the work by Savolainen (2007 and Savolainen et al. 2002), which suggests a date within the span of about 15,000 to 40,000 years ago, most likely ca. 15,000 years ago. There appears to be a basis here for calibrating the molecular clocks to genuine evidence of what actually took place. Given that the Bonn-Oberkassel dog is about 14,000 years old, and that domestication had to have started slightly earlier

---

[13] Unfortunately, an earlier version of this map (Morey 2006: 162) depicted Ust'-Belaia in Siberia well south and west of its actual location. As the present figure indicates, though, the site is north and east of Ushki-1. Like Ushki-1, "The dog was also present, as shown by a ritual burial" (Chard 1974: 58). An important point worth noting, given Ust'-Belaia's location, is that it is younger than Ushki-1, some 9,000 years old, though as J. W. Olsen (1985: 66) notes, the chronology there is problematical. Accordingly its relevance to the initial movement of dogs into the New World is ambiguous, especially when one considers Danger Cave in Utah, dealt with in Chapter 5 here.

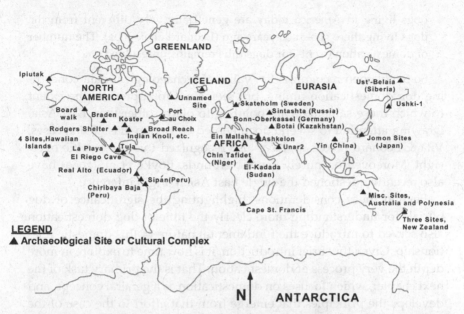

FIGURE 3.8. Generalized map of the world, showing locations of some known dog burials, highlighting their geographic extent. Only some that could be efficiently labeled are indicated. There are many others, introduced in Chapter 7, and inventoried in Table B.1. Adapted and expanded from Morey (2006: 162, Figure 1).

than that, 15,000 years ago is a reasonable estimate. In addition, harking back to the previous chapter, it is worth pointing out again that the Bonn-Oberkassel dog is surely derived not from one of the smaller, more southerly wolf subspecies, but from the basic Gray Wolf, *Canis lupus lupus*. From Sablin & Khlopachev's (2002: 797) work, it can be appreciated that the Eliseevichi skulls are also likely derived from *C. l. lupus*, if not being *C. l. lupus*. Possibly those skulls were curated and retained because they were of symbolic significance. To be sure, some dogs may have been derived from the smaller wolf subspecies, but not the earliest ones that are reported. This pattern highlights the probability that different wolf subspecies gave rise to dogs in different areas of the world. In fact, in commenting on Savolainen's genetically based position that there was one center of dog domestication, in East Asia, Vilà & Leonard (2007: 42) have suggested the following:

However, it is not clear that modern dogs can be used to infer patterns of diversity in the past. A study of ancient dog remains has shown that dogs living in America today are genetically very different from the dogs living there 500–1000 years ago (Leonard et al. 2002). The number of domestication events for dogs thus remains unresolved.

So Savolainen recognizes the value of archaeological data concerning dog domestication timing, but seems not to appreciate the extent to which those same data are pointing to places other than East Asia. Drawing attention to several documented early non–East Asian cases, Vilà & Leonard (2007: 46) have also recognized this geographic oversight. Moreover, on purely genetic grounds, Boyko et al. (2009) have also recently questioned the single East Asian origin inference.

Beyond those considerations, highlighting the significance of dog burials for understanding more clearly the time of dog domestication also served to introduce the fundamental nature of this domestic relationship. Given that brief introduction, it is now time to explore in more depth the very process of domestication. That is the primary task of the next chapter, which focuses on domestication as a general concept, and develops the principles that emerge from that effort to the case of the domestic dog.

DOMESTICATION OF DOGS
AND OTHER ORGANISMS

HISTORICAL PERSPECTIVES

In beginning a chapter that deals with the very process of domestication, it is important to establish its initial pretext. Specifically, for tactical purposes plants have a place in this treatment, simply because treatments of the domestication of both plants and animals are often bound, conceptually, by a common thread. That statement reflects my own primary background as an archaeologist, subjected to a genuine social science tradition of scholarship, in North American anthropology. Numerous passages from the literature clearly illustrate that conceptual framework. Because this is a volume about a particular domestic animal, it is useful to begin with several definitions, or characterizations, of animal domestication that have been offered over the years:

> I would define the essence of domestication as: the capture and taming by man of animals of a species with particular behavioural characteristics, their removal from their natural living area and breeding community, and their maintenance under controlled breeding conditions for profit. (Bökönyi 1969: 219)

> It seems reasonable to accept the fact that the events leading from animals that were wild to those that were finally domesticated would follow the process of capture, taming, and controlled breeding (but not necessarily conducted as a well-organized procedure). (S. J. Olsen 1979: 175)

> Domestication consists of isolating animals belonging to a species having particular characteristics (certain species will not be domesticated) from their free communities and raising them under the control and to the benefit of man. It is a result that the animals acquire the domestic

characteristics. (Gautier 1992a: 150; originally in French, English translation by the author; see also Gautier 1990, 1992b)[1]

Those were somewhat older definitions, but that this approach remains current in some quarters is indicated by the following passage:

> Of all the behavioral traits that that make certain animal species more attractive candidates for domestication than others (. . .), it is the possession of a more placid, tractable, less wary nature that is the single most important factor making certain individuals within these *target species* potentially more suitable domesticates. (Zeder 2006: 171, emphasis added)

In all these scenarios, different animals are targeted for domestication, and that state is brought about by human behaviors that are explicitly invoked, or strongly implied. The thread that these accounts hold in common is, of course, that domestication is a condition imposed on animals for the benefit of people. Domestication in this view is a human invention, a strategy devised by people to better their lives. At one level, such decidedly anthropocentric scenarios are perfectly understandable. The importance to our own lives of domesticates ensures that it is natural to regard domestication in such terms. One merely needs to contemplate the implications for humanity if our modern livestock or our bountiful corn crops were suddenly erased from human existence, to realize that they are presently not only beneficial, but also necessary. What gets lost in this approach, however, seems substantial, and is at least threefold. In the first place, it presumes that domestication is a purely human phenomenon. But it is not, as subsequent portions of this chapter clarify. Second, it loses sight of the fact that, from an evolutionary standpoint, the animals or plants benefit as much as we do. Finally, agriculture and animal husbandry weren't nearly as beneficial to people in the long run, from a certain vantage point, as we often presume. But before dealing more with those points, it seems useful to make a distinction between the processes often believed to have brought about domestic animals, and what characteristics are held to typify the animals, once domesticated.

---

[1] The original passage in French is as follows: "La domestication consiste à isoler des animaux appartenant à des espèces ayant des caractères particuliers (certaines espèces ne se laissent pas domestiquer) de leurs communautés en liberté et de les élever sous contrôle et au bénéfice de l'homme. Il en résulte que ces animaux acquièrent des caractères domestiques" (Gautier 1992a: 150).

Inferred Characteristics of Domestic Animals

Below are three definitions concerning what are held to be the characteristics of domestic animals.

> Domestication can be said to exist when living animals are integrated as objects into the socioeconomic organization of the human group, in the sense that, while living, those animals are objects for ownership, inheritance, exchange, trade, etc., as are the other objects (or persons) with which human groups have something to do (Ducos 1978: 54).

> A domestic animal is one that has been bred in captivity for purposes of economic profit to a human community that maintains complete mastery over its breeding, organization of territory, and food supply (Clutton-Brock 1981: 21).

> By a domesticate, I mean a species bred in captivity and thereby modified from its wild ancestors in ways making it more useful to humans who control its reproduction and (in the case of animals) its food supply (Diamond 2002: 700).

Again, it is about people, and in the case of people these definitions are perfectly reasonable. But at least one of the above writers, Jared Diamond, knows full well that cases of domestication are not restricted to people, a point highlighted shortly.

In any case, the traditional view highlighted above is characteristic of Anthropology, but it goes well beyond that particular academic circle. In fact, this traditional view of domestication is so prevalent that it is formally codified in the meaning that is commonly applied to the term. Consider one of the primary definitions of "domesticate" (the verb) offered by a standard dictionary: "to adapt (an animal or plant) to life in intimate association with and to the advantage of man" (G. & C. Merriam Co. 1974: 339). The more recent *Oxford American College Dictionary* (Putnam's Sons 2002: 400) deals with the same word comparably: "tame (an animal) and keep it as a pet or for farm produce." Plants, similarly, are cultivated for food. Biologist Susan Crockford has made a noteworthy suggestion on this basic topic: "the traditional definition of domestication we have been taught to accept as fact is no more than a deeply entrenched myth" (Crockford 2006: 40). Crockford was dealing with domestication as a general phenomenon in that statement, but she did deal especially with dogs in her work, and that aspect of her work warrants coverage at a later point in this chapter.

A Mechanistic Framework

For now, though, no compact definition of domestication appears here, but before identifying the view taken, it is useful to clarify the three-fold basis for my objections to the traditional viewpoint. First, perhaps the most fruitful way to view domestication is from a mechanistic framework. Such a framework emphasizes the fundamental nature of a domestic relationship, regardless of what particular organisms are involved. And it is useful to address that point by calling once again on the very same first dictionary that was just disapprovingly quoted for the definition of "domesticate." Here, however, the word is "domestic," as in the adjective form of the word: "of or relating to the household or the family" (G. & C. Merriam Co., 1974: 338). Similarly, the more recent Oxford dictionary defines it as "of or relating to the running of a home or to family relations" (Putnam's Sons 2002: 400). Obviously, these definitions are not about plants.

These definitions also presume that it is a human phenomenon, reflected by the word "family." Here, however, it is useful to regard the idea of household or family in more general terms. Genetically related kin may well comprise the group of individuals in question, but they need only be individuals living in close association with each other in a fashion that can genuinely be considered symbiotic to constitute family (e.g., Cohen 2002: 624). That is, both benefit in an evolutionary sense. And there are some well documented examples that do not involve people at all. The objective, of course, is to get back to people, and their associated dogs, but initially it is instructive to set the stage by covering some cases that involve animals (other than humans) and plants, or plant-like organisms, in fully agricultural relationships. Following that, attention then shifts to a conspicuous example that involves two different animals, but not people. This treatment begins with the former, animals and plants.

To do that, I re-create a version of one of my standard teaching ploys in an introductory college-level archaeology class that I have taught for a good many years. On the day when the designated topic is the origins of agriculture, the students hear that I want to describe for them the life habits and behaviors of animal X. They hear, in advance, that animal X does not represent people. First, this account goes, it lives in large groups and its members get most of their food by feeding on a particular kind of plant. Moreover, they don't just eat the plants, they grow them as well. They also construct special areas where they prepare beds for growing this plant. Not only that, they even fertilize these beds with processed plant debris and animal excrement. I continue by relating that they meticulously monitor the environmental conditions. They open and close special ventilation passages to regulate temperature and humidity. They monitor growth of their plants constantly, remove alien growth, and add chemicals that promote growth of the plants. The plants cared for this way regularly yield an edible growth structure, which is what animal X feeds on primarily.

At that point, I pose the sixty-four thousand dollar question: What am I talking about? Surely, they hear, this must be an intelligent, thinking, rational, planning animal that consciously anticipates the outcomes of its behaviors. Over many years of teaching, exactly one student has supplied the correct answer, so usually I answer the question myself: In fact, the answer is ants, and they're engaged in complex agriculture. As for the student who came up with the correct answer, she was an advanced student, bound for medical school the following year, and so had the relevant background in biology.

The plant in question is actually fungus, a plantlike organism. This complex symbiosis legitimately qualifies as domestication in the eyes of a major authority, Jared Diamond, whose work in part underlies the class content, and it is fitting to repeat here some of his own published words: "Molecular evidence indicates at least five clades of ant fungi, attesting to at least five independent *domestication* events, rather than the single event previously thought possible" (Diamond 1998: 1974, emphasis added). A companion piece to this 1998 paper, in the same issue of *Science*, by Mueller et al. (1998) is entitled "The Evolution of Agriculture in Ants." In that article, they also note that "Domestication of novel cultivars and switching between cultivars may be ancient themes in attine agriculture, which presumably began with an ancestral ant that facultatively associated with fungi" (Mueller et al. 1998: 2035).[2] Moreover, "Ant agriculture is highly specialized." (Diamond 1998: 1974), as the teaching ploy highlights, and ants constitute one of several varieties of fungus-farming insects (Mueller & Gerardo 2002).

---

[2] Interestingly, the documentation of modifications to the fungi has proven elusive: "modifications of the ant fungal crops are (surprisingly at least to me) not obvious, even though the ants themselves have evolved obvious modifications" (Diamond 1998: 1975). But Diamond goes on to indicate the likelihood of "chemical modifications of the fungi to match olfactory and gustatory capabilities of ants" (Diamond 1998: 1975).

In teaching, I then move to an animal-to-animal example and do so here as well. This is another kind of ant that lives in groups, whose members get most of their food from a different kind of insect called an aphid. But they don't eat the aphids. Aphids, the students hear, extract and feed on sap from several kinds of plants. They then produce as a byproduct, or waste, a sugar-rich liquid that passes on through their digestive tract. Ants feed on this liquid, sometimes called honeydew, which they entice the aphids to release by stroking them with their antennae. Aphids have evolved an organ for holding the honeydew droplet in place, apparently a specialization that facilitates the feeding behavior of the ants. These kinds of aphids have lost a variety of defensive structures that their close relatives have, because the ants protect and defend them. Ants literally herd the aphids around, sometimes keeping them temporarily in protected chambers in a nest. The aphids, the students hear, have been likened in print to little more than domestic cattle. Specifically, in his classic monograph, from which this account largely derives, insect specialist E. O. Wilson (1975: 356) wrote that "[certain] aphids have evolved to the status of little more than domestic cattle." In some cases, ants keep aphid eggs protected during the winter, and then take out newly hatched aphid nymphs to the roots of nearby food plants. They take them to the appropriate part of the plant, and they do it at the right time in the aphid's life cycle.

Does it seem meaningful, I ask the students, to try to explain these complex examples of agriculture and animal husbandry as inventions devised by one or the other organism, accomplished by means of one imposing conditions on the other? Did one domesticate the other? Which in each pair? Did aphids figure out a strategy for getting protection and reproductive assistance from ants, or did ants figure out a way of getting nutritious food from aphids? I suggest that if we're going to use the term "domestication" to describe what has happened to these organisms, surely we can only say that they domesticated each other. But even that sounds awkward because outside the realm of humans we automatically think in terms of evolution by natural selection, not in terms of different organisms literally accomplishing evolutionary ends for themselves. The basic point for the students is that talking about domestication as something that one organism actively does to another over evolutionary time only matches our intuition when we are considering ourselves. And that is why it's so hard to look at our relationship with corn plants through the same lens that we use to look at the relationship between ants and their fungus crops.

Artificial Selection

One symptom of the difficulty of viewing our own domestic relationships through the same lens that we use when viewing other organisms' domestic relationships is the widespread practice of often drawing a distinction between artificial selection, practiced by humans toward domesticates, and natural selection, characteristic of natural ecological relationships (e.g., Bökönyi 1969; Gautier 1992b). As Clutton-Brock has said,

> In the domestic animal or plant change occurs as a result of artificial selection by humans rather than by nature, and similarly reproductive isolation is maintained by the activities of man rather than by geographical barriers. (Clutton-Brock 1981: 10)

More recently, Raisor has offered a comparable view:

> Domestication differs from evolution in that humans, not nature, create different strains of a plant or animal through the careful selection of desirable traits. These traits are further continued by the reproduction of those animals or plants (Raisor 2005: 45).

When considering those passages, one can wonder just what is not natural about people. Surely the term "human selection" would make the same point, but without misleadingly setting people above nature. The first passage is from a prominent source, by a well-known scholar, and the tendency to regard selection that is under human control as artificial is deeply ingrained. Consider, for example, that in a recent edition of a prominent introductory archaeology textbook, the author writes of domestication as "manipulation of the reproduction of economically important plants or animals through **artificial selection** – that is, the directed breeding of plants and animals possessing characteristics deemed beneficial to human beings" (Feder 2007: 348, original bold emphasis). That passage occurs in a section entitled "Humans Taking the Place of Nature: Artificial Selection" (Feder 2007: 347). To be sure, humans are surely unique in many respects (Bingham 1999), but their participation in domestic ecological relationships is not one of them. Charles Darwin (1962: 114) himself used the term "artificial selection" when characterizing cases of domestication that involve people, and this fact may be regarded by some people as a warrant to invoke the concept in modern times. It should be remembered, though, that Darwin was writing some 150 years ago, much has been learned since then, and it is increasingly recognized that people are an integral part of nature (e.g., Pierotti & Wildcat 2000).

Moreover, the term "artificial selection," a personal pet peeve of mine, gets used sometimes in the case of dogs. As Coppinger & Coppinger (2001: 24) stated,

> Another problem in understanding domestic dogs is the tendency of researchers to see domestic animals differently from wild forms. They look at domestic species as if they had gone through some invalid kind of evolutionary process. Perhaps they feel that the dog's adaptation and evolution were not "natural," and so are inferior.

In a generally useful collection of pieces on the dog, at least three different contributors use the term 'artificial selection' (Clutton-Brock 1995: 15; Lockwood 1995: 132; Macdonald & Carr 1995: 214). Clearly, as for example when one is referring to the deliberate enhancement of aggressive behaviors in certain lines of dogs (e.g., Lockwood 1995), artificial is used in the sense of human selection. At face value, it does little harm in that context, but in other cases humanly directed change is often presumed to be operative, when that may not be the case. In commenting on the aforementioned collection, good on balance, my own particular objection to the general notion of artificial selection and dog evolution surfaced: "Perhaps it is counterintuitive to think of the human household as an ecological arena in which dogs play out intraspecific competitions. But that's exactly it" (Morey 1997: 88). That is, "The dog's natural environment is the human family" (Vilmos Csányi, as quoted by Douglas 2000: 24).

To move on, it is worth considering two related issues concerning my second and third indicated objections to the traditional view of domestication. My second point is that successful domesticates receive just as much benefit from the relationship as people. Consider corn plants. From a minor, wild weedy grass in Meso-America (teosinte), corn has come to be of worldwide importance. Yes, modern corn is dependent on us for its reproductive success. But worldwide, humans are quite dependent on corn as well. One can realistically say that corn plants have domesticated us, just as surely as we have domesticated them. Third, one must consider just what constitutes benefit, to people. To be sure, domestication-based foodstuffs, both plant and animal, support a burgeoning human population worldwide. It is indeed hard for us to imagine a life without our modern domesticates, on which we depend. But, regarding the final objection, is there a cost?

Is There a Cost?

Jared Diamond, whose work with ant agriculture we discussed earlier, has referred to human agriculture as "The Worst Mistake in the History of the Human Race" (Diamond 1987). More recently, Diamond explained why, while also making clear his view of why domestication could not have been a deliberate decision:

> Food production could not possibly have arisen through a conscious decision, because the world's first farmers had around them no model of farming to observe, hence they could not have known that there was a goal of domestication to strive for, and could not have guessed the consequences that domestication would bring for them. If they had actually foreseen the consequences, they would surely have outlawed the first steps towards domestication, because the archaeological and ethnographic record throughout the world shows that the transition from hunting and gathering to farming eventually resulted in more work, lower adult stature, worse nutritional condition and heavier disease burdens. (Diamond 2002: 700)

Quite simply, the initiation of some domestic relationships between people and other organisms does not seem quite so beneficial to us, in subjective terms of overall quality of life, when looked at more closely. With that issue in mind, it is worth pointing out that Susan Crockford (2006: 43) has suggested that "it's my firm belief that it was the animals that started it all, not us, and that the process must have been extraordinarily fast." In other words, domestic relationships work both ways, and the animals can benefit rather quickly, in evolutionarily time.

But the immediate question, for present purposes, concerns dogs and people. Just how did dogs benefit? Consider their ubiquity on this planet. Dogs now live in every part of the world where there are people, including, for different lengths of time, remote Antarctica (e.g., Bellars 1969). At the same time, their wolf ancestors, who once ranged over virtually the entire northern hemisphere, are now distributed only irregularly, having been exterminated from many areas. About 200,000 individuals made up their entire population, as of 2003 (Boitani 2003: 317). Because their numbers are limited, modern efforts are underway to have wolves recolonize areas where they once lived. Jon T. Coleman (2004) devoted an entire presentation to considering why wolves have been nearly extirpated in recent times, but are now protected, and are often highly regarded by people. He focused on North America, but the phenomenon is much more widespread. Ongoing studies of these modern wolves focus on different aspects of their lives, including, for example, their habitat preferences, landscape use patterns, and mortality characteristics (e.g., Darimont et al. 2003; Potvin et al. 2004, 2005; Chavez & Gese 2006; Oakleaf et al. 2006; Lovari et al. 2007). Related studies emphasize their hunting patterns (e.g., Kunkel et al. 2004; Smith et al. 2004; Wright et al. 2006; Barja & Rosinelli 2008), including the degree of foresight that they can display (e.g., Mech 2007), as well as their relations to modern human economic interests (e.g., Oakleaf et al. 2003; Theuerkauf et al. 2003; Mattioli et al. 2004; Blanco et al. 2005; Gazzola et al. 2005; Singh & Kumara 2006; Blanco & Cortéz 2007; Frame et al. 2008; Gazzola et al. 2008). Also studied is the manner in which different ecological factors, including the presence of other wolves themselves, can influence the genetic and/or social structure of wolf populations (Vilà et al. 2003; Pilot et al. 2006; Sidorovich et al. 2007). Yet other investigations focus on intraspecific social interactions within wolf packs (Cipponeri & Verrell 2003), and comparisons of the foraging ecology of modern and ancient wolves, as gauged by chemical signatures in bones (Fox-Dobbs et al. 2007). These variables are all relevant to monitoring the wolf's current tenuous lot in life.

Dogs, on the other hand, have a different lot in life. In terms of their sheer numbers, as of the beginning of the twenty-first century there were "Four hundred million dogs in the world – that is a thousand times more dogs than there are wolves" (Coppinger & Coppinger 2001: 21). And shortly before the end of the twentieth century, Lynette Hart (1995: 162) made the observation that

> Currently in the United States, more than 50 million dogs reside in roughly 38% of all households (Market Research Corporation of America 1987; American Pet Products Manufacturers Association 1988). Of these, the vast majority are kept as social companions.

In much more recent times Udell & Wynne (2008: 248) put the number at closer to 75 million. Indeed, "From a Darwinian perspective, wolves who took up residence with people a few thousand years ago made a smart move – at least from today's vantage point" (Morey 1994a:339). As for how people benefited in an evolutionary sense, there are several ways, of variable impact, and much of the next chapter is devoted to dealing with that issue.

DOMESTICATION AS EVOLUTION

For now, though, it is fitting to return to the viewpoint that is deeply ingrained into our way of thinking, as summarized earlier, where domestication is an invention designed by people of the past to improve their quality of life. Having repeated several prominent definitions of domestication that have been offered over the years, it is important to clarify, as effectively as possible, the perspective advanced in this volume. I begin by appealing to this statement: "domestication was and is evolution" (Rindos 1984: 1). David Rindos penned that conviction at the very beginning of a major volume on the origins of human agriculture, one in which he also highlighted the relevance of certain insects, especially ants, to this issue. Rindos's statement rings true, though at one level such a statement is an oversimplification. It remains to be determined just what kind of domestic relationship has evolved in any given case. But to break free of the traditional perspective, it does seem necessary, first, to conceive of domestication, involving any organisms, as an episode of biological evolution. In fact, meaningful precursors to the idea of domestication as evolution exist, as indicated by a question that was posed long ago by Charles Darwin's own cousin, Francis Galton: "Is it possible that the ordinary habits of rude races, combined with the qualities of the animals in question, have sufficed to originate every instance of established domestication?" (Galton 1865: 122). Galton answered this question in the affirmative, and while one must make allowances for his style of expression, this passage nevertheless emphasizes domestication as a natural result of traits shared by humans and other animals.

Domestication, in the view advocated here, is best regarded as the development of a symbiotic relationship between two organisms, with evolutionary benefits to both. In this relationship, both organisms are closely involved in each other's life cycles. This view does serve to indicate why simply referring to domestication as evolution is an oversimplification. It is the basis for the initial development of the symbiosis that varies from case to case. Regarding people, Ádám Miklósi (2007b: 207) has suggested that it could be "more advantageous to consider each domestication event (species) separately and investigate more closely the actual relationship between humans and the particular species in question." In some human cases, it clearly revolved around feeding strategies and still does. Corn and chickens are handy examples (not to mention aphids for ants), while other animals, like cattle, were used for food but also as beasts of burden. Additionally, they served as a source of raw material for making tools and fashioning personal apparel, in the case of sheep by means of their fleece (wool). The point is that many animals served a variety of primary roles for people, making their symbiotic association with people understandable, but this principle doesn't work as well for dogs. To be sure, dogs eventually were and are used for a variety of utilitarian purposes, including the purposes just indicated for other animals, and the next chapter highlights some of those roles. But that was not the initial basis for the development of this domestic relationship. Recall the dictionary sense of the word "domestic," as covered earlier, as of or relating to the family. I hold this as a social relationship, first and foremost, and the fact that dogs were so frequently buried at death much like a family member speaks eloquently to this principle. Wilks (1999) suggested this point as well, referring specifically to an 11,000- to 12,000-year-old joint human–puppy burial in Israel (see Chapter 7, specifically, Ein Mallaha). Kubinyi et al. (2007: 27) have also captured this point nicely:

> As far as we know, dogs were not domesticated for any direct benefit (e.g., food). As early dog fossils from burials indicate, dogs had a special, probably partly spiritual, relationship with humans from very early on.

To be sure, other animals were sometimes accorded ritualistic burials, but not nearly as frequently as dogs on a worldwide basis.

For much of the remainder of this chapter, the goal is to spell out how the domestic relationship between people and dogs likely got initiated. First, though, it is useful to call attention to a cogent point that has been emphasized elsewhere: "A vast range of human–animal relationships has existed throughout history, and the animals involved in many of these relationships cannot be easily categorized as strictly wild or strictly domestic" (Dobney & Larson 2006: 261–262). This point is surely true, and those authors highlight examples such as urban domestic pigs (Dobney & Larson 2006: 262) from different places in the world. These animals wander freely through towns foraging, returning to their owners in the evening. Their point serves to emphasize that the distinction between wild and domestic is not always clear. Surely, though, the primary issue is the degree to which the animal might be viable outside of the human-dominated setting. For dogs, there seems to be no point in questioning that they are now fully domestic, but initially their status was surely not so clear-cut. This is a useful point to bear in mind when considering the transition from wolf to dog, which is covered in the next section. And after focusing on how the domestic relationship between dogs and people got started, most of the rest of this volume then elaborates on how it has developed over time, in different places.

THE DOMESTICATION OF THE DOG

To begin, it is important to note that during this time frame (ca. 15,000 years ago), people were exclusively hunter-gatherers, with agricultural economies still in the future. Numerous investigators have recognized that, as hunter-gatherers, people surely had overlapping ecological niches with wolves (e.g., Zeuner 1963; Scott 1968; Clutton-Brock 1981, 1995: 8–10, 1999: 49; S. J. Olsen 1985; Fritz et al. 2003: 291). After making that point in a synthetic work (Morey 1990: 22), I continued in a closely related vein:

> Both were social species that hunted for many of the same prey items. Wolves, as opportunistic scavengers, may have learned to be aware of human hunting activities and to scavenge from human kills. Perhaps humans even learned to do the same with wolves. (Morey 1990: 23)

That this perspective, advanced a good many years ago, influenced the thinking of some others on this matter is indicated by considering the following passage:

> human hunter-gatherers and wolves existed in overlapping niches and were accustomed to contact with each other: both were social species that hunted for many of the same prey items. As opportunistic scavengers, wolves may have learned to follow human hunters and to scavenge from human kills, and humans probably learned to do the same with wolves. (Brewer 2001b: 21)

To be sure, wolves and people were often ecological competitors because they hunted for some of the same prey items. This fact may well condition the historically common perception in Western society, at least until recent times, that wolves are to be regarded as symbols of evil, and as such are highly feared (Lopez 1978; Fritz et al. 2003: 293–294). One need only consider that the fairy tales and folklore of Western society abound with evil wolf figures to appreciate the existence of that perception (e.g., Fiennes 1976: 175–190; Zimen 1978: 302–315). But this deeply conditioned fear is not likely to have characterized hunting-gathering people from thousands of years ago. One example of people who were recently hunter-gatherers, and who currently live in the same environment as wolves, is some Alaskan Eskimos. Significantly, they do not express misplaced fear of wolves, but general admiration for their intelligence, sociality, prowess as hunters, and purposiveness and individuality of their behavior (Stephenson & Ahgook 1975; Lopez 1978). These circumstances underscore that the recent near-extirpation of wolves is just that: recent, but in the past.

So, to follow a traditional scenario, given that many ancient hunting-gathering peoples surely had some regular contact with wolves, and probably did not have a misplaced aversion to them, it is logical to suppose that somewhere, maybe more than once, some young wolf pups were found and adopted by people (e.g., Clutton-Brock 1984: 204; Clutton-Brock & Jewell 1993: 24). As for the motivation in a specific case, that is a matter of speculation, but the idea that pet-keeping was the motivation has a significant pedigree (e.g., Zeuner 1963: 39; Reed 1969, 1984: Savishinsky 1983; Serpell 1989), a pedigree that extends at least as far back as Darwin's own cousin, Francis Galton (1973). To be sure, the very idea of pets may also reflect a common preconception among members of Western industrial societies, and there is no way to be sure that peoples of the past did such things. It can be established, though, that people in a variety of modern nonindustrialized societies do sometimes capture and maintain wild animals, including young wolves (Savishinsky 1983). Francis Galton (1973: 173–193) provided some examples as known from his era. Granting this probability for the past, then, the most important consequences for present purposes revolve around the immediate implications for the animals.

Indeed, that portion of the traditional scenario is an aspect of this issue that I have subscribed to in the past (e.g., Morey 1994a, 1995). At this stage, however, some modification of that scenario is in order. At issue is just what conditions would need to have been in place for people to adopt one or more wolf pups. That is, why might wolves have been receptive to human presence in general? Susan Crockford (2006) has made a noteworthy case regarding this issue, and it is worth exploring here.[3] Specifically, she suggests that subtle changes in thyroid hormone physiology, during what she calls "protodomestication" (Crockford 2006: 47), better conditioned wolves to life with people in

---

[3] Crockford's work plays a limited role in Chapter 8 as well, but for present purposes, the implications of her work for understanding the timing of canid domestication are most relevant. The initial expression of Crockford's case appeared several years before her 2006 book (Crockford 2000b, 2002), but is developed most fully in the later book.

TABLE 4.1. *Four hypothetical steps leading to the domestic dog, as envisioned by Coppinger & Coppinger (2001: 57)*[a]

Step	Event/Process
1	People create the new village niche.
2	Some wolves enter the new niche to gain access to food.
3	Those wolves are more tolerant of human presence.
4	Such wolves are at a selective advantage over the more wild wolves.

[a] This is an adaptation of their outline.

a human-dominated ecological setting: "I define protodomestication as a colonization process that occurred within human-dominated, or *anthropogenic*, environments" (Crockford 2006: 47, original emphasis). Moreover, as pointed out earlier, Crockford (2006: 43) thinks the animals started it and that the process was rapid. The theme of the animals starting it all, with additional elaboration of Crockford's work, appears before much longer, but for pragmatic purposes it first seems appropriate to call attention to another effort to model the domestication of the dog.

The Coppingers' Model

Table 4.1 is a synopsis of the domestication process as envisioned by biologists Raymond and Lorna Coppinger, in their book about the domestication of the dog and the animal's various roles among people (Coppinger & Coppinger 2001). Their first step, that people create a new village niche is only slightly problematical, a point highlighted later. Next, they have wolves entering that niche to gain access to food. What, specifically, is this new niche? "For simplicity, let's call this new niche the town dump" (Coppinger & Coppinger 2001: 59). In their view, dogs came into being largely by being specialized scavengers. The third step in their domestication process is that those wolves were more tolerant of human presence, and they model those animals as having reduced flight response with respect to people. Accordingly, "The wild wolf, *Canis lupus*, began to separate into populations that could make a living at the dumps and those that couldn't" (Coppinger & Coppinger 2001: 61). In their model, the size and shape changes shown by dogs are a consequence of the scavenging lifeway. The final step in their model is that those wolves were at a selective advantage over their more wild counterparts. Thus, the dog came into being.

To be sure, there are several useful aspects to the Coppingers' model, but some unfortunate missteps as well. One is represented by the following statement: "To me, dogs, like other domestic animals, are the products of serious agriculture" (Coppinger & Coppinger 2001: 283). In making that statement, they are clearly unaware of the 14,000-year-old Bonn-Oberkassel dog, from Germany, as simultaneously revealed by another statement: "But so far there is zero evidence for dogs before 12,000 B.P." (Coppinger & Coppinger 2001: 286). As discussed in Chapters 2 and 3, the Bonn-Oberkassel dog has been known about for some two to three decades, there was no agriculture going on at all at that time, and the identification of that specimen as a dog is convincing. Another problematical feature of the Coppingers' work stems from their conviction that dogs are simply the product of scavenging at human "dumps." To be sure, some wolves may well have found the dietary possibilities associated with a human settlement enticing, but that possibility doesn't account for the way that they rapidly came to be treated much like people. It seems seriously doubtful that people would routinely ritualistically bury mere dump scavengers when they die, but dogs, the immediate descendents of those putative 'scavengers,' are often buried at death, and they have been for thousands of years. Moreover, in bypassing Benecke's (1987) work with the Bonn-Oberkassel dog, the Coppingers simultaneously bypassed the fact that this animal, the earliest securely identified dog in the world, was a burial, part of a human double grave. Only three times in their book do the Coppingers acknowledge an instance of dog burial. That acknowledgment comes initially in their Chapter 3, where they refer to several dogs "found in Stone Age Natufian graves" (Coppinger & Coppinger 2001: 86). The reference to 'Stone Age Natufian graves' represents some well-publicized cases that are covered more fully in Chapter 7, here. Then much later, in their Chapters 9 and 10, they refer to one of the same cases again (Coppinger & Coppinger 2001: 277, 286). In sum, despite its several merits, the Coppingers' book reflects a lack of appreciation for the archaeological information that exists. It is also problematical on some biological grounds, as Susan Crockford, herself a biologist, notes. Specifically, writing about the Coppingers' book she says: "when pushed to describe the actual process of transformation, they suggest the literary equivalent of 'something happens'" (Crockford 2006: 42). Crockford set out to model what that something was.

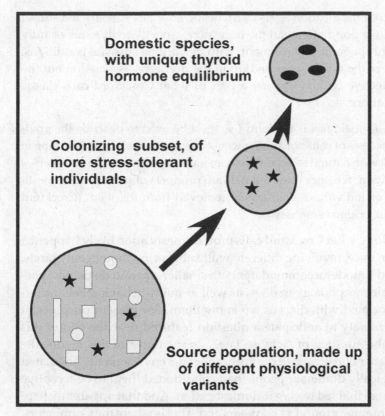

**Domestic species,
with unique thyroid
hormone equilibrium**

**Colonizing subset, of
more stress-tolerant
individuals**

**Source population, made up
of different physiological
variants**

FIGURE 4.1. Schematic depiction of the process of protodomestication, resulting in domestication, with a subset of the source population of animals being more stress-tolerant than the others, and coming to associate with people. This depiction is adapted from Crockford (2006: 48, Figure 2.4), and is explained more in the text.

Crockford's Model

In Crockford's model, individuals in all groups of animals are variable in their capacity to respond viably to different sources of stress, influenced especially by their individual thyroid hormone characteristics: "Individual variation in behavioural responses to stress is the key to explaining how wolves could have evolved rapidly into dogs without direct and deliberate human interference" (Crockford 2006: 48). Figure 4.1 provides a schematic illustration, adapted from Crockford's own illustration (2006: 48, Figure 2.4), of how this process of what she calls protodomestication leads to domestic animals. A restricted subset of the source population had the necessary capacity to live successfully in an anthropogenic environment, with its various stressful conditions. Those were the ones that ended up in the domestic association but the initial selective regime was *not* a part of what Crockford calls classic domestication:

> *Classic domestication* is the term I suggest be used to describe the gradual processes of conscious and unconscious human selection (working in concert with natural selection) that can modify any captive or commensal population, whether those animals are products of prior protodomestication or individuals deliberately removed from the wild. (Crockford 2006: 46, original emphasis)

With dogs, what Crockford calls protodomestication likely happened more than once, involving different wolf subspecies, though only rarely. Crockford's model accommodates both smaller size and distinctive paedomorphic morphology in dogs, as well as many behavioral characteristics associated with dogs as we know them. For present purposes, it is fitting merely to anticipate a question featured near the end of this chapter, the question of "why so late?" In the terms of this model, the canids were not ready until the anthropogenic environments populated by ecologically dominant people had conditioned them to be receptive to behaviors that led to classic domestication. And that apparently happened sometime around 15,000 years ago. This is where the Coppingers' (Coppinger & Coppinger 2001) idea about the new village niche, as covered previously (see also Table 4.1), is just slightly problematical. People were hunter-gatherers at that point, not settled agriculturalists, though as noted earlier, in the Coppingers' view dogs were the products of serious agricultural life among people. But as Crockford (2006: 47) points out, settlements didn't have to be absolutely permanent for anthropogenic conditions to have been obtained.

One noteworthy aspect of Crockford's thyroid hormone physiology model lies in its broad applications: "all of the physical and behavioural characteristics that change when wild animals become domestic are controlled by thyroid hormone" (Crockford 2006: 50). So this basic framework pertains to more than dogs, and Chapter 8 briefly deals with that aspect of it. For now, though, it is relevant to note that the Coppingers (Coppinger & Coppinger 2001) were surely on target in suggesting that reduced flight response in some wolves was likely an important factor. But it was Crockford who devised a framework to account for how and when that factor came into play. To be sure, Crockford's specific mechanism, involving thyroid hormone physiology, continues to be in need of empirical backing, as pointed out by Richardson (2007). But whatever the specific physiological mechanism turns out to be, the overall concept that a more stress-tolerant subset of the original population was the founding stock seems emminently reasonable. Moreover, recalling the earlier comment about animals starting it all, Crockford (2006: 47) notes that colonization of an anthropogenic environment was a choice that the animals could make. And that point returns the focus to the pet-keeping scenario, one that I continue to favor in modified form. To begin, the circumstances modeled by Crockford may have been instrumental in facilitating the introduction of these pups into human society. Such dog–wolf pups would already have been different, more conditioned to living with people than were truly wild ones, and the next logical step is to consider what would have happened with such pups.

Initial Domestic Dog–Wolf Pups

It has been demonstrated that wolf pups taken at an early age and reared by people can, to some extent, be tamed and socialized (e.g., Fentress 1967; Pulliainen 1967; Woolpy & Ginsburg 1967).[4] As Scott & Fuller (1965) showed, the most crucial social bonds of a dog or wolf's life are formed when the animal is three to eight weeks old, an interval called the critical period. More recently, Jensen (2007b: 70) has subscribed to that point.[5] Consequently, when raised by people from an early age, that is the context in which they will form their primary social bonds. As part of a broader mammalian pattern, young brains are the most susceptible to guiding influences, and Michael Fox's (1971) work

[4] "To some extent" is the operative concept. As James Serpell (1995d: 258) has emphasized, "no wolf is ever as tameable or trainable as a domestic dog." Chapter 8 in this book explores just why that is the case.

[5] As others have subsequently pointed out, Scott & Fuller's (1965) original model has been refined over the years. For example, many workers, including both Serpell & Jagoe (1995: 80–84) and, more recently, Ádám Miklósi (2007a), favor the term "sensitive period" for "critical period." Thus, Miklósi writes that "*sensitive period* might be a more appropriate term because the time boundaries involved are more varied than originally assumed" (Miklósi 2007a: 209, original emphasis). Moreover, "Available evidence suggests that the sensitive period for socialization is much shorter in wolves" Miklósi (2007a: 209). This factor underscores that wolves would need to have been adopted at very young ages. In fact, Kubinyi et al. (2007: 28), in a paper with Miklósi as a coauthor, indicate, based on modern experimental work, that wolf socialization to people really has to begin before day 10 to be maximally effective.

documented that pattern with respect to dog brains. But one point that should not be lost sight of is that the mere rearing of tamed wolf pups did not automatically create the animals known as dogs. It was only a beginning. For one thing, the animals had to sustain their initial toehold in the domestic niche, a multifaceted accomplishment. The bulk of the remainder of this chapter explores how they did that, but initially just two points serve as a frame of reference for what follows. First, recalling the second sense of the word "domestic," as emphasized earlier, this was about dogs becoming much like family members. Second, though this factor is not explored until later in the volume, beyond their greater tolerance of stress, dogs changed in some notable ways from wolves in their behavioral capacities. For now, though, recalling that these wolves would already have had a head start, how did the earliest dog/wolves maintain their toehold in the new domestic niche with people?

The initial rearing of tame wolf pups, even ones with a head start, did not ensure their successful integration into human society. But that initial rearing would mean that the young animals did not have to be tamed, or confined to prevent their escape. Given that circumstance, it is noteworthy that wolves and people are, overall, remarkably compatible in terms of their hierarchical social organization, especially basic pack (family) structure and communication systems (Scott 1968; Clutton-Brock 1977; Feddersen-Petersen 2007: 108–110). Those factors predisposed our stress-tolerant wolves to domestic life with humans. At the same time, individual wolves are highly variable in temperament and behavioral tendencies (A. Murie 1944: 25; Stephenson & Ahgook 1975; Sullivan 1978; K. MacDonald 1987; Clutton-Brock 1999: 51; Packard 2003: 55–57), and some individuals would not have the necessary temperament to mesh with a society in which humans were dominant. Those individuals that could not fit in were likely either disposed of, or forcibly driven away, to fend for themselves. As Serpell (1995b: 258) put the matter, the route to a genuinely domestic relationship "could have been achieved 'unconsciously' merely by killing or driving away temperamentally unsuitable individuals." So let us assume that a few individuals did fit tolerably well into human society.

These young domestic wolves also had to eat. Wild wolves are carnivorous, and one or two large ungulate species make up their principal prey in most areas where they live today (Peterson & Ciucci 2003: 104). Deer are primary examples in many temperate regions, and animals like caribou and moose are often taken in arctic or subarctic regions. Occasionally, however, smaller animals such as mice, rabbits, or birds are also taken (Mech 1970; Pulliainen 1975). In the wild, juvenile wolves also grow up on a diet of animal products, including, as they are weaned, regurgitated stomach contents provided by one or both parents (Mech 1970). Eventually, the juveniles are taken to kill or rendezvous sites where they eat unprocessed meat. At a later point they begin to accompany adults on hunts, and they learn some of the skills necessary for adult life.

The young domestic wolf, though, grew up quite differently. Humans are, of course, omnivorous. As hunter-gatherers, people surely did not provide a growing wolf with an optimal diet. Such wolves were probably provided with a diverse array of food, consisting mostly of plant products and meat scraps, some of which may have been no longer considered acceptable as human fare, due to spoilage or other factors. Quite simply, a finicky wolf would have stood little chance making it. Moreover, their survival skills beyond that initial stage needed to be fundamentally altered. For example, there were no experiences available for learning the group hunting skills of wild wolves. For the young wolves, successful solicitation of food from people was surely an important skill, along with some scavenging. They likely supplemented these activities with the individual hunting of small animals such as rodents, whose capture did not depend on the more refined group-oriented tactics used by wild wolves for dispatching larger prey. In short, their omnivorous diet would have continued into adulthood, and only individuals that could adjust successfully to these conditions could potentially contribute to future generations. One can appreciate that, under such conditions, smaller body size itself may well have been favored, due to the reduced nutritional requirement of smaller animals. And this is potentially not a trivial point, since "Body size is manifestly one of the most important attributes of an organism from an ecological and evolutionary point of view" (Werner & Gilliam 1984: 393).

Captive wolves provide a crude model for this situation (see Stockhaus 1965; Clutton-Brock 1970: 305; Epstein 1971: 83–86) and tend to show developmental changes in morphology that are roughly consistent with some of the changes shown by dogs.[6] But those basic traits are clearly inherited in dogs (e.g., Degerbøl 1961: 41; Stockhaus 1965; Lawrence 1967; Epstein 1971: 86). As a consequence, developmental

---

[6] One should recall, especially from Chapter 2, that the reservations expressed by Ádám Miklósi (2007a: 103) about the early canids from Eliseevichi-1 in Russia, as reported by Sabin & Khlopachev (2002), included this very possibility.

alterations at an individual level do not account for basic dog morphology, distinctive at an early point. One instead must invoke secure reasons that the basic traits in question would come to hold selective advantage, and thus be favored in an evolutionary sense. At this juncture, it is appropriate to return to some points raised only cursorily in the previous chapter.

Beyond needing to eat, and adjust successfully to the social dynamics of human society, new domestic wolves needed to reproduce in order for the new line to be perpetuated and have a chance to sustain itself. A young domestic wolf would not likely be afforded an opportunity to reproduce with a wild wolf, as wild wolves do not readily accept outsiders into their pack (see Mech 1970: 51–56; Mech & Boitani 2003: 1–2). Occasionally, however, for reasons that are not well understood, strange wolves will be allowed to join, at least temporarily, a pack that already has a breeding pair. Such individuals are sometimes known as adoptees (Mech & Boitani 2003: 2). Regardless of this poorly understood exception among wolves, the point for present purposes is that new domestic wolves/dogs, having grown up in human society, almost certainly would need to accomplish reproduction in the context of that setting. Among wild wolves, breeding opportunities depend on social competition within a pack (Peterson 1977: 80–85; Jenks & Ginsburg 1987; Schotté & Ginsburg 1987; Moehlman 1989: 154; Mech & Boitani 2003: 3–6). Again, the dynamics of life in human society would complicate matters, but the relevant social factors had to be negotiated in that setting. Obviously, they were.

Ecological Colonization

Early in the previous chapter, I suggested that with the colonization of a new ecological niche, a domestic association with humans, early sexual maturation was at a premium as dogs underwent rapid population growth. According to Richard Lewontin (1965: 78), "Colonization is the establishment of a population of a species in a *geographical* or *ecological* space not occupied by that species" (original emphasis). Lewontin went on to note that it is of maximum general interest to consider colonizing episodes of a given species, rather than to consider colonizing species per se. Not long after Lewontin wrote those words, J. P. Scott (1968) pointed out that dogs could be regarded as a species that entered a new habitat, where they rapidly spread to fill a new ecological niche. Put simply, they were colonizers, ecologically poised for rapid population growth. Moreover, following up on Lewontin's additional point, it was an episode of colonization among wolves that then evolved into what are now recognized as dogs.

Wild wolf populations are regulated primarily through adjustments in rate of reproduction, and by juvenile mortality (Pimlott 1975: 284; Fuller et al. 2003: 183–186). Fall is an important seasonal juncture for young ones, and their mortality is often related to malnutrition (Van Ballenberghe & Mech 1975: 57; Mech 1977; Fuller et al. 2003: 176). Rausch (1969: 119) once reported that in Alaska, pups may comprise as much as 60 percent of a wolf population at any given time. Thus, it can be appreciated that juvenile mortality is a significant variable and is related to the pattern whereby wolves generally live in close balance with their important resources. Consequently, in terms of population levels, mortality tends to be density-dependent in wolves. But one hallmark of a colonizing episode is the absence of density-dependent mortality patterns Lewontin 1965: 78), and a colonization episode entails rapid population growth.

In my earlier synthetic treatment of dog evolution, the points being raised here were juxtaposed against the concepts of r- and K-selection (Morey 1990: 31–34), originally developed by MacArthur & Wilson (1967), and subsequently expanded by others. I bypass that step here, the focus being exclusively on linking some important life history traits with ecological parameters, an endeavor that draws freely from some of the literature of that era on r- and K-selection.[7] In fact, even before MacArthur & Wilson's (1967) work, it was recognized that selection in rapidly growing populations will lower the age at first reproduction (Cole 1954; Lewontin 1965). That point played a role in some of the subsequent literature produced during what might be called the r-K era (e.g., Meats 1971; Giesel 1976; Stearns 1976; Gould 1977: 326). The reason is, as Lewontin (1965) pointed out, that small absolute alterations in developmental rates have approximately the same impact as large changes in fertility. For dogs, if they could articulate a rational evolutionary logic to what their best strategy was at this point, it would be,

[7] As expanded from the work by MacArthur & Wilson (1967), the r/K dichotomy predicted a conceptual division of life history traits into two groups. Specifically, r-strategists would be characterized by early maturity, large numbers of offspring, little or no parental care, and large reproductive effort. K-strategists, on the other hand, would have delayed reproduction, small numbers of young, iteroparity, parental care, and small reproductive effort. Intensive scrutiny of this concept resulted in a good many objections, though, in broad outline, I (Morey 1990: 32–33) contrasted wolves as K-strategists, evolving as r-strategists, when associated domestically with people.

to phrase matters in shamelessly vernacular fashion, "Have as many viable pups as you can, as fast as you can, and get them out there!"

In that light, it is significant that, as noted in the previous chapter, wild wolves reach sexual maturity at about two years of age (A. Murie 1944; Young & Goldman 1944; Novikov 1962; Mech 1970; Pulliainen 1975; Kreeger 2003), whereas dogs may breed at six months to a year (Scott & Fuller 1965; Fox 1978; Clutton-Brock 1984). One complication here is that even though six to twelve months reflects the onset of puberty in most dogs, in larger breeds especially this may not occur until into the second year (Hancock & Rowlands 1949; Andersen & Wooten 1959; Johnston et al. 1982). In addition, patterns in modern dogs do not establish, by definition, that the same was true of dogs several thousand years earlier (e.g., Price 1984: 22). In any case, modern dogs appear to be progenetic (accelerated onset of sexual maturity) with respect to wolves, but the morphological consequences are more directly observable with dogs of the past. As already noted, progenesis is expected to result in size reduction and paedomorphosis, and the reality of those two traits of dogs, relative to wolves, was covered in the previous chapter.

As always, complicating factors abound. If progenesis alone underlies size reduction and paedomorphic morphology in dogs, one would expect a clear correlation between adult body size and the onset of puberty in dogs. There is no strict correlation between adult size and the onset of sexual maturity in dogs (Houpt & Wolski 1982: 133), though smaller breeds do generally attain puberty earlier than larger breeds. And likely correlated with that pattern, larger breeds tend to take slightly longer than smaller breeds to reach about 90 percent of their adult weight (Kirkwood 1985:102, 105). Given that these are only tendencies, it is noteworthy that Sutter et al. (2007) invoke a particular genetic mechanism (the *IGF1* gene) as being a strong determinant of body size in mammals. They specifically advance a case that this mechanism underlies the small size of many modern breeds of dogs. Moreover, they suggest that this mechanism was likely present before the emergence of recognized breeds. This finding may well be correct, though it doesn't apparently pertain to initial modest size reduction or paedomorphic morphology relative to wolves, given that initial size reduction in dogs is functionally linked with paedomorphic morphology, as already covered here. However, Sutter et al. don't even address the issue of paedomorphism at all, and there is no indication of how the specific genetic mechanism that they have surmised could possibly bear on that pattern. In short, even though the new information is certainly welcome, despite the inevitable complications, a basic pattern exists, and dogs have apparently undergone progenetic heterochrony with respect to wolves. Their basic morphology, as covered in the previous chapter, is consistent with that physiological process, and the morphological outcome is anticipated by long-standing theoretical predictions, as covered more fully in this chapter. Given that pattern, before offering a summary assessment of the fundamental nature of this domestic relationship, it is appropriate to return briefly to a question anticipated earlier in this chapter and address it more fully now. It is a question that concerns the content of the previous chapter, but in the context of what has been covered in this chapter.

WHY SO LATE?

Given the genetics-based work that suggested an antiquity for dog domestication tens of thousands of years earlier than the archaeological record indicates (e.g., Vilà et al. 1997), this is a legitimate question. Having dealt with the substantive issues concerning the timing of canid domestication in the previous chapter, that is not the focus here. However, it is worth pointing out that others are also deeply skeptical of the timing of canid domestication as suggested from genetics-based inferences that are centered on modern animals:

> I believe that mtDNA studies are an exciting area of evolutionary research. Sooner or later they might reveal a breakthrough on dog genealogies and ages, even if so far they have just added to the confusion. But at the present time, I think this information about dogs, wolves, and coyotes, as reported and interpreted, has done some real harm. (Coppinger & Coppinger 2001: 293)

I agree, and will take the general timing as inferred here (ca. 15,000 years ago) and elsewhere (Crockford 2006; Morey 2006; Savolainen 2007) as a given and deal only with the issue of why it didn't happen sooner. Certainly both wolves and our species have been in existence for much more than 15,000 years. Crockford's model should apply, one would think, to those earlier animals as well. For example, at Klasies River Mouth in South Africa, skeletally modern hominid remains have been found that date in the vicinity of 100,000 years ago (Singer & Wymer 1982; Grün et al. 1990; Rightmire 1991). Nearby, also in South Africa, a path of preserved footprints, clearly made by people with anatomically modern walking apparatus, is dated to about 117,000 years ago

(Bower 1997). And there is one case of modern-looking hominid fossil remains from Africa that may even approach 200,000 years old (McDougall et al. 2005). Of course, skeletally modern does not necessarily indicate behaviorally modern, and here we run into the inevitable complications. At one level, one can't really say that a person who lived 200 years ago was truly modern behaviorally. At a more realistic level, though, humans who were engaging in some incredibly modern-like behaviors, minus technology of course, existed for several thousand years before there is any credible evidence of dog domestication. A handy example is some of the late Pleistocene peoples of Europe, who executed some truly stunning artwork in caves, 17,000 to 31,000 years ago (e.g., Vouvé et al. 1982; Ruspoli 1987; Chauvet et al. 1996; Clottes 2003). Lascaux and Chauvet Caves, both in France, contain several hundred paintings and/or engravings of many different kinds of animals, often rendered in great detail and recognizable to species. From the art depicted in the books cited earlier, there are no images that are even suggestive of dogs. The Coppingers have also made that point: "The Chauvet and Lascaux cave paintings are significant for dog investigators for several reasons. Most notably, no domestic animals are pictured, including dogs" (Coppinger & Coppinger 2001: 285). In one case, Lascaux, I can verify that reality, having actually been there and witnessed this stunning art in person.[8]

At face value, this is an absence of evidence argument, but it meshes with others. That is, from this ancient time frame, there is no compelling

[8] In the same study in which they held that dog domestication dates back to more than 30,000 years ago (see Chapter 2), Germonpré et al. (2009: 481) offered the following statement:

> Other evidence for an early date for the domestication of the dog was found in Chauvet cave, France. Here, occurring in the deepest part of the cave, a track of foot prints from a large canid is associated with the one of a child (Garcia, 2005). Torch wipes made by this child were dated at c. 26,000 B.P. Based on the short length of medial fingers in the footprints the canid track was interpreted as being made by a large dog (Garcia, 2005).

The association referred to in that passage is spatial only. As Clottes (2003) noted regarding this circumstance, there is "no proof that they existed at the same time" (Clottes 2003: 40). Moreover, according to Clottes (2003: 35), the same part of the cave had footprints from other animals, namely bear and ibex. Beyond the fact that the canid and human footprints cannot be held as literally occurring at the same time, given the difficulty distinguishing large dogs from wolves based on their major bones, it strains credibility enormously to think that footprint impressions showing the "short length of medial fingers" could suffice.

skeletal evidence of dogs, no contextual evidence (e.g., burials), and no suggestive artistic depictions of them. And one should realize that the earliest securely established dog in the world, from Germany, is from the same general region as those French caves. But the same absence of evidence is true of other regions in the world, as well. What is the soundest conclusion? There were no dogs. We return to the issue of the coexistence of wolves and people for thousands of years prior to domestication, and R. K. Wayne, in continuing with a genetics-based case for a more ancient date of dog domestication, has written as follows: "This origin is much older than suggested by the archaeological record, a possibility at least not contradicted by the fossil record, since wolves and humans coexisted for as much as 500,000 years (Clutton-Brock 1995)" (Wayne et al. 2006: 282). True, but this seems like a disingenuous statement. The source cited, Clutton-Brock (1995), indicates only that bones of wolves have been found in some *hominid* (premodern human) sites as old as 400,000 years. The wolves were surely using those places at different times than the people, though such a pattern does speak to the likelihood of periodic contact, presumably what Wayne et al. wished to highlight. But the fossil record of that era does not at all imply domestication.

So, let us return to the original question: Why so late? As noted, dogs were apparently the first animals to enter into a domestic association with people. As for the specific timing, some 15,000 years ago, this was approaching the end of the Pleistocene Period, and, following Crockford (2006), but to put it more simply than she does, the world was filling up with people. Animals that could adjust to anthropogenic environments were at an advantage. From those circumstances came the dog.

Certainly there continue to be many important and unanswered questions about canid domestication. But I believe there is a compelling answer to the question of just what is the fundamental nature of the human–dog domestic relationship. It is with my answer to that question that this chapter comes to an end.

THE HUMAN–DOG DOMESTIC RELATIONSHIP: JUST WHAT IS IT?

As suggested earlier, this was first and foremost a social relationship, with dogs often being regarded as family members, as reflected by their burial at death, all over the world (Morey 2006). In suggesting that

FIGURE 4.2. The essence of the dog–human domestic relationship as portrayed in the 1990s, in *American Scientist* (Morey 1994a: 346; 1995: 150).

point, it is useful to touch base again with their wolf ancestors and call on Mech & Boitani (2003: 6), using their own words:

> The wolf and the wolf pack are as closely linked in the human mind as a child is linked to a family, and rightly so. The human family is a good analogy for the wolf pack. The basic pack consists of a breeding pair and its offspring, which function in a tight-knit unit year-round.

From their statement, in conjunction with some of the other points developed here, it can surely be appreciated that, in a real sense, the human family is a dog's "pack."[9] And it seems fitting to highlight an effort to capture photographically the principle of dogs being regarded

---

[9] Given that principle, it seems fitting to anticipate now a point developed more in subsequent chapters, beginning with the next one. That point is the mutual influence of dogs and people on each other. More specifically, Donna Haraway made this point in summary form rather effectively: "No wonder dogs and people share the distinction of being the most well-mixed and widely distributed large-bodied mammals. They shaped each other over a long time" (Haraway 2003: 118–119).

as family members, thereby encapsulating the incredibly strong social bond between dogs and people (Figure 4.2). It is a photograph of my wife, Beth, and our dear departed dog, Jingo, published over a decade ago in an article that tried, for the nonspecialist reader, to make clear part of the content of this chapter (Morey 1994a: 346; Morey 1995: 150). At any rate, that is one way of trying to capture the point graphically. Perhaps James Serpell captured it best verbally: "The dog–human relationship is arguably the closest we humans can ever get to establishing a dialog with another sentient life-form" (Serpell 1995c: 2).

5

THE ROLES OF DOGS IN PAST HUMAN SOCIETIES

JUST AS DOGS FULFILL A VARIETY OF FUNCTIONAL ROLES IN TODAY'S societies, they played a wide range of roles in past societies, aside from their role as social companions. The purpose in this chapter is to identify some of the roles that have been at least suggested on logical grounds, and in some cases inferred with great reliability. One should not lose sight of the difference between plausible supposition and secure inference, and it is important to identify which applies best in any given case, though the boundary is not always fully distinct. Quite obviously, one role for which the term "secure inference" definitely applies is that dogs were often objects (or subjects) of deliberate burial at death, and one can reliably say that they fulfilled a mortuary role. That is also an example of a role that they played in past societies that they continue to play in today's societies. In fact, it turns out that the roles played by dogs in the past were much like the ones they now play, allowing for the different twenty-first century setting of today's societies, with all its technology, transportation and communication capabilities, and so on.

DOGS AS A FOOD SOURCE

One of the roles that dogs clearly played in the past seems, at first, to run directly counter to their role as social companions, or friends, a role highlighted by their frequent burial at death. Quite simply, dogs sometimes served as dietary fare. In fact, that reality poses a particularly intriguing paradox that emerges when one looks closely, for they played both roles, in some cases within the same society. As to dogs serving as dietary fare, without additional complications, obviously one cannot travel back in time and personally witness dogs being eaten. But secure evidence that some were eaten has been found in different parts of the world. What counts as secure evidence in this case can be one or a combination of factors. For example, bones of dogs that occur in particular settings, such as what appear to be cooking facilities, are suggestive. Even more suggestive is if the bones have been burned, or at least exposed to intense heat. It is worth pointing out that these are the same kinds of criteria by which bones of other animals, such as deer in North America, are evaluated. In fact, that has long been true. Consider, for example, the following passage from William Webb's report on Indian Knoll in Kentucky, an Archaic Period site dating between roughly 5,500 and 4,500 years ago (see Morey et al. 2002):

> From the general excavation of this midden, a large amount of animal bone was taken from the debris. This mass of bone revealed that deer were eaten by the hundreds. Such bones occur in quantities on occupation levels, about camp fires, or scattered through the midden. Along with the bones of deer and other animals, undoubtedly used for food, were found the bones of dogs. These dog bones, occurring scattered in the midden and about fireplaces, in precisely the same situation as the bones of deer, wild turkey, and fish bones, seem to suggest that the dogs were also eaten on occasion by these people, and the bones thrown out in the debris along with those of other animals. (Webb 1946: 156–158)

Webb made that observation in the midst of a section of his report that was devoted to presenting information on the dog burials at Indian Knoll, of which he and his crew reported finding twenty-one in all. Indian Knoll, then, is an example of the paradox referred to above. Moreover, this situation also highlights how the roles played by dogs in the past could be consistent with those that they play today. For example, on the one hand, dogs are sometimes (illegally) raised as a food source in modern Korea (Korea Animal Protection Society 2003). At the same time, Korea is also a place where dogs sometimes are treated quite reverently at death. A quoted passage from the English language newspaper *Korea Herald* served to highlight this situation recently (Morey 2006: 169), and serves that purpose here as well:

> Recently, a dog funeral service that oversees the cremation and burial of dogs is gaining a reputation in Gangnam and other parts of Seoul as more people come to regard the pet as an actual member of the family. The funeral procedures are no different from those of a human funeral. The dog is dressed in ramie burial clothes and laid in a small coffin made of paulownia wood. After the body is cremated in a special pet

crematorium, the ashes are laid to rest in a dog charnel. (*Korea Herald* 2000)

One wonders if the same people who treat their dogs in this way also sometimes eat them. Sandra Olsen has expressed the paradox of some dogs' roles quite appropriately:

Even among cultures that consume dogs, some individuals may be kept as pets, named, and even breast-fed by humans (Serpell, 1995[d], Titcomb 1969). Within a society, dogs may be divided into different functional groups, such as those that are considered edible and those that are companions. (S. L. Olsen 2000: 71)

Returning to Webb's work, his era predated modern approaches to analysis, and although modern approaches are more refined, they include the principles that he invoked. Today, in conjunction with such patterns or separately, one evaluates indications that the animals in question were systematically rendered by the people, in ways consistent with one or more butchering activities. One reliable indication is the scoring of bones by stone or metal tools, leaving what are known as cut marks, a consequence of skinning and/or dismembering parts of an animal. The analysis of butchering patterns, as inferred from the frequency and placement of cut marks on bones, has become a routine activity (e.g., Guilday et al. 1962; Binford 1981: Chapter 4; Noe-Nygaard 1989; Lyman 1994: 297–315). With dogs of the past, such evidence has been found in numerous places, representing many different time periods and many different geographic areas of the world. Figure 5.1 is an example of a cut-marked dog bone from the Qeqertasussuk site in Greenland, dating to about 3900–3100 B.P. (Grønnow 1994: 201). This example, along with two others, was shown previously in an entire presentation about dogs from that time period in the North American Arctic (Morey & Aaris-Sørensen 2002: 51).

In some instances, ethnohistoric data play a useful role. For example, in writing about some of the classic Aztec centers in Mexico, Elizabeth Wing consulted written accounts, and drew attention to a case in which an historic observer

describes the market-place at Alcoman in 1539, where over 400 dogs were for sale at higher prices than other meat. The dog meat was consumed for special feasts . . . small dogs were fattened for consumption in the home and for sale at the market at Tlatelolco. (Wing 1984: 228)

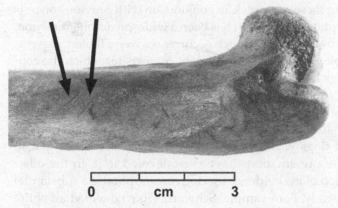

FIGURE 5.1. The proximal end of a cut-marked dog femur from Qeqertasussuk, a Paleoeskimo site in Greenland, dating between about 3900 and 3100 B.P. The arrows point to two conspicuous cut marks, of four, on the lateral face of the bone. See text for additional discussion regarding the significance of cut marks. Photograph by Geert Brovad, Zoological Museum, University of Copenhagen, Denmark. This image was previously published in the journal *Arctic* (Morey & Aaris-Sørensen 2002: 51, Figure 5).

Significantly, Tlatelolco became the location of "the largest and most active marketplace in the Valley of Mexico" (Berdan 1982: 10). But in other places in the same region, ethnohistoric data just help confirm what is evident archaeologically. Several years prior to her 1984 piece, Wing (1978) dealt with animal remains from three sites in Veracruz, Mexico, dating in the range of about 2300–1700 B.P., and an earlier site, San Lorenzo, dating to 3200–2900 B.P. From these sites, Wing indicated that "Dogs may be considered the basic animal protein, supplemented by snook, a seasonally available aquatic resource" (Wing 1978: 39). She estimated, based on the quantities of animal remains, that dog meat may have constituted anywhere from 11 to 53 percent of the diet, depending on the site.

Ethnohistoric information also goes hand in hand with archaeological data elsewhere. For example, in the 1800s, two Presbyterian missionaries, John Dunbar and Samuel Allis, lived intermittently with the Pawnee people of Nebraska. Both wrote accounts of their experiences and in one account, Dunbar wrote of the different foods eaten by the Pawnee, noting that "A fattened dog also constituted a most delicious repast" (Dunbar 1880: 323). Using archaeological evidence, Snyder (1991) has documented the use of dogs as food at several protohistoric Pawnee sites. The point of these examples, in conjunction with previous ones, is that the consumption of dog flesh has been a widespread phenomenon.

To highlight how widespread such occurrences were, Table 5.1 assembles a few notable examples from different contexts for which the consumption of dogs has been reliably inferred, based on archaeological evidence. A survey of the Table 5.1 sample reveals some noteworthy patterns. For one thing, these few settings come from widely separated locations in the world, representing different time periods. In addition, consumption of dogs was a conspicuous enough occurrence that in some cases entire presentations have been devoted to it. In the other cases, presentation of the evidence for dog consumption is a substantial part of a larger study. For example, Schwartz (1997) devoted an entire chapter (Chapter 3) of a book on dogs in the early Americas to "The Edible Dog."

The overall point of this section is that the consumption of dogs was a widespread practice, one that still occurs in some places today. Evidence for it in modern times is direct, and evidence for it for the past is secure. It is indeed a noteworthy paradox of the dog–human relationship, that animals so revered that they often receive ritualistic burials at death can also be regarded as little more than a food source, sometimes within the same society. But rather than dwell further on this paradox, it is appropriate to shift attention to yet another sort of role, one that required that they be alive.

TRANSPORTATION USES

As with the consumption of dog flesh, it is not possible to observe directly that dogs were used for transportation purposes. It is sometimes inferred that such has happened, however, and at the outset it is necessary to draw a distinction between different kinds of transportation. There are two primary categories to be considered here, and they are best taken in turn.

Individual Dogs

As considered presently, transportation refers to dogs individually moving goods for people, and that task clearly places a serious limit on the size and weight of what a dog can transport. When dogs are used in teams, a transportation strategy considered next, there are still limits, but they are much less restrictive. In terms of individual dogs, there are

TABLE 5.1. *Examples of archaeological contexts that have yielded reliable evidence of dog consumption by people*[a]

Site/Region	Time Period(s)	Reference(s)
Britain	Several contexts	Harcourt 1974: 171–172
Ban Chiang, Thailand	ca. 5500 B.P. to present	Higham et al. 1980: 159; J. W. Olsen 1985: 69
Qeqertasussuk, Qajâ, Greenland	Paleoeskimo (3900–3100 B.P.)	Morey & Aaris-Sørensen 2002: 50–52
Eretria, Greece	Helladic through Hellenistic Periods (3600–2150 B.P.)	Chenal-Velarde 2006
San Lorenzo (Veracruz, Mexico)	Olmec (ca. 3200–2900 B.P.)	Wing 1978
Durezza Cave, Austria	Late Iron Age (ca. 2700–2450 B.P.)	Galik 2000: 136
Cuello, Belize (Central America)	Pre-Classic Maya (ca. 2600–1800 B.P.)	Clutton-Brock & Hammond 1994
Siracusa, Sicily	2550–2400 B.P.	Chilardi 2006: 34–35
Santa Luisa, Chalahuites, Patarata (Veracruz, Mexico)	Late Formative/Early Classic Maya (ca. 2300–1700 B.P.)	Wing 1978
Levroux, Côte-d'Or, France	Gallic (ca. 2200–1900 B.P.)	Horard-Herbin 2000: 119
Belgium and Romania	Three Roman Period Sites (fourth century B.C.–A.D. fourth century)	Tarcan et al. 2000
El Riego Cave, Tehuacan Valley, Mexico	Venta Salada Phase (ca. 1300 B.P.–1521 A.D.)	Flannery 1967: 168
Clachan, Nuvuk, Beulah, Central Canadian Arctic	Thule Period (ca. 850–550 B.P.)	D. A. Morrison 1983: 236–238, Tables 22, 23
Grasshopper Pueblo, Arizona (American Desert Southwest)	Mogollon Tradition (ca. 700–600 B.P.)	J. W. Olsen 1990: 109–115
Several Sites in South Dakota (northern plains, North America)	1000 B.P.–160 B.P.	Snyder 1995 (esp. Chapter 7, analysis of cut marks on bones)
Gray, Packer, Nebraska (central plains, North America)	Protohistoric Pawnee	Snyder 1991
Mobridge, South Dakota (northern plains, North America)	Protohistoric Arikara	Parmalee 1979: 211

[a] Entries are generally arranged from earliest to latest, with the exception of the first (Harcourt 1974), which covers multiple settings.

two primary modes: (1) when some kind of pack is physically attached to the animal, and the dog literally carries the load; and (2) when an individual dog pulls a load that is attached to a line or harnessing device of some kind, and the dog pulls the load by that means.

In North America, among the historically known Assiniboine peoples (see Denig 2000), dogs sometimes wore packs made up of two skin pouches cinched around their middle. Walker et al. (2005: 84) have provided a photographic illustration of the use of this pack arrangement among the Assiniboine. Also well known from ethnohistoric sources, at least in North America, is the use a travois (Figure 5.2). Most often, a travois was made from two poles that were crossed and then tied together at one end. Two other cross pieces were placed across the dragging end of the poles, and a load might be placed on those cross pieces, or a woven device on a hoop frame would serve as a platform for the load. Snyder (1995: 198–199) has described such a device in those terms. As she noted, Kurz (1937: 293) once estimated that a dog hauling a load of 70 English pounds (32 kg), could do so for 30 to 40 English miles (ca. 48–64 km) per day. In this capacity, a dog might haul any number of things, ranging from meat at kill sites to loads of firewood or household goods, for example during a residential move.

The North American Plains

Samuel Allis, in writing about his experiences among the Pawnee in Nebraska, indicated that dogs might haul up to 100 English pounds (45.5 kg), "according to the size of the dog (Allis 1887: 140). Henderson (1994) worked with a modern animal, a husky, taken to be adequately representative of the capability of a prehistoric dog of the northern plains, and evaluated its efficiency with a travois. In a series of winter trials he determined that this ca. 55 lb (25 kg) dog, hauling a load of about 31 lb (14 kg), could go for up to 17 miles (27 km) under favorable weather conditions. As for more direct evidence of construction and use of the travois, Buffalo Bird Woman, a Hidatsa (North Dakota) woman, born about 1840, related such information in 1913 to Gilbert Wilson, who transcribed it for posterity (G. Wilson 1924: 216–221). Figure 5.2 is an illustration in which, among those people, a dog is pulling an unloaded travois. According to Buffalo Bird Woman,

> One of the chief uses for dogs was to carry the wood gathered for fuel. In our family, sometimes my sister, Cold-medicine, or my two mothers, or

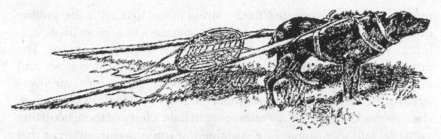

FIGURE 5.2. Depiction of a North American plains dog pulling an unloaded travois. This image is used as part of the educational mission associated with the Knife River Indian Villages National Historic site in Stanton, North Dakota. As such, this would represent a Hidatsa or Mandan dog. This illustration is part of the Gilbert L. and Frederick N. Wilson Papers and is jointly owned by the Minnesota Historical Society in St. Paul, Minnesota, and the Division of Anthropology at the American Museum of Natural History in New York City. It is used here by permission.

all four of us would gather wood. We always took the dogs and travois with us. (G. Wilson 1924: 206)

It is interesting to note that, in a major ethnography of the Pawnee people of Nebraska based primarily on native informants in Oklahoma in the 1920s and 1930s, Gene Weltfish wrote "In the Pawnee language, the word horse means 'super-dog,' the only pack animal the Pawnees had before the impact of European culture" (Weltfish 1965: 143). According to Allen, in a classic work, "In the plains country from Saskatchewan to the Mexican boundary, the *travois* was in general use" (Allen 1920: 453, original emphasis). Clearly, dogs made a substantial contribution to North American plains peoples' transportation needs in the historic past.

If dogs were used in a similar fashion prehistorically, then one might logically anticipate some kind of discernable evidence suggesting that function from the archaeological record. The most promising source of such evidence is from the skeletal remains of the dogs themselves. Quite simply, the physical strain of carrying or hauling substantial loads is likely to leave behind indications of that kind of strain in some portions of the anatomy, especially involving the articulating parts of bones (e.g., Davis 1987: 162–165). In fact, such evidence comes from a variety of contexts around the world, dating to different times. Not surprisingly, given what was just covered in identifying some examples, it is useful to begin with the North American plains.

In the state of South Dakota, especially at two sites in the Missouri River Valley, Snyder (1995) encountered vertebral anomalies in dog remains that she attributed to the stress associated with the pulling of travois. One site was the prehistoric Sommers site, most likely dating in the time frame of about 1000–900 B.P. (Snyder 1995: 58). The other site was the Larson site, post-dating Euro-American contact, and apparently occupied continuously from about 350–250 B.P., or 1650–1750 by the Christian calendar (Johnson 2007: 191–197). In both cases, she observed osteophytic development in the vicinity of the articulating vertebral surfaces, suggesting sustained, if intermittent, strain in that region. Again, for both cases, she summarized the situation as follows:

> Degenerative changes to the vertebrae from the Sommers site, plus similar osteophytic reactions on articulating thoracic (Figure 46) and lumbar (Figure 47) vertebral units from the Larson site (39WW2), may well be related to chronic or intermittent stress to the vertebral column associated with travois pulling. (Snyder 1995: 216)

Beyond vertebrae, she covered a case from another site in the area where there was "extensive osteophytic development around the margins of both the scapular glenoid fossa and humeral head" (Snyder 1995: 221). She went on to note that these kinds of traumas are represented in canid bones from several different sites in this region, of the same general time period. Accordingly, then, this is a situation in which relevant ethnohistoric accounts, in conjunction with direct archaeological evidence, combine to make a strong case for dogs in this region being used to haul loads. Moreover, they did so both prehistorically and historically.

The Southeastern and Midwestern United States

In the southeastern United States, the terrain is such that hauling loads by travois is less likely than in the Great Plains. In keeping with that point, "The use of the dog travois is not described in the ethnohistoric Southeast" (Warren 2004: 5). Accordingly, archaeologically derived information is a necessity in this setting. For example, in what might be regarded as a pilot study, Diane Warren (2000) studied dog remains from Archaic Period sites in the Southeast, with a specific focus on paleopathologies. In that study she examined 194 dogs, most from burial contexts, from 12 different sites in Kentucky, Alabama, and Tennessee. Overall, Warren found that vertebral paleopathologies in the form of marginal osteophytoses were prominent, increasing in frequency as one moves down the vertebral column. For example, Warren (2000: 113, Table 8) found that on the initial cervical vertebrae, the frequency of osteophytic development ranged from only 0 to 16.7 percent. Some Kentucky dogs and the small Tennessee sample showed no osteophytic development at all, whereas the Alabama dogs had the highest frequency. On the lumbar vertebrae, however, the corresponding frequencies were 22.2–92.9 percent. Interestingly, the Tennessee sample had the lowest incidence (22.2 percent), while nearly all the Alabama specimens showed lumbar paleopathologies. Regardless of these disparities, the data reveal a consistent pattern of vertebral paleopathologies, increasing in frequency as one moves down the vertebral column. As Warren notes, though, there are difficulties in ascribing functional significance to vertebral paleopathologies, as dealt with again later in this chapter.

In 2004, Warren presented a much more comprehensive and wider ranging investigation, involving the study of 455 dogs from 44 different archaeological sites (Warren 2004). In addition to Archaic Period dogs from the Southeast, she also considered later Woodland, Mississippian, and Protohistoric Period dogs, including 27 from the state of Illinois, all of which are Woodland in age. Some of the expansion of the series came from adding bones from different non-Archaic sites or site components to the original series in her earlier work. Nevertheless, the majority of dogs studied are Archaic in age, a pattern that is not surprising considering that most of the dog remains studied are also from burials. That, quite simply, is why parts of their skeletons, frequently the whole skeletons, were so often available (see Chapter 3).

Warren (2004) devoted an entire chapter (Chapter 7) to investigation of the overall pattern regarding vertebral marginal osteophytoses: "In all groups examined, marginal osteophytosis was rare in the cervical region. [and] increased drastically beginning with the TV1/2 [first and second thoracic vertebra] joint" (Warren 2004: 146). The frequencies tended to decline slightly through TV8/9 but then rise again through the upper lumbar vertebrae. Another peak in frequency tended to occur at the joint involving the last lumbar vertebra. Overall, Warren found that some of the documented patterns mirrored those observed in frequencies of vertebral spinous process fractures (covered in her Chapter 6). Significantly, she indicated that "To the extent that pack carrying contributed to the spinous process fractures, this provides support for an association between vertebral marginal osteophytosis patterning and the use of dogs as beasts of burden in the prehistoric Midwest and Southeast" (Warren 2004: 153).

In short, nothing definitive can be said, but these paleopathologies likely reflect the use of dogs for bearing burdens. As indicated by her passage quoted earlier, Warren also focused on fractures or misshaped development of the vertebral spinous processes as a likely consequence of carrying packs. Given similarities in the patterns of representation of marginal osteophytoses and spinous process fractures of the vertebrae, it is useful to highlight the latter by means of noting a more recent study.

Specifically, at Dust Cave in Alabama, two dogs, from among four that were deliberately buried, exhibited antemortem damage to the spinous processes of the thoracic and lumbar vertebrae (Walker et al. 2005). One of these, along with a third dog, exhibited caudal curvature of the scapular spine on both scapulae. As with Warren's (2000, 2004) work, there are no directly relevant ethnohistoric data, given that the Dust Cave dogs are from the Southeast and are Archaic in age. As with Warren's work also, this is a situation in which the most likely possibility involves dogs carrying individual loads, affixed to their backs. Given that possibility, those bones were analyzed by veterinary faculty at the State University of New York at Delhi, and their conclusion was that "the damage to the vertebral dorsal spinous processes and scapular spines was probably due to weight bearing down on that area of the dogs' backs" (Walker et al. 2005: 88). Similarly, writing about comparable pathologies on dog vertebrae from the late prehistoric McCulloch site in southern Missouri, Darwent & Gilliland (2001: 159) indicated that "At this point we leave the origin of the bent spines open, but note that heavy load bearing on the back of this small dog could have contributed to, if not actually caused, the pathology." Thus, even though one cannot state definitively that these dogs carried packs, one can state that they surely had an unusual amount of weight on their backs, consistent with packs.

To return briefly to Warren's (2004) work, some of the suggestive patterning she found involved temporal changes in the representation of vertebral pathologies. For example, in comparing the overall Archaic series to the small Woodland series, she found that mid-lumbar marginal osteophytoses in the Woodland series were reduced in frequency compared to the Archaic series. At the same time, "Lower thoracic and upper lumbar pathology may conversely have *increased* over time from the Archaic to the Woodland and beyond" (Warren 2004: 149, original emphasis). She suggested some change over time in the use of dogs as beasts of burden, and at least one apparent change ran counter to an expected result. Specifically, she had hypothesized that "As sedentism increased from the Archaic to the Mississippian and beyond, the use of dogs as beasts of burden is expected to have declined" (Warren 2004: 143). What she found, though, recognizing the inherent ambiguity posed by small post-Archaic samples, was that the use of dogs as beasts of burden may have increased rather than decreased after the Archaic. According to her, "Possibly this reflects increased burden loads or carrying distances occurring as dogs were used more frequently by larger, more sedentary populations for the collection of firewood and other resources" (Warren 2004: 246).

So there are a variety of complications and uncertainties posed by Warren's thorough study, but it does seem clear that archaeologically observable paleopathologies can provide some genuine insights into the use of dogs as beasts of burden.

The North American Arctic

In the North American Arctic a source of considerable uncertainty over the years has been the role, or even the presence, of dogs from early times, several thousand years ago (Morey & Aaris-Sørensen 2002). Dogs were an integral part of life in that region during the last several hundred years, but what the situation with dogs was prior to that time, during what is known as the Paleoeskimo Period, has been unclear (see Chapter 6). Charles Arnold (1979) worked with remains from the Lagoon site, on Banks Island, located in the High Arctic, between mainland Canada and Greenland. This site, with radiocarbon dates placing it at about 2400 B.P. (Arnold 1979: 264), yielded, among other things, three canid vertebrae – the twelfth and thirteenth thoracic and the first lumbar. Arnold suggested that a domestic dog was represented based not on traditional morphological criteria, but on the pronounced osteophytic development and flattened spinous processes on them, consistent with use of a pack or harness. Similarly, Morey & Aaris-Sørensen (2002: 50) noted that one of the Paleoeskimo dog vertebrae that they documented from Greenland exhibited "an abnormally shortened, flattened spinous process." Osteophytic development and/or flattened spinous processes on dog vertebrae, regardless of the region, are both highly suggestive, but inconclusive, evidence that dogs were being used to carry or haul goods.

Dogs in Teams

More compelling would be evidence that such pathologies occur with unusual frequency among dogs that are known to be used in such ways. With such a possibility in mind, it is relevant to draw attention to some information about the British Antarctic Survey's dog sledge teams, operative in the modern era. The standard sledge team consists of nine animals, hauling a maximum load of about 120 pounds (54.4 kg) for variable distances, with a round trip of some 900 miles (1,450 km) not being unusual (Bellars 1969: 2). In its lifetime, a good dog might cover at least 8,000 miles (12,875 km), and dogs are usually "incapable of further work by the time they are 8 years old" (Bellars 1969: 2). Due to their stressful working conditions, these dogs developed high frequencies of osteoarthritis in their joints, and in fact "osteoarthritis is the reason for destruction for nearly 72 percent of males and 52 percent of females that survive beyond 5 years of age, excluding deaths due to accidents" (Bellars & Godsal 1969: 15).

It is probably misleading to compare the consequences of travois pulling directly, previously highlighted for the North American plains, with the hauling of sledges by the Antarctic dogs. In the first place, the travois usually involved a single dog hauling a load, whereas the Antarctic dogs work in teams of about nine. In addition, the terrain of the North American plains is markedly different from the Antarctic. Quite obviously, the climate is dramatically different, too. In combination, one result is that the most frequently observed skeletal anomaly in the Antarctic dogs involves the limb joints, not the vertebral column, as was the case in the North American plains (Snyder 1995). Specifically, "26 out of 34 dogs examined *post mortem* had erosions of the articular cartilage of hip and shoulder joints" (Bellars 1969: 15, original emphasis). With that pattern in mind, one should recall that Snyder (1995) did encounter shoulder joint anomalies in dogs from several sites in her North American plains setting. So while the two situations do not correspond directly, they do share some genuinely common ground. It is also worth repeating from an earlier discussion that situations where dogs likely carried packs individually (e.g., Warren 2000, 2004; Walker et al. 2005), rather than hauled loads, generally resulted in dogs with misshapen vertebral processes due to the weight of the loads, rather than the systematic strain of pulling.

In all such cases, though, it should not be assumed that observed skeletal anomalies that are typically associated with the strain of pulling or the direct weight of loads automatically signify those activities. That is, considering the pattern of vertebral anomalies, degenerative pathologies in that part of the anatomy sometimes occur in both wild and domestic canids that have never served in such a capacity (e.g., E. C. Cross 1940; Harris 1977; Palmer 1993). Accordingly, in an archaeological case, a single isolated specimen may be considered suggestive, but far from conclusive. A series of specimens from the same setting constitutes a stronger case. Relevant ethnohistoric data, if available, can strengthen the case further, as in the North American plains example covered here. These considerations highlight the rationale behind making the point at the beginning of this chapter that there is, in principle, a difference between plausible supposition and secure inference, but that difference is not always clear-cut. The North American plains case can be considered secure. The cases covered from the North American southeast are compelling by virtue of there being a broad series, though relevant ethnohistoric data are generally not available. The Paleoeskimo cases from the Arctic are basically just logical supposition, but the same is not true of later arctic situations.

In prehistoric times, Thule peoples rose to prominence in the North American Arctic, first in the westerly regions (e.g., northwestern Alaska), from where they rapidly spread eastward, all the way to Greenland (summaries in Maxwell 1985; Schledermann 1996, Chapter 9), some 700–800 years ago (see Chapter 6). A conspicuous part of the Thule peoples' dramatic success was the routine use of sleds, each pulled by a team of dogs, to move both people and goods over substantial distances. Precisely because that role receives much more attention in the next chapter, including the security of that inference, it is best left at that for now, and just one final point on that particular topic warrants identification here. That point is simply that most modern Inuit peoples of the far north are directly descended from Thule predecessors and those people routinely used (and use) dogs in teams to pull sleds across the icy landscape. Moreover, this is a situation in which, even if archaeological evidence was ambiguous, which it is not, the historical link is so direct that ethnographic analogy alone would be compelling.

DOGS USED IN HUNTING

Just as one can't literally document dogs being used for transportation prehistorically, one can't document their use in hunting, either. Their common role for this purpose in modern times does render it logical, however, to suppose that they were used similarly before recorded history. Clutton-Brock & Jewell (1993: 23) graphically portrayed the possible derivation of different kinds of dogs, including two kinds of hunting dogs, labeled as "scent hounds" and "sight hounds" (see also Bubna-Littitz 2007: 96). They were concerned with modern breeds that can be assigned to each category and their probable derivation from different geographic varieties of wolves. Regardless of the accuracy of their specific inferences, it is a useful way to conceive of different kinds of hunting dogs.

Writing about indigenous peoples of the North American plains region, Waldo Wedel once made the following observation:

> The success of this seasonal hunting technique in the capture of incompletely fledged cygnets and goslings, *especially with the help of a dog*, has been historically documented by Meriwether Lewis during the passage of The Corps of Discovery through the Montana region in the summer of 1805 (Lewis and Clark 1904, vol. 2, p. 255). (Wedel 1986: 24, emphasis added)

Thus, some historically known North American peoples used dogs as aides in hunting, at least sometimes. Also in the North American plains, Buffalo-Bird-Woman, the North Dakota Hidatsa woman, has described the use of a travois to Gilbert Wilson (1924), as mentioned earlier. In the same work, he transcribed an account by her relating a story about a hunt, on which dogs accompanied the hunting party (G. Wilson 1924: 231–262). From her account, it is not clear if the dogs actually played a substantial role in assisting with the hunting itself, but a total of twelve dogs accompanied them on the hunt. Since bison were the main target, dogs may well have played only a minor role, if that. However, dogs clearly provided some transportation assistance. Specifically, Buffalo-Bird-Woman indicated how certain of the dogs carried what are known as bullboats by travois (G. Wilson 1924: 231). According to authorities, the Hidatsa bullboat was

> Created by stretching a buffalo hide over a frame of tough but flexible willow boughs, it was propelled by one oar. To venture into deep water in a bullboat has been likened to going to sea in a tub. (Ahler et al. 1991: 16)

The buffalo hide was affixed with the hair-bearing side toward the water, and a bullboat could carry up to half a ton, but weighed only about 30 English pounds (13.6 kg). Additionally, they could be built in one day, and cottonwood oars were used for navigation.

In general, bearing in mind Ahler et al.'s (1991) comment about deep water navigation, bullboats were used for transportation across or down a river, or perhaps a standing body of water. One or more dogs could accompany people in a bullboat. Thus, as said, it is not clear if dogs were used for direct hunting assistance in this case, but they clearly provided transportation assistance to the hunting party. It is also worth noting that the bullboat, used for water navigation of modest scope, has an Arctic analog with much greater capability called the umiak (see Chapter 6).

Dogs in Hunting: A Genuine Impact?

In some scholars' estimation, dogs may have had more than a casual influence on the hunting practices of some people. As an example, Juliet Clutton-Brock (1984) wrote about dogs among Old World hunting-gathering peoples. She was specifically focused on Mesolithic Period peoples, or Natufian to use the terminology of the Near East. Roughly speaking, Mesolithic in Europe corresponds to the earlier portion of the Archaic in the New World, in both time (about 10,000–5,000 years ago) and in comparable lifeways. Clutton-Brock focused on a change in hunting technology associated with the Mesolithic. Specifically, she called attention to a change from short range attacks on animals to using the flight capability of arrows armed with what are known as microliths. As Clutton-Brock (1984: 204) wrote,

> I should like to put forward the theory here that this change of weapon was associated with different hunting methods resulting from the world-wide spread of the domestic dog as a hunting partner that could track down wounded animals and could retrieve game from difficult terrain such as undergrowth or water. Washburn and Lancaster (1968: quoting Lee) described how one bushman with a trained pack of hunting dogs brought in 75 percent of the meat of a camp, whilst six other resident hunters in the group, without dogs, brought in only 25 percent of the meat.[1]

Clutton-Brock's suggestion seems especially appropriate to the following situation. The early Mesolithic site of Star Carr, in England, has been well known for many years, being the subject of a comprehensive study by J. G. D. Clark (1954). That site yielded, among other things, a rich assemblage of microliths of the kind Clutton-Brock referred to

[1] In more recent times, Jeremy Koster (2008) has drawn attention to the role of dogs in the hunting activities of the indigenous Mayangna and Miskito peoples in Nicaragua. Though the dogs sometimes identify unprofitable prey types from an optimal foraging perspective, on balance the dogs can be worthwhile to the hunting of some animals by these people.

in her piece. Moreover, as suggested by J. G. D. Clark (1954: 103), and inferred by later workers as well, microliths were used in archery, as in the flight of arrows (e.g., Andresen et al. 1981). Microliths were hafted either singly at the end, or accompanied by a series hafted along the shaft of an arrow. The latter sort of arrangement has been reconstructed and illustrated based on a specific case from Mesolithic Denmark, dating to about 8,500 years ago (Aaris-Sørensen & Petersen 1986; Aaris-Sørensen 1988: 167–172).

Star Carr dates to about 9,500 years ago, and its special relevance to this work is that it yielded some cranial bones identified as a domestic dog by Magnus Degerbøl (1961). As of 1987, those bones represented the earliest domestic animal known from Britain (Davis 1987: 175). So Clutton-Brock's (1984) suggestion, while unverifiable, gains credence in a noteworthy case. Star Carr, it should be noted, yielded remains of a variety of terrestrial mammals that were hunted, including some that were relatively fleet of foot, such as red deer and roe deer. Dogs may well have turned out to be of genuine assistance in tracking down those animals when they were wounded. Naturally, the dogs may also have been of assistance in locating them during hunting activities. At any rate, Star Carr represents a case in which the role of dogs in influencing human hunting strategies can readily be envisioned. At the same time, it is worth stressing again that this was, first and foremost, a social domestic relationship, as emphasized in the previous chapter.

A case that is consistent with that statement comes from another Mesolithic site, Vedbaek, in Denmark, dating in the range of 7300–6500 B.P. Vedbaek included a Mesolithic cemetery (Albrethsen & Petersen 1976) with the remains of at least twenty-two people. As with Star Carr in England, microlithic technology is characteristic of the Danish Mesolithic, as indicated by the Danish case noted previously (e.g., Aaris-Sørensen & Petersen 1986). At Vedbaek, these people were remarkably effective in their quest for food. Though there was clearly a coastal emphasis, they obtained animals from virtually every conceivable environment. Moreover, the Vedbaek assemblage, like Star Carr, included many microliths as well (P. V. Petersen 1977). So given their clear terrestrial hunting prowess, Clutton-Brock's (1984) suggestion about the importance of dogs in hunting at this time seems thoroughly reasonable. What is especially worth drawing attention to is that two dog skulls were identified from Vedbaek (Degerlbøl 1946), and one, a complete skull, is from a documented burial (Nielsen & Petersen

1993). Moreover, that skull was measured in my own earlier synthetic treatment of dog evolution (Morey 1990: 96). Someone who is deeply familiar with the circumstances of this situation has left no doubt of his inference regarding the special association between dog and hunter in that setting:

> The Stone Age dog's role as hunting partner was touched on in the introduction. They probably also served as guard dogs and maybe as well for carrying and pulling loads, and finally for the sake of completeness it should be mentioned that they also have served as human dietary fare now and then. We have many examples of dog bones broken for marrow, and skulls with cut marks and chopping marks from our Stone Age sites, but the two Vedbaek skulls show neither kind of mark indicative of killing, skinning, or later consumption, so how they finally passed away must remain unknown. (Aaris-Sørensen 1977: 176; originally in Danish, English translation by the author)[2]

Thus the Vedbaek dogs were apparently not dispatched by people, at least not in a manner that has left any clear evidence behind.

Pleistocene Extinctions: Did Dogs Have an Impact?

In an unusual piece, Fiedel (2005) has argued that dogs may have played a conspicuous role in the well-known extinction of late Pleistocene animals in North America. In an explicitly speculative essay, Fiedel suggested that when dogs surely accompanied people to the New World after about 13,500 years ago, by way of the Bering Land Bridge, they likely provided hunting assistance as well as transport capability to those early Paleoindian peoples. By doing that, Fiedel suggested, they facilitated peoples' rapid colonization of the continent and may have played a role in the extinction of North American megafauna. One should understand that Fiedel's scenario draws in part from the previous modeling, without dogs, of Paul S. Martin, known for what is

[2] The original passage in Danish is as follows: "Stenalder-hundens rolle som jagtpartner blev berørt indledningsvist. Som vagthunde har de vel også tjent og måske ligeledes som last- og traekdyr, og endelig skal det for fuldstaendighedens skyld naevnes, at de også nu og da har tjent som menneskeføde. Vi har adskillige eksemplar på marvspaltede hundeknogler og kranier med skaerre- og hugspor fra vore stenalderbopladser, men de to Vedbaek-kranier viser ingen spor efter aflivning, flåning eller senere fortaering, så hvordan deres endeligt forløb må forblive ubesvaret" (Aaris-Sørensen 1977: 176).

commonly termed the "overkill hypothesis" (Grayson & Meltzer 2003: 585).

The crux of Martin's argument was that people emerged from the Old World into a new continent where the fauna was initially naïve regarding human hunting practices. As a consequence people basically swept through North America, quickly driving large fauna like mammoths extinct through wasteful hunting practices. Martin developed this scenario in a series of publications (e.g., P. S. Martin 1967, 1973; 1984; Mosimann & Martin 1975), and the model has been quite influential. Grayson & Meltzer (2003) presented evidence against this model, emphasizing that the North American animals went extinct, as did many in the Old World, for causes that likely had little to do with human hunting practices. Fiedel & Haynes (2004) subsequently objected to their piece, but the pros and cons of that exchange need not concern us here. Nor need it be of concern here that Martin has not ameliorated his position in any significant way (P. S. Martin 2005), though its flaws continue to be articulated (Grayson 2006). As for what is of present concern, on the one hand, it seems fortunate that that Fiedel labeled his essay as speculative. On the other hand, his essay unfortunately reports some incorrect information that should be identified.

For example, one would think that it would be to his advantage to highlight accurately what the earliest known dogs in North America are. In that vein, he offers the following statement: "The oldest securely identified domesticated dogs in the continent are the three that were intentionally buried about 8500 RCBP within Horizon 11 of the Koster site, in southern Illinois (Struever and Holton 1979: 210; Morey and Wiant 1992)" (Fiedel 2005: 13). Although a Middle Archaic dog burial was known from the 1970s (F. Hill 1972), the definitive work on the three early Archaic Koster dogs, establishing both their taxonomic identity and age of about 8,500 years old, was not forthcoming until Morey & Wiant's later (1992) work. But Fiedel's statement is incorrect in that, quite simply, the Koster dogs are not the oldest securely identified dogs from North America. Although they are the oldest known established dog *burials* from North America, the oldest securely identified dog from North America is represented by specimens from Utah, dating in the range of 10,000–9000 B.P., as reported by Grayson (1988: 23). Fiedel notes this case in the following manner: "A *possible* early dog (*c*. 9000–10,000 RCBP) was identified in Danger Cave, Utah (Grayson 1988)" (Fiedel 2005: 13, emphasis added). One wonders why this dog was designated as being merely a "possible early dog." Grayson, an experienced zooarchaeologist, expressed no reservations about its taxonomic identity: "The two *Canis familiaris* mandible fragments (UU-23684; DC-87) came from a single element and were readily reassembled" (Grayson 1988: 23). He also provided an illustration of one fragment, from two different views, and labeled as the caption to Figure 5: "Danger Cave *Canis familiaris* mandible UU-23684 from stratum DII" (Grayson 1988: 23). Grayson's identification is unequivocal, and in the very paper on the Koster dog burials that Fiedel (2005) calls attention to, Morey & Wiant (1992: 225) wrote: "At present, specimens from Danger Cave, Utah, dating between 9000–10,000 B.P., are the *oldest well-documented* remains of domestic dog from North America" (emphasis added).

Since dogs earlier than Koster would be relevant to Fiedel's case, one wonders about his handling of the Danger Cave material. Danger Cave has the earliest securely documented dog remains from North America, and while it is certainly legitimate to suspect that some dogs were present in Paleoindian times, if so, they haven't been securely identified. Certainly, one cannot credibly advance a case for them having played any more than a largely invisible role at that time. In that sense, North American Paleoindian dogs, if present, are something like Dorset dogs in the North American Arctic, covered in the next chapter. That is, by certain elements of logic they should be there, but they haven't been securely identified yet.

THE USE OF DOG PRODUCTS

Eating dogs, or using them for transportation and/or for hunting purposes, are clearly consequential uses of the animals. In this section of the chapter, the objective is to highlight some of the other ways that dogs, or more properly, the products of dogs, have been used. Very likely, nothing covered in this section qualifies as a consequential use, in a selective sense, but what I summarize here constitute a few examples of the other useful roles that can be attributed to dogs, all of which characterize certain other animals as well.

Dog Wool?

Obviously, dogs are not generally thought of as animals that provide wool. On the northwest coast of North America, though, some apparently did. At about the time of European contact, about 1790, a particular kind of dog was known as a wool dog. This circumstance has been covered by Crockford (1997: 3; Crockford & Pye 1997), and more recently by Barsh et al. (2006). Regarding these wool dogs, the following description tells the basic story: "Wool dogs were said to have been deliberately bred and sheared like sheep for their wooly fur, which was woven into blankets" (Crockford & Pye 1997: 149). These dogs were apparently restricted to a small area of southern British Columbia and northern Washington State, and other dogs were larger village or hunting dogs. Regarding these wool dogs, now apparently extinct, there is even a direct archaeological indication that such a function really existed (Schulting 1994). Again, this use may not have been too consequential, in a selective sense, but it was a pragmatic use for a dog product.

Dog Bones Used for Tools or Ornamental Purposes

That dog bones were used to make tools sometimes is really no surprise. Bones from many different animal species were used similarly. Often enough, the functional nature of the tool can be inferred, sometimes with great reliability. In other cases, such an assessment is difficult, if not impossible. Since nothing about the use of dog bones for tools sets them apart, only one example is mentioned here, because it is directly relevant to the next chapter on dogs of the Arctic. That example is from a late prehistoric settlement in northeast Greenland, where Glob (1935: 76) noted that "A piece of dog humerus bears the marks of much wear at the middle and may have been used as a handle for a towing line." This use clearly reflects the use of a dog bone for a basic functional task, and the use of their bones in such ways, including some extensive and distinctive modifications to them, plays a conspicuous role in the next chapter.

Dog bones have also been used to make ornamental objects. For example, at Qeqertasussuk, the Paleoeskimo site (ca. 3900–3100 B.P.) in west Greenland with the illustrated cut-marked dog bone (Figure 5.1), one item recovered is what Grønnow & Meldgaard (1988: 437) illustrate and describe as a dog bone needle case. Also in west Greenland, in what are referred to as the mummy caves at Qerrortût, dating to within about 150 to 200 years ago, there were a number of mummified human bodies, sometimes accompanied by mammal teeth with holes drilled through them. In all, the situation was described as follows: "Pierced teeth to the number of 17, vis a bear incisor, four *dog* incisors, a seal canine, ten fox canines, and two fox molars" (Mathiassen & Holtved

1936: 107, emphasis added). In this case, dogs were clearly not singled out more than some other animals. In fact, modified or not, animal parts of a variety of species are often placed with human burials in different settings (O'Shea 1984: 304–316), but the use of dogs' teeth in this fashion merited a brief note now. Bearing in mind that Mathiassen & Holtved (1936), quoted previously, did not indicate a suggested role for those teeth, the meaning of some similar burial accompaniments elsewhere is less ambiguous. Specifically, in dealing with dogs' teeth found with human burials from Archaic Period sites in Kentucky and Alabama, Haag (1948: 121) wrote as follows: "Further, there are several instances of the use of dog teeth for necklaces." Moreover, it is worth highlighting a case from Awatovi, a village in Arizona dating from the early 1200s A.D. until its destruction in 1701. At that site, "There is evidence that limb bones of adult dogs were used as sources for bones tubes or 'beads' by the inhabitants of the village" (S. J. Olsen 1976: 104). Once again, it seems that dog bones were being used for personal ornamental purposes.

Dogs as Objects of Artistic Expression

Beyond the use of dog bones or teeth for ornamental and/or pragmatic purposes, dogs also functioned sometimes as objects of human artistic expression. Moreover, they still do. For the past, though, Marion Schwartz (1997) devoted an entire chapter (Chapter 5) to the topic in her major work on dogs of the early Americas. In that chapter, and in a later presentation (Schwartz 2000), she emphasized artistic renditions produced by Mayan peoples of Mesoamerica, and Mississippian peoples of late prehistory in North America. In both cases she provided graphic examples. The use of dogs in this capacity plays a much more restricted role in this book because it has been dealt with ably before, and the overall focus here precludes all but the most cursory treatment of this aspect of the journey of the dog. Nevertheless, it is worth sharing a particular example that meshes nicely with one of the themes increasingly developed as this volume proceeds. That theme is how dogs have come to be regarded like people in many different settings. Figure 5.3 is a ceramic dog from prehistoric Mexico, from the Colima province, and dates to between about 2300 and 1800 B.P.[3] What makes

[3] Given the proportions of the dog depicted, one might recall Elizabeth Wing's (1984: 228) passage, quoted earlier, about dogs being fattened for consumption at the Aztec center of Tlatelolco in the early 1500s (A.D.). The Colima region of West Mexico is quite near the location of Tlatelolco (essentially present-day Mexico City), and while

this piece so appropriate for this volume is that the dog is anthropomor-phized, by virtue of the mask depicted as affixed to its face. This image also anticipates a theme developed most clearly in Chapter 7, namely the perceived spiritual qualities of dogs. That is, writing about a similar piece from the same general location and time fame, Peter Furst suggests that "the human mask may represent the animal's *tonalli*, its inner essence, soul, or life force, which is here given a human guise" (Furst 1998: 186, original emphasis).[4]

THE USE OF DOGS BY ARCHAEOLOGISTS

Finally, the intention here is not to be annoyingly trite and smug in pointing out that archaeologists sometimes use dogs (their bones), too. In fact, dog remains provide a kind of information that one might not initially suspect. Specifically, chemical signatures in dog bones have been used to arrive at general dietary inferences, about both people and dogs. As an Old World example first, working from an assumption that dogs ate generally similar foods as their people, with respect to predominantly marine versus nonmarine foods, Noe-Nygaard (1988) compared $\delta^{13}C$

this effigy is earlier in time by at least several hundred years, it renders reasonable a supposition that perhaps the dog depicted represents one fattened for that purpose. This is only a possibility, but it is reasonable to suspect so. Strengthening that case is the fact that this general region, including the modern state of Colima, may well be "the place of origin of the Aztecs, as well as of their ancestors the Toltecs" (Mountjoy 1998: 251).

4 It also seems fitting to (foot) note the continued occurrence of dogs in artistic expressions. There are many examples, and only a few are identified here, beginning with one that focuses on archaeological expressions. Specifically, the National Museum of Anthropology, in Roma, Mexico, maintains a collection of ceramic dog figures from the Colima Province, some reminiscent of the one shown in Figure 5.3, and of comparable age. More to the point for present purposes Carolyn Baus de Czitrom (1988) has assembled a booklet that displays images of many of these dogs. One that is especially similar to the one in Figure 5.3, also from the Colima region, is shown in Furst (1998: 185, Figure 26). That piece is part of the Proctor Stafford Collection, and is maintained in Los Angeles County Museum of Art. Close to modern times, an exhibition at the Bruce Museum, in Greenwich, Connecticut, features dogs as expressed in paintings, especially with people, from as far back as the Renaissance period of several hundred years ago. This exhibit also has some photographic images from recent times. Again more to the point for present purposes, the images of this exhibition have been published in a large format book (Bowron et al. 2006). In keeping with the theme of that exhibition, Eichhorn & Jones (2000) have devoted a book to studio portraits of dogs with people. Silverman (1985) and Hall (2000) have also assembled images of dogs in photographs. Finally, for this brief diversion, Brackman & Brackman (2002) have produced a brief booklet that assembles mostly photographic images of dogs in North America, especially as they appear with people.

FIGURE 5.3. A ceramic dog from the Colima province of Mexico, dating between about 2300 and 1800 B.P. Note the human mask affixed to its face (see fn 3). This piece is in the Gilcrease Museum in Tulsa, Oklahoma, and the image of it above was provided by the Thomas Gilcrease Institute of American History and Art in Tulsa, Oklahoma. Photograph by Shane Culpepper, Gilcrease Museum.

isotopic ratios in dog and human bones. The range of these ratios can be taken as a general indicator of the marine versus nonmarine origin of most foods eaten. She was working with several late Mesolithic/early Neolithic sites in Denmark and found that dogs and people shared similar ratios, the values of those ratios depending on proximity to coastal areas. Shortly after that study, Clutton-Brock & Noe-Nygaard (1990) generated comparable data on dog bones from Star Carr and nearby Seamer Carr in England. Both were situated off the coast, but the chemical signatures indicated a predominantly marine diet. From this unexpected result, they concluded that the peoples responsible for these sites probably lived for most of the year in immediate proximity to marine resources (on the coast), and those two sites were likely temporary camps where the dogs died. Their results were apparently corroborated by Schulting & Richards (2002), though S. P. Day (1996) raised some technical objections, with interpretive implications, to Clutton-Brock & Noe-Nygaard's (1990) original study. Subsequently, Dark (2003) raised some technical objections to Schulting & Richards's (2002) more recent study as well.[5]

Regardless of these technical disputes, it is noteworthy that the same basic analytical approach has also been used productively in the New World. As one example, Burleigh & Brothwell (1978) used stable carbon isotope ratios from hair and bone collagen samples from archaeological dogs from Peru and Ecuador to arrive at an inference that maize had formed a part of their diet. Though skeletal remains of dogs from Peru, dating within the range of about 3000–700 B.P., were later studied from the perspective of morphometric variation (Brothwell et al. 1979), the dog remains studied by Burleigh & Brothwell (1978) included one from Ecuador in the vicinity of 5,000 years old, and a later one from Mexico (ca. 1700 B.P.). Sporadic evidence of maize in the coastal Peru region comes in at about 3800 B.P., and for the dogs, dating somewhat later than that, they suggested the possibility that "maize formed between 20–60% of the diet of these dogs" (Burleigh & Brothwell 1978: 358). Overall, they estimated that, after about 2850 B.P., "maize was a significant part of the diet of these animals" (Burleigh & Brothwell 1978: 360). They also raised, as a matter of incidental interest, the question of whether these dogs were perhaps being "deliberately fattened for human consumption" (Burleigh & Brothwell 1978: 358). In light of Wing's (1984: 338) observation about dogs at the Aztec marketplace at Alcomon, quoted previously, it is reasonable to suspect so.

Elsewhere, the Josey Farm site in Mississippi (U.S.) dates to between about 1300 and 1840 A.D. (Hogue 2003: 186), and a nearby site (22OK904) dates within the latter portion of the same time span. There is no question of maize cultivation in this general temporal and regional context, but three dog burials, two from Josey Farm, the other from 22OK904, provided a way to evaluate the applicability of this basic analytical technique. One result was that "Carbon isotope analysis clearly shows that

---

5 More recently, Anders Fischer and colleagues have conducted comparable investigations using bones from 75 people and 27 dogs from some two dozen sites in Denmark. Comparisons were broad, involving both Mesolithic and later Neolithic contexts, representing both coastal and inland contexts. Significantly, they concluded that "the consumption of marine food was clearly much less important in the Neolithic than during the Middle and Late Mesolithic" (Fischer et al. 2007: 2147). Moreover, "Middle and Late Mesolithic humans and dogs found in the Danish interior have all consumed significant amounts of marine food [and] Most probably the Middle and Late Mesolithic population in the coastal zones of Denmark moved seasonally between the seashore and the interior" (Fischer et al. 2007: 2147). Again, dog bones (as well as human bones) have turned out to be useful to archaeologists.

the proto-historic dog diets are similar to the human diet" (Hogue 2003: 189). The significance of this result is that there was no direct archaeological evidence of maize agriculture, at least at Josey Farm, and this study indicated its role. Therefore, this approach can clearly be useful as a means of assessing empirically the relative role of agricultural products in settings where their general use is inferred or known, but direct archaeological evidence is lacking.

CLOSING THOUGHTS ON THE PAST USES OF DOGS

The ways in which dogs have been used that have been covered are undoubtedly only a small subset of the ways that existed, albeit some of the conspicuous ones. Beyond the utilitarian roles highlighted, the frequency with which dogs appear in artistic representations, both past and present, signifies the substantial role that this animal can take in the human psyche. In concluding this chapter, one can legitimately wonder if dogs of the past fulfilled certain roles that they sometimes play today. For example, were dogs of the past used in racing, as they are sometimes today? Similarly, were there prehistoric versions of dog shows, in which dogs were displayed and competed for honors, based on their appearance and manner of comportment? Did dogs serve a warning function regarding unwelcome intruders? Answers to such questions are elusive, and it is hard to know what kind of archaeological evidence might suggest such roles for the past, if they existed. Except for the last possibility, it is legitimate to suspect that they did not play these specific kinds of roles until recent times, but they likely played some distinctive ones in the past, tailored to the lifeways of certain groups of people. Raising this probability in closing this chapter only serves to emphasize that people are creative, and just as we are creative with dogs in our own societies today, there is no reason to think that past peoples were any less so.

6

DOGS OF THE ARCTIC, THE FAR NORTH

Oh, Greenland is a dreadful place,
A land that's never green,
Where there's ice and snow, and the whalefishes blow,
And the daylight's seldom seen, brave boys,
And the daylight's seldom seen

Colcord 1938: 148

THE ARCTIC AS A REGION

Geographically, the lands that comprise the Arctic represent but a small subset of the total land surface of the world. Similarly, the people who have inhabited this region represent only a small subset of the world's human population. Those people, however, have lived in some quite distinctive ways, a pattern that is not surprising given the distinctive environmental challenges they regularly faced. The passage at the beginning of the chapter, the last verse of a traditional eighteenth-century British whalers' song known as "Greenland Fishery," captures something of the perception commonly held by many European-Americans about arctic environments. Especially intriguing, given the purposes of this volume, is a real contrast in the roles played by dogs in those environments at different times. Specifically, they sometimes played a conspicuous and even vital role among people in some contexts, but only a minor role, if any, under other circumstances that entailed similar or even more challenging environmental conditions. The purpose in this chapter is to describe these situations and explore the different roles that dogs played. Given the ephemeral presence of dogs at some times, in contrast to their clear importance at others, "from a theoretical perspective, tracking the evolution of the canid–human domestic relationship in the Eastern Arctic becomes an even more intriguing challenge" (Morey & Aaris-Sørensen 2002: 53). In keeping with that statement, the primary focus here is the eastern North American Arctic, as defined below, simply because I have the most background in that region. This treatment does not, however, ignore arctic areas in other parts of the world.

As an initial frame of reference, Figure 6.1 presents a generalized map of the North American Arctic. The easternmost region of northeast Siberia is also depicted, technically a part of Asia, not North America. Its relevance to the North American Arctic is clear in several ways, though, and for this volume its key role in the peopling of the New World is an obvious one. At later junctures in this chapter, views of more specific portions of the North American Arctic will be important, including certain site locations. But for present purposes, it is sufficient simply to note that, by convention, the North American Arctic is often separated into three broad divisions: (1) the eastern Arctic, consisting of Greenland and northeastern Canada, including the islands of the Canadian archipelago between Greenland and mainland Canada; (2) the western Arctic, comprised of parts of Alaska, especially northwestern coastal Alaska, along with extreme northeastern Siberia, and (3) the central arctic, the expanse in between the eastern and western arctic. There are no definitive boundaries separating these areas, just the broad, vaguely defined transitional zone, the central Arctic. Harp (1984) employs these basic divisions in outlining the course of work in the North American Arctic after about the mid-twentieth century. At any rate, these broad divisions of the North American Arctic structure much of the content of this chapter. Moreover, the eastern Arctic, the main focus here, is considered distinctive enough as a region that others (e.g., Maxwell 1985; Helmer 1994) give it separate treatment.

As indicated earlier, one of many intriguing aspects of life among past arctic peoples is variation in the role that dogs played. That is, at some points dogs played virtually no role, if any, while at others they played a discernable though apparently minor role, and at yet at others they clearly played a major role, which extends up to the historic period. But this pattern does not follow a general temporal sequence, a sequence that would be intuitively expected. That is, the period in which they were largely absent occurred between the other two periods.[1] This

---

[1] It is well worth pointing out that this disparity is accompanied by the likelihood that earliest (Paleoeskimo) peoples and latest (Thule/Inuit) peoples originated from

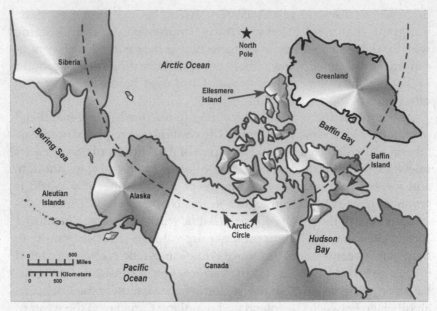

FIGURE 6.1. Generalized map of the North American Arctic, with the eastern end of Siberia included.

account, however, does follow a temporal sequence and begins with the early period when they did play a discernable role. In the eastern Arctic, this includes what is known locally as the Saqqaq culture, a Greenland variant of the Arctic Small Tool Tradition (see Fitzhugh 1984; Maxwell 1985), the latter spanning a period of time from about 4500 to 1000 B.P. (Maxwell 1985; Helmer 1994: 18; Møbjerg 1998). In western Greenland, Saqqaq spans the period from about 3900 to 2700 B.P. (Grønnow 1994: 201).

EARLIEST PALEOESKIMO DOGS

Figure 6.2 presents the basic time frame to be followed in this chapter.[2] The complexes that are of primary concern are pre-Dorset

> largely unrelated genetic sources, a probability highlighted by recent mtDNA work (Gilbert et al. 2008). In short, there is no meaningful indication of historical continuity between the two groups, so their practices regarding dogs arose independently. But the authors of that study indicate that how Dorset people, the intervening group, fit into this mtDNA picture is still not known. A consideration of the roles of dogs among all these groups comprises the bulk of this chapter.

[2] This figure was adapted from a previous version (Morey & Aaris-Sørensen 2002: 45, Figure 2), and it contains some detail that is not of major concern here. Specifically, the

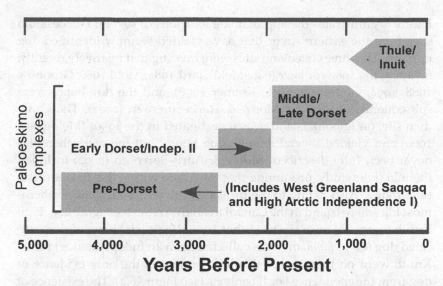

FIGURE 6.2. Simplified timeline of major cultural complexes associated with the North American eastern Arctic. Adapted directly from Morey & Aaris-Sørensen (2002: 45, Figure 2).

(including Saqqaq), Dorset, and Thule. By convention, the first two are regarded as Paleoeskimo, while Thule is sometimes regarded as Neoeskimo (e.g., D. Morrison 1994). As indicated on this timeline, Thule, an archaeological designation, and Inuit, referring to recent or modern people, are the same basic phenomenon for present purposes. Regardless, just less than 5,000 years comprise the entire sequence. One might recall from the previous chapter the suggestion made by Charles Arnold (1979) that dogs likely were present in early Paleoeskimo times, based on vertebral pathologies found on canid vertebrae. Arnold felt, in fact, that his conclusion had broader implications:

> This conclusion lends some measure of support to the idea that domestic dogs are an integral part of cultural adaptation to the arctic, and as such will probably be shown to have had a widespread distribution in Paleoeskimo cultures. (Arnold 1979: 265)

Further work has not supported this idea. To see why, it is best to begin with the earliest dogs that are securely documented. Specifically,

pre-Dorset and Early Dorset spans include units designated as Independence I and II. These are High Arctic units, named for a particular fjord (Knuth 1966/67), and receive only brief attention in this presentation.

Figure 6.3 illustrates the approximate locations of several Paleoeskimo sites from the eastern Arctic that have yielded securely identified dog remains. The three Greenland sites were investigated relatively recently, meaning the 1980s or later (e.g., Meldgaard 1983; Møhl 1986; Grønnow 1988, 1994; Gotfredsen 1996; Kramer 1996), and the dog bones were subsequently studied by Morey & Aaris-Sørensen (2002). The Canadian site, on Igloolik Island, was investigated in the 1950s (Meldgaard 1960) and yielded several definite dog bones, but the work there has never been fully described. Morey & Aaris-Sørensen (2002) included them in their study, presuming that they date within this Paleoeskimo time frame. Also dating to the pre-Dorset time frame, from northernmost Ellesmere Island in the Canadian High Arctic (see Figure 6.3), Eigil Knuth (1967: 32) once reported that "one diggit [sic] bone of domesticated dog (*Canis familiaris*) was collected from an Independence I ruin." Knuth went on to note that this lone bone was the only evidence of dog from Independence I or II contexts (see Figure 6.2). The existence of that single dog bone has not been verified. Some other suggestive but unconfirmed cases that might represent dogs have been pointed out by Morey & Aaris-Sørensen (2002: 47).

One of the noteworthy features regarding the dog bones from the three Greenland sites is how infrequent they are, compared to other animals that are represented. There are a total of 79 dog bones from those sites, from among some 200,000 examined animal bones in all (Morey & Aaris-Sørensen 2002: 49). All told, those bones accounted for eight individuals. Moreover, from stratified sequences with some time depth, the dog bones were not scattered continuously through the deposits, occurring almost exclusively in the earliest or latest deposits at the sites. It is clear that while some dogs were present, at any given time there could only have been one or a few. Figure 5.1 highlighted the one role of these dogs that could be reliably ascertained. That is, at least some of them apparently were eaten. In short, what other roles they may have played is not clear, and the occasional dog can hardly be considered to have been an important subsistence resource, except maybe at a time of genuine food shortage. It is entirely reasonable to suspect that these few early dogs were used as pack animals, as suggested by Arnold (1979), or as aides in hunting. Both suggestions have been made at various times for dogs of this time period (e.g., Maxwell 1984: 361; 1985: 89; Dumond 1987: 92; McGhee 1996; Gotfredsen 2004: 167). Regarding the latter suggestion, in a thorough and compelling work,

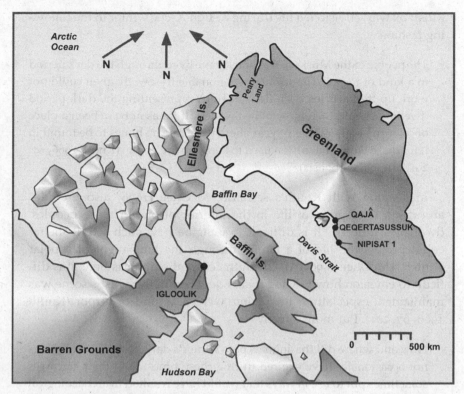

FIGURE 6.3. General map of the eastern Arctic, showing the approximate locations of four Paleoeskimo sites that have yielded identified dog bones. Adapted from Morey & Aaris-Sørensen (2002: 45, Figure 1).

McGhee (1996: 56–57) provides an envisioned scene in which a group of Paleoeskimo hunters in High Arctic northern Greenland prepare to dispatch some muskoxen, in defensive formation surrounded by dogs. As McGhee (1996: 57) notes in the explanation of this scene, "Muskoxen are vulnerable to human hunters accompanied by dogs. Almost an entire herd might be killed before breaking from its defensive formation." In keeping with such a scenario, this region (and time) has been referred to as "the Musk-ox way," and refers especially to the Independence I and II complexes, as indicated on Figure 6.2 (Knuth 1966/67).

If the few Paleoeskimo dogs that may have been there were used in such a way, an intriguing question concerns how they passed the long winter night in the High Arctic. Knuth, who championed the term

Musk-ox way, envisioned life during a High Arctic winter in the following fashion:

> The people of the Musk-ox way must have lived through the dark period in a kind of torpor. The women were unable to sew, the men could not work up flint articles. Rich finds at ruins representing the dark period dwelling should therefore not be awaited. If it has not *also* been a place of sojourn at other times of year, there would be no bones to be found in it for these would have been used to keep the fire going. (Knuth 1966/67: 201, original emphasis)

In keeping with Knuth's scenario, McGhee (1996) also provides an artist's conception of life in the interior of one of these peoples' dwellings. Indeed, "It is difficult to imagine how such a frail shelter could protect the life of a family through the long night of a polar winter" (McGhee 1996: 63). Given the current emphasis here, it is difficult to envision how the dog that is depicted in the artist's scene was maintained, especially if the people were in a kind of torpor (Knuth 1966/67: 201). Put more directly,

> How and where did the animals pass winter's dark period? Would dogs not occasionally have expired during these times, with their carcasses sometimes put to use in ways that would have resulted in archaeological evidence that they were regularly present? (Morey & Aaris-Sørensen 2002: 53)

The purpose in raising those questions was not to suggest that these peoples absolutely never had dogs. But there is no evidence that they routinely did, and "there is no archaeological evidence to support an assumption that dogs were a regular part of Paleoeskimo life" (Morey & Aaris-Sørensen 2002: 53). Moreover, there apparently is no evidence that has emerged since that study that would call that statement into question. For example, subsequent to that work it has been pointed out, based on a comprehensive review of patterning among animal bone assemblages across the eastern Arctic, that dogs "are exceedingly limited and found in isolated patches until Neoeskimo migration into the area around 1,000 years ago when dogs became a widespread part of arctic culture" (Darwent 2004: 65). So there were a few dogs during earliest Paleoeskimo times, but to attribute a major role to them during this span is far-fetched. Dogs were an important part of life in the Arctic much later, but before they assumed that distinction they were apparently of even more marginal importance to people, if playing any role whatsoever.

DORSET DOGS (?)

As explained by McGhee (1996: Chapter 6), the Paleoeskimo complex known as Dorset was named in 1925, based on artifacts found near Cape Dorset on Baffin Island, between mainland Canada and Greenland. Emerging at roughly 2,500 years ago, it was associated with colder conditions, and "The Dorset people apparently thrived during a period when the climate was rapidly cooling, eventually reaching conditions that were significantly colder than those of today" (McGhee 1996: 117). Dorset people predominated in the eastern Arctic until Thule times, and lived under some of the harshest conditions ever known in the Arctic. And they did not merely live for they apparently thrived, as McGhee noted, for a long time. They are, for example, responsible for some of the most vibrant and intriguing artistic output that is known archaeologically, as highlighted by McGhee (1996). In addition, they thrived while giving up or never developing certain technologies that other arctic peoples have found important or, in certain cases, vital to their well-being. For example, they apparently gave up the bow and arrow, known among earlier Paleoeskimos, and seem to have lacked the efficient boats and open water hunting gear associated with later peoples. For present purposes it is worth emphasizing that they had few, if any, dogs, and so apparently lacked the assistance of dogs in hunting or, in archaeologically tangible terms, transportation. That is, there is no indication of dog sleds, so essential to later peoples, as emphasized in the next major section.

Dogs for Help in Hunting: Where Are They?

For the present, however, their success in hunting, apparently without the aid of dogs, has occasioned comment. Specifically, in writing of Dorset people, Moreau Maxwell has related that

> Middens contain remains of caribou, musk-ox, polar bear, and other inland mammals, birds, varieties of seal, walrus, narwhal, and beluga [see Table 6.1]. To simply list these animals, however, glosses over amazing feats of human skill and endurance. Hunting polar bear with trained dogs and a high-powered rifle, or walrus from a wooden schooner,

TABLE 6.1. *Examples of Thule/Inuit sites, or sites with Thule components, that have yielded trace buckles or occasionally whip shanks, used for team dog-sledding. The sites are primarily in the eastern Arctic, especially Greenland; however, one case (Kittigazuit) is from the western Canadian Arctic.*

Site, Country	Reference
Kittigazuit, western Canadian arctic (coastal)	McGhee 1974: 42 (Table 1)
Three.Sites, Baffin Island, Canada	Schledermann 1975: 180–183, 194–195, plates 21, 22, and 28
Brooman Point, Bathurst Island, Canada	McGhee 1984: 120–121, plate 11
Dead Man's Bay, Clavering Island, northeast Greenland	Larsen 1934: 111
Several settlements in the Kjempe and King Oscar Fjord regions, northeast Greenland, including Geographical 2	Glob 1935: 43, 52, 63, 70, plate 6
Igdlutalik, west Greenland	Mathiassen 1934: 71, 86, plate 1
Inugsuk, west Greenland	Mathiassen 1930: 204, plates 21 and 23
Qeqertarmiut, west Greenland	Mathiassen 1931: 67, 142, plate 2
Utorqait, west Greenland	Mathiassen 1931: 22
Sermermiut, west Greenland	Mathiassen 1958: 43
Misigtoq, Kangartik, southeast Greenland	Mathiassen 1933: 69, plate 6
Ruinnaesset, southeast Greenland	Mathiassen 1936: 38 (Figure 21), 41

exposes one to a certain amount of danger. It was more dangerous still in the pre-rifle hunting days when walrus were pursued by a crew of harpooners from a large umiak. But to hunt these dangerous animals *without dogs* or large boats and within the force range of a man's arm – 20 m at best – is incredible. (Maxwell 1985: 129, emphasis added)

As an aside, the umiak mentioned by Maxwell is the more sophisticated version of the North American plains bullboat that figured in the previous chapter. The frame of an umiak was usually made of driftwood or whalebone supports that were lashed or pegged together, with bearded seal skins stretched over that framework. Saladin d'Anglure (1984: 486) has provided a photograph of a group of Inuit people in one at Wakeham Bay, Quebec, in the early 1900s. In a classic monograph on "The Central Eskimo," Franz Boas (1964: 119–121) provided details of its construction, as known in the late 1800s. Among Thule peoples (see below) dogs could be transported in umiaks, alongside people. As for where they are (dogs) in Dorset times, for any purpose, the only adequately documented case at present is a single skull from early Dorset deposits, dating to about 2200 B.P., from the Nanook site on Baffin Island (Cleland 1973). That single case occurs as one of tens of thousands of animal bones that have been examined from Dorset contexts across the eastern Arctic.

Thus, it does appear that there was only an occasional dog among Dorset people, and those few played an indiscernible role, at most, in helping Dorset people hunt. The overall success of Dorset hunters, in spite of that fact, and in spite of their other limitations, is genuinely amazing, as emphasized previously by Maxwell. In fact, the near-absence of dogs in Dorset times is often regarded as one of several remarkable oddities of these people (e.g., Dumond 1987: 97; Maxwell 1985: 129; McGhee 1996: 145–146). Earlier Paleoeskimos did not have many dogs, but they did have enough that their inconsistent presence can be documented. Remarkably, it does appear that Dorset people largely gave up dogs, like a number of other things that their inventory lacked. It is not clear why these patterns came about. But as puzzling as it is how Dorset people managed and even thrived without dogs for hunting, it is at least as intriguing to ponder this absence as it pertains to their basic transportation capabilities.

Dogs for Help in Transport: Where Are They?

Among pre-Dorset Paleoeskimos, the few dogs present were sometimes eaten, as noted, but the possibility that they served as individual pack animals is also quite reasonable, given evidence in the form of vertebral paleopathologies, as covered. But strong contrasts exist between the earlier Paleoeskimos, Dorset people, and especially the people who followed, the Thule, precursors of the modern Inuit. Among Thule peoples, dogs were not only present in substantial numbers, they played a vital role in transportation, and gear associated with the role of dogs in transportation is routinely recovered from Thule archaeological sites. Robert McGhee has made this point effectively, and in direct contrast to Dorset people:

> There is no archaeological evidence, for example, to indicate that the Dorset people used dogsleds. We can be sure that at least some Dorset groups must have had dogs. . . . But there is none of the complex harness tackle used by Inuit dog handlers and no other evidence that dogs were

used to pull sleds. In fact, an amusing Inuit tale of the Tunit [Dorset] relates that these people dressed in bearskins, a material that slides almost frictionless over sea ice; when a man wished to travel, he hooked his dog to the hood of his parka and was pulled along sledless across the ice! (McGhee 1996: 145–146)

At least two points are worth noting about McGhee's passage. First, he, like many others, feels sure that there must have been some dogs among Dorset people, though he knows how meager the evidence is. Second, modern Inuit people, who continue to make use of dogs for transportation (see discussion that follows), are apparently befuddled by the Dorset situation as well. This amusing Inuit tale reflects both their amazement and their own conviction that Dorset people must have had dogs. As the next section makes clear, dogs were an integral part of life among Thule people, a role transplanted, in modified form, into the modern era. Thus it is apparently hard for Inuit people, too, to conceive of life in the Arctic without dogs.

To be sure, Dorset people did have sleds, a point that has been made clear by Maxwell (1985: 152–153). But the presence of sleds does not automatically signify that dog teams were used, a point illustrated previously by McGhee's observation that Dorset sites lack any recognizable harness tackle that signifies team dog-sledding. This is an important point to bear in mind when considering the Siberian case discussed later. For the moment, though, to offer a quick answer to the question posed in these last two brief (sub)sections, I do not know where the Dorset dogs are. They apparently were not there, except once in a great while. As for why, the Dorset pattern needs to be understood in its larger context:

> there is no need to generate a special explanation for the virtual absence of evidence for dogs from Dorset contexts. . . . Given overall limitations to their mobility and associated subsistence production, we find nothing odd about the Dorset pattern. (Morey & Aaris-Sørensen 2002: 53)

So, unlike succeeding Thule people, Dorset people do not seem to have pursued a way of life for which dogs would have been a real asset. In this volume about dogs, it may seem odd to have invested even this much verbiage in a people that basically did not have the animals. But it is a matter of intriguing contrasts, one with earlier Paleoeskimos, among whom there were clearly some, though not many, dogs, and another much more major one, with the Thule people.

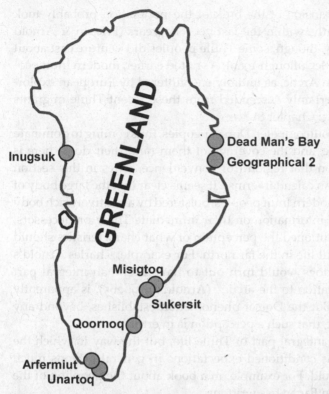

FIGURE 6.4. Map of Greenland, showing the approximate locations of several Thule sites from which bones are illustrated in this chapter.

THULE DOGS

The Thule Tradition originated slightly more than 2,000 years ago in the western Arctic (Dumond 1984: 101–105; Hoffecker 2005: 134), and its peoples developed a maritime focus (Maxwell 1985: Chapter 8; Schledermann 1996: Chapter 9). Figure 6.4 is a generalized map of Greenland, showing the approximate locations of several Thule sites that are highlighted in this chapter. The Thule Tradition was given its name by Danish archaeologist Therkel Mathiassen (1927) who identified three key transportation possessions, among the different traits he listed, that made Thule peoples so remarkably successful: dogsleds, kayaks, and umiaks. By longstanding convention, from their area of immediate origin in the western Arctic, beginning some 1,000 years ago, Thule peoples spread rapidly eastward through the Canadian Arctic and into Greenland. Recent information suggests, however, that the eastward expansion of the bulk of the population probably took place more recently, within the last 700–800 years (Friesen & Arnold 2008). Apparently, though, some Thule peoples did venture east about 1,000 years ago (McCullough 1989). As noted earlier, modern Inuit peoples of the eastern Arctic, as initially encountered by European explorers, are almost certainly descended from these recent Thule migrants (see Maxwell 1985: Chapter 8).[3]

Thule peoples outcompeted Dorset peoples, thus coming to dominate the eastern Arctic, and the coverage of them (and their dogs) here is largely focused on that region (for convenience, dates in this section appear in Christian calendar terms). It seems clear that the large body of information on modern Inuit peoples, bolstered by a relatively rich body of archaeological information on their immediate Thule predecessors, has strongly conditioned the perception of what characteristics should reflect a successful life in the far north. For example, Charles Arnold's suggestion that dogs would turn out to have been "an integral part of cultural adaptation to the arctic" (Arnold 1979: 265), is apparently a case in point. But the Dorset phenomenon establishes, beyond any reasonable doubt, that such a perception is in error.

Dogs were an integral part of Thule life, but the way in which the Thule pattern has conditioned expectations in general extends much further than Arnold. For example, in a book about the peopling of the New World, Brian Fagan has written:

> A burial of a domesticated dog dated to about 11,000 years ago is the earliest record of this vital animal in the far north. Perhaps this was the point at which dog-sled travel became a viable means of getting around on arctic ice, a technology that was to have momentous consequences in later millennia. (Fagan 2004: 97)

The dog burial that Fagan is referring to is at Ushki-1, from extreme northeast Asia, highlighted in Chapters 2 and 3. It seems as though the Thule pattern conditioned him to anticipate evidence of dog-sledding at this time, and it is unfortunate that he apparently did not

3 Historically, some sites and associated complexes subsumed here under cultural historical category of Thule have been regarded as distinctive enough to warrant a separate designation, though clearly linked to Thule (e.g., Inugsuk). The use of one general label for these later people does not at all imply that Thule should be regarded as some kind of a homogeneous entity. To the contrary, there is much regional variation, and the use of one inclusive label implies only that different peoples regarded as Thule were broadly similar in some fundamental ways. That is to be expected, given that they were all surely descended from the same basic stock.

consult the relevant literature. Specifically, in the 1990s Pitul'ko & Kasparov (1996) reported on the Zhokov site in Siberia, documenting a fragment of a wooden sled runner and some domestic dog skeletal remains. Accordingly, they sought to make a case for early dog traction at that site. The first point worth noting is that the time frame, about 8,000 years ago, is younger than Ushki-1. More compelling is the fact that none of the recovered objects indicate the kind of equipment that typically occurs in Thule contexts where dog-sledding occurred, for example trace buckles (see discussion in the next section).

Recognizing the limitations of what they found, and that compelling evidence of dog traction comes from a later time frame, Pitul'ko & Kasparov (1996) nevertheless suggested the possibility of dog traction. However, they concluded that portion of their larger study by stating that "Even so, it is impossible to prove, at this point, that dog traction was actually practiced in the High Arctic as far back as 8000 B.P." (Pitul'ko & Kasparov 1996: 14). As suggested earlier, sleds alone are not compelling evidence of dog traction. So, Fagan was off base, and there is still no convincing evidence of team dog-sledding in the Arctic prior to the general Thule time frame. At that juncture, it did indeed start to become important, and it is now appropriate to explore directly how Thule people utilized their dogs. Robert Park has provided a good opening framework statement:

> The importance of the domestic dog to traditional Inuit culture cannot easily be overestimated, since so many aspects of their organization for mobility and subsistence were closely tied to the use of dogs. (Park 1987: 184)

Dog-Sledding Evidence

Accordingly, it is fitting to begin with the abundant direct evidence of dog-sledding among Thule peoples in the eastern Arctic. Figure 6.5 is an image of several bone trace buckles from a Thule site on Baffin Island, Canada, as reported by Schledermann (1975: 68–95). Such items were part of the harnessing gear used for dogs in sled teams, as known from the historic Inuit. They are routinely reported from Thule sites in the eastern Arctic, along with sled parts such as documented by Pitul'ko & Kasparov (1996), as summarized earlier. But again, Pitul'ko & Kasparov recovered no such harnessing gear from their Siberian site. Table 6.1 assembles several examples of Thule sites, from High Arctic

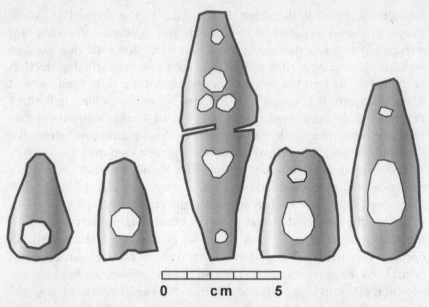

FIGURE 6.5. Schematic representation of several bone trace buckles, as fastened to the end of a dog trace, part of the harnessing system used for dogs that were members of dog sled teams. These items are from the B-1 site on Baffin Island, Canada, dating from about 700 B.P. to the early period of Euro-American contact (Schledermann 1975: 68–95). The elongated central piece was evidently unfinished, and was "in the process of being shaped into two trace buckles" (Schledermann 1975: 120). These images are adapted from photographic images in Schledermann (1975: Plates 21 and 22, pages 181 and 183).

Canada and Greenland that have yielded trace buckles and/or whip shanks, associated with team dog-sledding. There are undoubtedly many more, but these few serve to demonstrate the widespread occurrence of such paraphernalia, and they simultaneously attest to Therkel Mathiassen's pioneering role in archaeological research in Greenland. For present purposes, he and several colleagues documented dog-sledding gear from sites in most areas of Greenland, including into the southeastern region (see Table 6.1). Only from the most southerly region, for example the Julianehaab District of the extreme southwest, were no trace buckles reported. However,

> They have had fairly large numbers of dogs. We have no evidence of sledge-traction, but the dog sledge must have been known to the Eskimos who migrated round to the East Coast, where natural conditions were more suitable for it. (Mathiassen & Holtved 1936: 88).

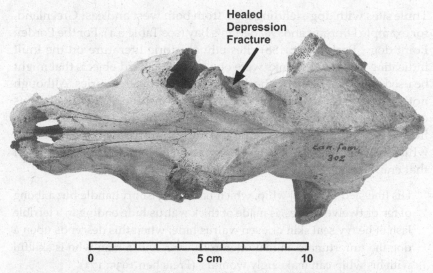

FIGURE 6.6. A dog skull from the Inugsuk site, in west Greenland, highlighting the healed depression fracture (arrow) in the right frontal bone. Photo by Geert Brovad, Zoological Museum, University of Copenhagen, Denmark.

In his classic work on the North American Arctic in the late 1800s, Franz Boas (1964: 124) illustrated both trace buckles, and the manner in which the associated harness was put on a dog.

Aside from trace buckles, which attest to the reality of dog traction gear, the presence of whip shanks, found at some Thule sites, calls attention to another dimension of the disciplinary management of dogs by Thule and Inuit people. That is, bones of Thule dogs sometimes exhibit pathologies, clearly stemming from traumatic injuries, consistent with the animals sometimes having been severely disciplined as part of their management. Park (1987: 186–187), in the same article that stressed the importance of dogs in Inuit/Thule life, has provided three examples from the Porden Point site, on Devon Island in High Arctic Canada. Figure 6.6 is an example of a dog skull from Inugsuk, in west Greenland, from deposits dating in the time range of about 1200–1500. The healed depression fracture, pointed out, is reminiscent of skull no. 3 from Porden Point, Canada, illustrated by Park (1987: 187, Figure 4). In fact, out of thirteen more-or-less complete dog skulls from Inugsuk, five others also showed comparable healed injuries. Inugsuk (Figure 6.4) was excavated by Mathiassen (1930) and yielded both sled parts and trace buckles (see Table 6.1), attesting to dog-sledding in that context. Moreover, there were dog skulls with healed injuries from several other

Thule sites with dog-sledding gear, from both west and east Greenland, for example Utorqait and Dead Man's Bay (see Table 6.1). For the Porden Point dogs, Park (1987: 188) cites ethnohistoric literature on the Inuit indicating that whip shanks were one of several hard objects that might be used to deliver blows to the dogs, resulting in such injuries. Although not dealing with the use of whips in that fashion, Peter Freuchen's own words from the Fifth Thule expedition of the early 1920s leave little doubt about their periodic use in a more conventional fashion while the dogs were working among some Canadian Inuit peoples of that era:

> His [the sled driver's] whip, which has a very short handle but a thong of ten or twelve metres, is made of thick walrus hide ending in a terrible lash of heavy seal skin or even walrus hide; when this descends upon a dog the unfortunate animal often rolls over, and a man who is skillful with his whip can make ugly wounds. (Freuchen 1935: 173)

Thus it seems clear that Thule/Inuit dogs were sometimes disciplined, rather harshly, when they did not perform as members of a sled team in optimal fashion. Having them perform optimally was important because "The dog is indispensable, especially in winter when hunting journeys are made by sledge" (Møhl 1979: 387).

Pragmatic Uses of Thule Dog Bones

Beyond what happened to them when they were alive, when they were dead, from whatever causes were responsible in a given case, the bones of Thule dogs were sometimes put to use in some distinctive ways, as first illustrated here by evidence from the Qoornoq site in southeast Greenland (see Figure 6.4). The work at Qoornoq was included in Gotfredsen et al.'s (1992) basic summary of evidence for Thule occupations throughout the larger project area, in Skjoldungen Sound. I participated in excavations there during the summer of 1990, and the dog bone portions shown in Figure 6.7 (there were several more) were recovered at that time. These bone portions are associated with three different winter houses there. Figure 6.8 is one of the Thule house structures prior to any excavation work, while Figure 6.9 is the midden area in front of a house during excavation. Collectively, the houses apparently date to the 1700s, although one may be slightly more recent.

As for the bones themselves, these five clearly reflect a distinctive manufacturing process that I, for lack of a better term, referred to as

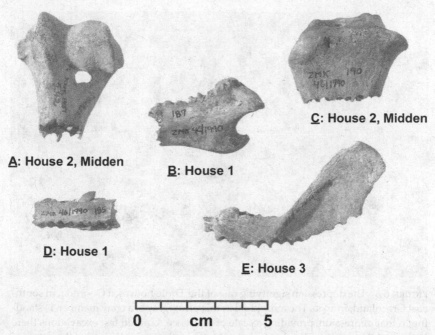

A: House 2, Midden

B: House 1

C: House 2, Midden

D: House 1

E: House 3

0   cm   5

FIGURE 6.7. Dog bones modified by the drill-and-snap technique, explained in the text, from Qoornoq, in southeast Greenland. A, right distal humerus; B, left mandible; C, right proximal tibia; D, left mandible; E, left mandible. Photo by Geert Brovad, Zoological Museum, University of Copenhagen, Denmark.

drill-and-snap. One can see on these pieces the contiguous sets of partial drilling holes, in some cases circumscribing the shaft of a limb bone (e.g., Figures 6.7A and C). Most likely the holes were made with a bow drill, as explained by Maxwell (1985: 282). Apparently in such cases the bone was separated into two pieces at the point of the drilling rows, and the piece that remained, in the form of those illustrated, was discarded. The piece that was not discarded was then modified into some kind of a recognizable implement, one that likely retained no direct evidence of this drill-and-snap procedure. Either that, or in a given case, it may have been set aside for such modification but was not further worked, for unknown reasons. The question, of course, is what was made from the piece that was not discarded, when that happened. Because such pieces are not likely to retain any recognizable evidence of the drill-and-snap process, their identity must be inferred on other grounds. Many years ago, Mathiassen (1936: 39, Figure 22) illustrated a seal mandible from a particular house at the Thule site of Ruinnaesset in southeast

FIGURE 6.8. The depression signifying one of the Thule houses at Qoornoq, in southeast Greenland, in 1990. The individual in this photograph, a crew member, is standing in that depression, providing a scale of reference. Limited test excavations there took place that season. The photograph is by the author, from the end of the entrance passage. The relevance of this picture is explained in the text.

Greenland that had been subjected to this drill-and-snap procedure. He in fact reported that there were eighteen such cases

> from which a length of the teeth had been drilled away (Figure 22.8) (similar specimens were found in the other houses, as stated); the piece drilled off has apparently had some purpose or other, but I am unable to say what. (Mathiassen 1936: 43)

Certainly these pieces were different from any of the three dog mandible portions in the illustration here, and the one he illustrated is most like the piece that would have been left from the creation of the one in Figure 6.7E. In short, this piece could be an example, as suggested earlier, of one that was set aside for further modification, but that modification never took place. Granting that possibility, and accepting Mathiassen's suggestion that such a piece from a seal jaw would have been the object of the process, it is possible to make some well-founded inferences as to what was made. However, doing so calls for a return to Qoornoq.

FIGURE 6.9. Excavation of the midden area in front of a Thule house at Qoornoq, southeast Greenland. The photograph, like the previous one, was taken by the author in 1990.

In two different presentations, Gotfredsen et al. (1992, 1994) have offered a compelling indication of what was likely made. In the first article they provided a photograph of a large piece, like that illustrated by Mathiassen, and the smaller piece evidently generated by the drill-and-snap procedure, with the margin originally bearing the drilling indentations subsequently smoothed over. They explained that the smaller piece was used to fashion a fishhook, and in the second article, they also provided a photographic image of an archaeologically recovered composite fishhook from a Thule setting in the project area. Along with that they described two related items, also recovered archaeologically, with direct reference to the illustration:

> Three items that document fishing. Upper: loop of baleen that may have been used for fish line. Middle: Fishhook bored from seal jaw with an inset barb. Bottom: Lower jaw of a hooded seal [*Cystophora cristata*] from the midden in front of House 2 at Qoornoq. From this jaw is produced a fishhook like the one shown. Drilled lower jaws can be found much more often than fishhooks themselves and are therefore an important piece of

evidence for these fishing tools. (Gotfredsen et al. 1994: 52, originally in Danish, English translation by the author)[4]

In their model, the main portion of the fishhook was made from the smaller of the two pieces of bone that resulted from drill-and-snap, whereas the barb of the hook, inset into it, was made from a different piece. The reasonableness of this reconstruction is further suggested by the fact that Mathiassen himself has illustrated such composite fish-hooks, as recovered from Thule sites. In fact, he provided a photograph of a broken fishhook with inserted bone point (Mathiassen 1936: 37, 38, Fig. 21), from Runnaesset, the same site from which he illustrated the larger drilled seal jaw piece. Moreover, he indicated, as quoted earlier, that he did not know what was fashioned from the drilled-off smaller piece of seal jaw. Though not definitive, Gotfredsen et al. (1992, 1994) seem to have figured that out, and it is likely the very sort of fishhook that Mathiassen himself illustrated. Moreover, in the general text of Gotfredsen et al.'s (1994) presentation, they wrote:

> For certain fish bones, it cannot be excluded that they come from stomach contents of birds, big fish, or seals; but to judge from the abundant fishing tackle, documented in the form of drilled seal and *dog jaws*, sinkers, and small knots of baleen for fish line, the fishery has quite obviously played an essential role. (Gotfredsen et al. 1994: 47, emphasis added, originally in Danish, English translation by the author)[5]

The noteworthy point here is that drilled dog jaws are being invoked as part of Thule fishing gear. And in fact, the main portion of the fishhook illustrated by Mathiassen, noted earlier, appears more robust than one might expect from a seal jaw. Thus, it might be a dog mandible, although what was done with the other skeletal elements is less clear. Regardless,

---

4 The original passage, a figure caption in Danish, is as follows: "*Tre genstande som documenterer fiskeri. Øverst: Løkke af hvalbardehår, som kan have været brugt til fiskeline. I midten: Fiskekrog udboret af saelkaebe med isiddende modhage. Nederst: Underkaebe af klapmyds fra møddingen foran anlaeg 2 på Qoornoq. Af denne kaebe er fremstillet en fiskekrog som den viste. Udborede underkaeber findes langt oftere end selve fiskekrogene og er derfor et vigtigt bevis for fremstillingen af disse fiskeredkaber*" (Gotfredsen et al. 1994: 52, italics in original).

5 The original passage in Danish is as follows: "For visse fiskeknogler kan det ikke udelukkes, at de stammer fra maveindhold af fugle, store fisk eller sealer; men at dømme efter de mange fiskeredskaber, dokumenteret i form af afborede sael- og hundekaeber, fedtstenssynk og små knuder af hvalbarde til fiskeline, har fiskeri dog helt klart spillet en vaesentlig rolle" (Gotfredsen et al. 1994: 47).

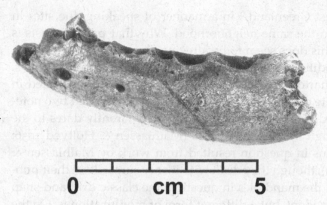

0   **cm**   5

FIGURE 6.10. Portion of a left dog mandible from the Unartoq site, extreme south-west Greenland, with drill-and-snap separation clearly evident at the posterior (right) end. Photo by Geert Brovad, Zoological Museum, University of Copenhagen, Denmark.

the next logical question is whether this technique is evident elsewhere, or is specific to Qoornoq.

To answer that question, another Thule site in the immediate project area, known as Balders Fjord (Gotfredsen et al. 1994), also yielded several examples of drill-and-snap dog bones. Beyond that project area, a left mandible portion from a dog, from the Unartoq site in extreme southwest Greenland (see Figure 6.4) is illustrated in Figure 6.10. According to Mathiassen & Holtved (1936: 56–64), who excavated and reported on this site, it was occupied from the mid- to late 1700s to the mid-1800s. The piece shown exhibits the drill-and-snap alteration, circumscribing the entire corpus, or body, of the mandible. Though Mathiassen was at a loss to determine what was made from the resulting piece of seal jaw, in his case described previously, it is legitimate to wonder what was made from the opposite piece here (a fishhook?). But the presence of this piece indicates that the drill-and-snap technique was not unique to the Skjoldungen project area. Moreover, when studying the collections at the University of Copenhagen's zoological museum, I recorded four drill-and-snap dog bones from Misigtoq (see discussion that follows), in southeast Greenland (Mathiassen 1933: 16–26). Included were a proximal right femur, a proximal right ulna, a proximal right metacarpal, and a distal right humerus. These pieces were reminiscent of the ones from Qoornoq, illustrated as Figure 6.7A and C.

Notably, examples of the basic drill-and-snap procedure that I know about, all of comparable age, are restricted to southeast Greenland or extreme southwest Greenland.[6] In a manner of speaking, the sites in question are all in the same neighborhood. Why that pattern exists is perplexing, but this does seem to be a regionally restricted expression of Thule bone modification technology, at least as applied to dog bones.

Beyond the Unartoq mandible, noted previously and illustrated in Figure 6.10, the site of Arfermiut (see Figure 6.4) also yielded two noteworthy dog mandibles (Figure 6.11). This site apparently dates to the latter 1600s through much of the 1700s (Mathiassen & Holtved 1936: 90). The specimens in question resulted from work by Mathiassen & Holtved (1936: 69), though they dealt with it only cursorily in their published report. On the mandibles in question, the classic drill-and-snap technique is not evident, but a different form of modification is. On the upper one (Figure 6.11), the mandible is complete, but one can clearly see a longitudinal groove incised in the bone, running from the anterior end to just behind the last visible tooth (right to left). A comparable groove is present on the other side of the specimen as well. On the lower one, the lower portion of the mandible has been broken away, perhaps related to an original incision, appearing much like that on the upper piece. In fact, there are still faint incision marks visible on that lower piece, as indicated on the illustration. Examples of longitudinally incised dog bones, including limb bones, were also recorded from several other sites in southeast Greenland, though their significance is unclear.

Put in broader context, even when surveying no more than the distinctive manufacturing processes used by Thule people on dog bones, and the probable diversity of resulting tool forms, one can readily appreciate Moreau Maxwell's words:

> Attempting an inventory of Thule material culture becomes a seemingly endless task of trait listing. These were perhaps the most gadget-oriented people in prehistory, nearly as much so as we are today, and monographic reports on Thule excavations can consequently make for hard sledging. Furthermore, with the blessings of permafrost we know virtually as

[6] Technically speaking, an exception to this generalization occurs in the form of several specimens from what is known as Sandells Vinterhus (Winterhouse), dealt with by Sandell & Sandell (1991). These bones showed classic drill-and-snap alteration from a setting that is late in the Thule sequence (ca. 1800–1850) and is in east Greenland, consistent with the generalization. But it is well north of the Misigtoq setting, being located in the Scoresby Sound region, closer to Geographical 2 than it is to Misigtoq (see Figure 6.4). It seems likely that this late (final?) Thule settlement in northeast Greenland was initiated by more southerly peoples who moved north.

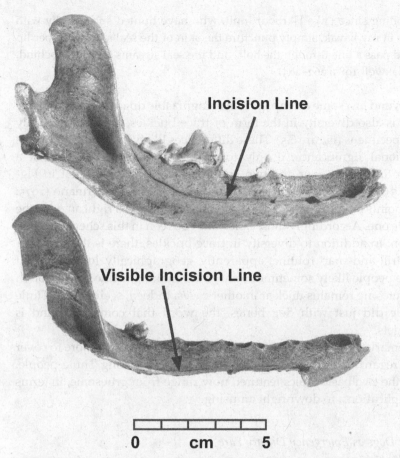

Incision Line

Visible Incision Line

0   cm   5

FIGURE 6.11. Two right dog mandibles from the Arfermiut site in extreme southwest Greenland, with visible incision lines, in the form of grooves in the bone surface, pointed out. Photo by Geert Brovad, Zoological Museum, University of Copenhagen, Denmark.

much about their possessions as we do about those of modern Inuit. For example, I once found in a northern Ellesmere Island dwelling for a single family, occupied for only one or two winters, three distinctively different toggles. From ethnographic analogy these three, of quite different shapes, were used to tow dead seal behind kayaks, one being inserted through the septum of the nostrils, a second through the skin of the neck, and the third through skin at the navel. No one, archaeologist or Inuit, has yet been able to tell me why one would need such different towing devices or under what conditions they would have been used. This is particularly baffling since Lake Harbour Inuit, who have hunted successfully with me in my kayak, simply puncture the skin of the seal's tough upper lip and pass a line through the hole, and the seal streams smoothly behind. (Maxwell 1985: 262–263)

Beyond just some examples from recognizable dog bones themselves, there is also diversity in the form of trace buckles, evident from only five specimens (Figure 6.5). These differences likely had some unknown functional significance, if only to the one who made an item at a given time. Moreover, the two specimens farthest to the left in Figure 6.5 appear to have only one hole each, but as Schledermann (1975: 120) points out, the other hole in each one is drilled at a right angle to the visible one. Accordingly those holes can't be seen in this schematic illustration. In addition to diversity in trace buckles, there is the noteworthy drill-and-snap routine, apparently geographically localized, that Thule people likely sometimes utilized for making fishhooks, although the outcome remains unclear in other cases. Indeed, seeing what Thule people did just with dog bones, the word that comes to mind is "gadgets."

Remarkably, and all gadgets aside for now, there is yet more to cover with regard to the representation of dog bones among Thule people. And the two basic topics featured now range from gruesome, in terms of implications, to downright amusing.

Thule Dogs as Emergency Dietary Fare

As already covered, especially in the previous chapter, dogs have often been consumed by people, sometimes rather routinely. At other times, though, their consumption stemmed from dire circumstances. In the eastern Arctic, part of a global climatic shift sometimes designated as the "Little Ice Age" (Maxwell 1985: 304) was pronounced shortly after about 1400, extending well into the 1600s. This episode was associated with even colder conditions, though they were ameliorating somewhat by the late 1600s. In Greenland it was clearly correlated with the onset of hard times for people, as manifested in several ways. In addition, adverse climatic conditions were especially pronounced in east Greenland, as opposed to west Greenland, where conditions were generally more favorable, due especially to the influence of prevailing ocean currents. At one point the interpreter and explorer Knud Rasmussen, who was born in Greenland and spent considerable time there, highlighted that environmental reality with the following story about a man from east Greenland:

All those who have the slightest to do with Greenland know the legend of the grand old hunter from Aluk, the old man whose heart broke with the joy of recognition when, after many years of absence in parts that were richer and more luxuriant on the west coast, he came back to the old settlement and one morning saw the sun rising out of the sea; so great was the love which bound him to East Greenland's apparently barren and stern coast. (Ostermann 1938: 8)

This difference was undoubtedly most pronounced in the more northerly regions. Figure 6.4 indicates the approximate locations of two Thule sites in northeast Greenland, Dødemandsbugten (Dead Man's Bay), and a site designated as Geographical 2, on Ella Island, immediately adjacent to Geographical Society Island (Glob 1935: 24). Dead Man's Bay spans a period of time beginning in the 1500s, and extending to sometime in the 1700s. Temporally, Geographical 2 is comparable to much of the span of time represented at Dead Man's Bay, though perhaps not to its earliest occupations. For present purposes, Dead Man's Bay is the most compelling, and it is appropriately named: "the name sounds mournful and sinister, yet it is a very apt one" (Larsen 1934: 5). As Helge Larsen reported, his crews excavated twenty-three of the forty-three recognized house ruins, and the fate of certain dogs is downright gruesome. Writing of House 3 (ruin group 1) at that site, he stated:

Still, one thing indicates that hunger has ravaged this house, and that is the presence of a disproportionately large number of dog bones, both in the house and in the passage, whereas there were relatively few bones of other animals. The possibility is, then, that the people have eaten their dogs owing to lack of other food, and later have starved to death. (Larsen 1934: 18)

Beyond those dog bones, several bones attributable to at least seven people, including three children, were found in the house or in the passage leading into it. I had a chance to examine the Dead Man's Bay dog bones in 1991. The bones from House 3, commented on by Larsen, experienced atrocious preservation conditions, such that the surfaces were so fragile and abraded away that even assessing relative maturity of the dogs in question was highly problematical. I strongly suspected that there was some gnawing damage on those bones (see discussion that follows), but given their badly degraded surfaces I was unable to

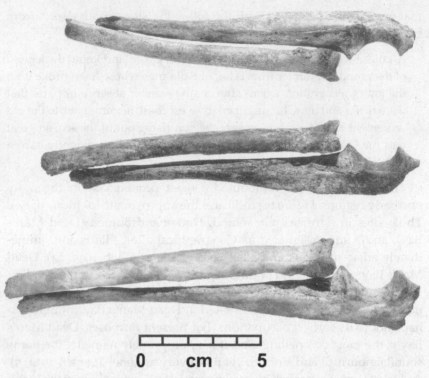

0   cm   5

FIGURE 6.12. Articulating gnawed dog ulna and radius pairs from the Dead Man's Bay site in northeast Greenland. The upper two pairs are from right (top) and left (middle) forelegs, possibly from the same individual. The lowermost pair is from a left foreleg. Photo by Geert Brovad, Zoological Museum, University of Copenhagen, Denmark.

evaluate a case for (or against) that based on the physical evidence. But other settings at Dead Man's Bay were more cooperative. Any evidence of gnawing damage to the bones that might be discernable was recorded, on all the Greenland archaeological dog bones that I studied in Copenhagen, including numerous examples from Dead Man's Bay.

Figure 6.12 shows three pairs of dog forelimb bones, an articulating ulna and radius in each pair, recovered from two different houses at Dead Man's Bay. What can be seen immediately from these bones is that the proximal ends of the bones (to the right on the illustration) are mostly intact, whereas the distal ends are lacking their ends, or epiphyses. The gnawing damage on the proximal ends of the lowermost pair is striking in its unusual nature. In general, gnawing damage to animal bones is not uncommon and is often inflicted by scavenging dogs

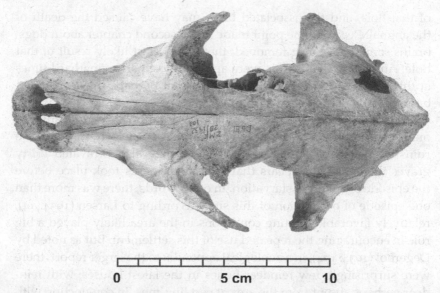

FIGURE 6.13. A complete dog cranium from the Dead Man's Bay site, in northeast Greenland, with a large hole in the right rear portion of the braincase (upper right in photo). Photo by Geert Brovad, Zoological Museum, University of Copenhagen, Denmark.

themselves. Previous personal experience with that very phenomenon comes from work in eastern North America (Morey & Klippel 1991). Gnawing canids, as well as some other animals, typically leave a series of recognizable tooth marks, in the form of relatively distinct shallow pits and furrows on the bone surface. In the case of the lowermost Dead Man's Bay dog bones, the tooth marks are considerably less distinct on an individual basis, appearing somewhat worn down. Given that starvation conditions apparently occurred there, an intriguing possibility to account for the appearance of the lowermost pair of bones is that perhaps some desperately weakened people were, as a last resort, accessing edible tissue from the joint areas of the front leg of a dead dog. The tooth marks on those bones seem consistent with what would be left if the starving people were using mostly the surfaces of their molars and premolars in an effort to chew on them.

Another Dead Man's Bay dog bone specimen helps shed further light on difficult circumstances there. Figure 6.13 is a skull of a dog from that site, apparently from a slightly earlier house. Its most noteworthy characteristic, given the current subject, is the large hole in the rear portion of the cranial vault. Impact marks are evident in the vicinity of that hole, and the associated blow may have caused the death of the animal. Recalling the point made in the second chapter about dogs' brains sometimes being removed, that is the most likely result of that hole. And given that people were apparently experiencing hard times at this settlement, it is worth suggesting that perhaps they consumed brain tissue in a time of need.

Several aspects of the Dead Man's Bay situation are significant. For one thing, in addition to the twenty-three (out of forty-three) house ruins excavated by Larsen and his crew, they also excavated thirty graves there, and it appears that some interments took place before the episode of apparent starvation. In other words, there was more than one episode of occupation of this site. According to Larsen (1934: 14), relatively favorable hunting conditions in the area likely played a big role in encouraging the repeated use of this settlement. But as noted by Degerbøl (1934: 174), in a zoological appendix to the larger report, there were surprisingly few reindeer bones in the latest houses, with reindeer perhaps absent from the area at certain times. In conjunction with that, a disproportionately large number of hare bones at times implied less than optimal hunting conditions. In all, Larsen (1934: 171–172) concluded that the dead people there were not from the final occupation, and that the last ones there didn't leave because of dire hunger, given the presence of many animal bones in the latest ruins.[7] Perhaps their leaving was due to poor reindeer hunting at that time. Whatever the truth, the events at Dead Man's Bay pose a situation "bearing witness of the death and extinction of a small tribe [and their dogs?] that once made a struggle for existence there" (Larsen 1934: 5).

In any case, gaining access to especially nutritious dog tissue is also suggested by a case from the site of Geographical 2, on Ella Island (see Figure 6.4). Though roughly contemporaneous with Dead Man's Bay, imminent starvation does not seem to have been a factor there. In fact, in

---

7 Recalling that there were many graves at the site, some people who starved were likely interred eventually. Given that likelihood, along with the fact that people occupied this site after the apparent starvation episode, it is not surprising that winter houses were generally lacking in human skeletal remains. An exception is House 3, Ruin Group 1, where there were "cranial bones of three adults, an individual of 1–7 years, and of three children . . . in all seven individuals. In addition we found a radius and a tibia of an adult and a femur and tibia of a child" (Larsen 1934: 17). It seems likely that this house was not reoccupied and so was incompletely cleared of human skeletal remains, whereas perhaps human remains were often removed from other houses upon subsequent occupations. Alternatively, some people may have survived.

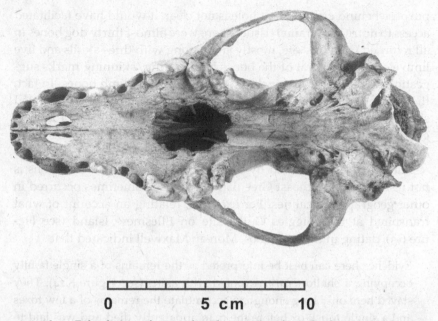

FIGURE 6.14. Palatal view of a complete dog cranium from the Geographical 2 site, in northeast Greenland, with a large hole in the rear portion of the palate, likely to facilitate extraction of the cranial tissue. Photo by Geert Brovad, Zoological Museum, University of Copenhagen, Denmark.

summarizing the overall situation there, Glob noted certain similarities to Dead Man's Bay, but reported the following:

> A common feature of the houses in which these human jaws were found [at Geographical 2 and three nearby sites] is that they contain a large number of artefacts, often ornamental things, but never such objects as would lead to the assumption that the inhabitants have died there of hunger or poisoning. (Glob 1935: 83–84)

Glob was familiar with the approximately contemporaneous work at Dead Man's Bay, referred to the work at that site several times, and rather clearly his assessment of the situation at Geographical 2 was influenced by his familiarity with Dead Man's Bay. But that the inhabitants at Geographical 2 likely did not starve there does not mean that all was good. Figure 6.14 is a dog skull from one of four winter houses at Geographical 2. There is no indication of how the animal died, but one can clearly see a large hole through the rear portion of the palate. While the purpose behind creating that hole is not clear, it would have facilitated access to nutritious cranial tissue. There were almost thirty dog bones in all returned from this site, mostly jaws, along with three skulls and five limb elements. Several of the bones bore cutting/skinning marks suggesting human butchery, including one of the few limb bones. In fact, there are cut marks anterior to the hole in the palate of that illustrated skull. It appears that some dogs there were eaten, if not in a time of genuine famine, perhaps when the circumstances of life were less than optimal. Rather clearly, times were tough then in northeast Greenland.

But use of dogs as food by Thule peoples in difficult situations is not restricted to northeast Greenland, and also sometimes occurred in other geographic localities. For example, relating an account of what transpired at the Ruggles Outlet site on Ellesmere Island (see Figure 6.3), dating in the late 1400s, Moreau Maxwell indicated that

> Evidence here can best be interpreted as the remains of a single family occupying a shallow *qarmak* roofed with skins (see Figure 8.24). They stayed here only long enough to accumulate the remains of a few foxes and a single musk-ox before the man apparently died and was laid to rest under a pile of rocks outside the house. From the evidence on the sleeping platform, the widow, left alone, ate their three sled dogs and died, two whalebone combs still in her hair. The significance of these remains is that they constitute the complete inventory of a single family during a short period of time. (Maxwell 1985: 305, original italics)

This scene took place during a time of pronounced hardship among Thule peoples in the northern parts of the eastern Arctic. It is unclear why the man died, but an occurrence such as this serves as a reminder that, like Dead Man's Bay, difficult circumstances among people in general sometimes entailed scenes of nothing less than life or death importance for individual people.

Fun with Thule Dog Bones

Quite obviously, the people just covered above were not having any fun with Thule dog bones at those times. But from a slightly later time frame, perhaps in conjunction with improving climatic conditions, especially farther south, some people apparently did just that. Figure 6.15 is a photograph of two left dog mandibles from the Misigtoq site (see Figure 6.4), as reported by Mathiassen (1933: 16–26). The occupation at that

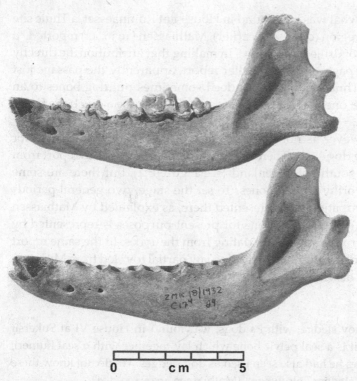

FIGURE 6.15. Two left dog mandibles from the Misigtoq site in east Greenland, both components of a toy dog sled arrangement, as explained by Mathiassen (1933: 100), and in the text. Photo by Geert Brovad, Zoological Museum, University of Copenhagen, Denmark.

site that yielded these bones spans the period from the late 1600s until about 1800. On each mandible, a clean hole is clearly visible, drilled through the upper portion of the ascending ramus. As for what that hole likely signifies, Mathiassen's own words serve best:

> Pl. 8.11 is half a dog mandible drilled through the upper angle; several of these were found, and one of the Greenlanders working for me said that as children they used to play sledges with two of them, a stick being pushed through the two rami, which acted as upstanders. (Mathiassen 1933: 100)

By "half" he clearly means right or left. Mathiassen illustrated only one, and while only two are shown in Figure 6.15, that site yielded two others. And that is not the only site where such items were found. For example, referring to what was later found in House 1 at Ruinnaesset, a Thule site in the same region (discussed earlier), Mathiassen (1936: 42) reported "a [toy] sledge of drilled dog's jaw." In making that attribution, he directly appealed to page 100 of his earlier report, apparently the passage just quoted. So Thule peoples did indeed sometimes put dog bones to an amusing use, one that is also fitting, since in this case a dog bone forms the dog sled.

That, however, is not the only creative use of bones they had for representing dog sleds. There are no modified dog bones to report from Sukersit, in southeast Greenland (see Figure 6.4), but there are some other noteworthy animal bones. To set the stage, two general periods of Thule occupation are represented there, as explained by Mathiassen (1933: 33–41). The relevant one for present purposes is represented by three house ruins, apparently dating from the 1700s. In the same report in which he drew attention to the dog jaw partial toy sled from Misigtoq, Mathiassen described a distinctive find from House 6 at Sukersit as follows:

> Another toy sledge, with its dogs, was found in House VI at Sukersit (Fig. 34); it is a seal pelvic bone which lay together with 9 seal humeri; these bones he had also seen used as dog sledges. We do not know these forms of "sledges" elsewhere. (Mathiassen 1933:100–101)

"He" is the Greenlander who weighed in on the meaning of the dog jaws covered previously. Mathiassen (1933: 100) provided a photograph of this particular set of seal bones, and Figure 6.16 is a recent photograph of this distinctive set of bones. They are arranged in this figure much as they were when illustrated in Mathiassen's report. What is especially worth calling attention to here is that the outermost "dogs," on the wings of the simulated dog team, are immature seal bones, whereas the others appear more mature. If this is how they were originally arranged, as seems likely, those young "dogs" likely symbolize young and inexperienced animals, placed where they could best learn the necessary skills from both the human sled driver and the rest of the dog team. At least that is certainly an intriguing possibility.[8]

[8] I am indebted to my colleague and coauthor, Kim Aaris-Sørensen for making this suggestion to me. I confess that I had not even noticed this aspect of the seal bones until he pointed it out to me. I then looked at Mathiassen's picture of the set and could see for myself that it was true.

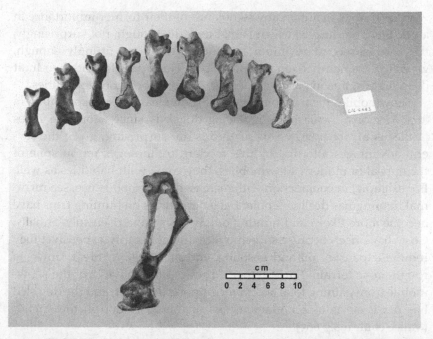

FIGURE 6.16. Nine seal humeri and one seal pelvic bone from the Sukersit site in east Greenland, forming a simulated dog sled team. As originally documented and explained by Mathiassen (1933: 100–101), and now in the text, in this toy dog sled team, the humeri (upper) served as the dogs, while the pelvic bone was the sled. Photo by Mikkel Myrup, Greenland National Museum and Archives, Nuuk.

RECENT INUIT DOGS

The preceding accounts relating the uses of Thule dog (or seal!) bones for pragmatic or amusement purposes often relied substantially on ethnographic analogies with recent Inuit people. And in certain cases, of course, Inuit people themselves provided an assessment of the purpose of a given modified bone, or a set of bones. Recalling a point suggested earlier, the connecting link between Thule peoples, known archaeologically, and modern Inuit peoples, known historically, is so direct that ethnographic analogy is valid in this case. At the same time, modern conditions of life have clearly altered the ways in which these northern peoples now make their livings. Cities, modern technology, economy, educational systems, transportation innovations, and so on are all part of the modern scene. Given all that this chapter has covered, especially the clear importance of dogs in Thule life, one must immediately wonder if dogs maintain any semblance of their former importance in arctic life. The simple answer is that they do, though, not surprisingly, they are of secondary importance now. And here, fittingly enough, given that the comparison is between past Thule peoples and their Inuit descendants, this in part is a reference to transportation capabilities.

Snowmobiles have become an important means of transportation in arctic regions, increasingly replacing dogsleds since about the 1960s (Vallee et al. 1984: 664). This pattern is not surprising, given the several advantages afforded by this modern technology. Yet, in spite of the appeal of modern snowmobiles, they come with liabilities as well. For example, in conjunction with increased snowmobile use, sea mammal hunting has declined, potential emergencies on hunting trips have become more likely, and hunting outlay costs have risen substantially. Costs have risen because snowmobiles are monetarily expensive, they require expensive oil and gasoline, and are liable to break down in rough arctic terrain. Put together, "These factors have kept the snowmobile from gaining total acceptance among the Inuit, and the dog sled has persisted as a secondary means of transportation in the Arctic" (Vallee et al. 1984: 665).

That statement was made especially about modern Inuit in Canada, but applies just as well to other areas, including Greenland. For example, in an issue of *Naturens Verden* magazine that features Grønnow & Meldgaard's (1988) popularized account of work done at the Paleoeskimo Qeqertasussuk site in west Greenland, there are two color photographs of dog teams pulling sleds. One picture is on the cover of that issue, and a team of dogs pulls a laden sled up an icy slope, guided by the driver, on foot for the ascent. The other is on the last page of the piece (Grønnow & Meldgaard 1988: 440), where a team of dogs pulls a sled with two riders down the streets of Christianshåb, also in west Greenland, past the local museum.

In fact, twenty years after Valle et al.'s (1984: 665) observation about the Canadian Inuit, in Greenland there is sometimes a conflict perceived between users of snowmobiles and those who continue to use dog sled teams. Moreover, the difficulty there tends to fall along ethnic lines, involving recent sociopolitical circumstances:

The conflict is often communicated as a conflict between Greenlandic cultural activities (dog sledge) and Danish recreational activities (snowmobile). . . . Generally speaking, the lack of communication between the two groups increases the feeling of many dog sledge drivers that snowmobile drivers are anonymous and alien to the landscape. (Sejersen 2004: 78–79)

Also, in northwestern Alaska dog sled races are not uncommon among some peoples, as a recreational expression of the traditional importance of dog traction. But "Its usage as the principal means of winter transport was eclipsed by the adoption of snowmachines beginning in the mid-1960s" (Sprott 1997: 79). Nevertheless, dog sled races continue to be held there, and also in Greenland, where they may involve both native Inuit people and others. In fact, a recent one was reported by Irene Jeppson (2008) in Greenland's largest newspaper, *Sermitsiak*. In that report, she related that "It took one hour and 43 minutes for Jens Ole Jensen and his dogs to travel the 40 kilometer long distance to the Greenland championship in the dog sled race" (Jeppson 2008, originally in Danish, English translation by the author).[9] And in that same piece, the winner, Jens Ole Jensen is reported as saying, "I know that my dogs are strong, so I was sure the whole time that I could lead the course" (Jeppson 2008, originally in Danish, English translation by the author).[10] So the advent of modern technologies in the Arctic has clearly impacted the traditional importance of dogs in arctic peoples' lives, but in spite of the major changes ushered in by modern conditions, dogs continue to play a discernable role in transportation, including recreational transportation, in some regions.

Importantly, activities of recent Inuit peoples also make clear how Thule peoples almost surely used their dogs for specific kinds of hunting tasks, involving sea mammals. For example, the Copper Eskimo people are associated with the vast, poorly defined region known as the Central Arctic. Among those people, the following account indicates one role: "From December until May breathing-hole sealing was the chief activity in most of the Copper Eskimo area. This method was carried out with a number of hunters and their specially trained dogs who located the breathing holes" (Damas 1984: 398). But by no means was seal breathing hole hunting known only from this

---

[9] The original passage in Danish is as follows: "I time 43 minutter tog det for Jens Ole Jensen og hans hunde at køre den 40 kilometer lange straekning til GM [Grønlands mesterskab] i hundeslaeddevaeddeløb" (Jeppson 2008).

[10] The original passage in Danish is as follows: "—Jeg ved, at mine hunde er staerke og derfor var jeg sikker på, at jeg hele tiden kunne føre løbet" (Jeppson 2008).

area. In fact, a comparable description comes from farther east, in Quebec:

> In March with the birth of ringed seals winter taboos ceased and a period of active seal hunting began. With dogs, the small birth shelters and breathing holes were sought out, or seals whose curiosity was aroused ambushed at the floe edge. (Saladin d'Anglure 1984: 498)

While it obviously is not possible to verify directly that preceding Thule peoples operated with this approach, it is likely that they did.

CLOSING PERSPECTIVE ON THULE/INUIT DOGS

When put in broader perspective, the roles played by dogs among Thule peoples, the immediate predecessors of the Inuit, encompass the entire realistically imaginable spectrum of roles for those times. At one end of that spectrum, dogs sometimes served as an emergency food source, under conditions that led to eventual death by starvation. At the other end of that spectrum, dog bones were sometimes fashioned into children's toys, toys that directly called attention to one of the routinely important pragmatic roles played by living Thule dogs, their role as members of dog-sled teams. That particular role is part of the central portion of the spectrum just alluded to. The same is true of the frequent use of their bones for utilitarian purposes, though the specific function sought from a given modified dog bone may remain enigmatic (e.g., some of the drill-and-snap pieces). In short, dogs regularly played a conspicuous role in the life of Thule peoples. Given that their roles when alive were sometimes correlated with frankly harsh discipline, as emphasized earlier, one is reasonably led to wonder if Thule peoples, or even arctic peoples in general for that matter, really ever seemed to like their dogs in the sense of a social bond, a major theme of this volume. With that vague and thoroughly subjective issue in mind, it is worth pointing out, as indicated in Chapter 3 (Figure 3.8), that there are dog burials known from different arctic or immediately sub-arctic regions in the world. These settings range from northern Alaska, to Greenland, and to Siberia. In addition, the northern Alaska site is one highlighted previously (Morey 2006: 164) as where "a dog was individually buried in a deliberately prepared log tomb." And in a Thule-style stone-built grave of uncertain age at the Greenland site there was "a central chamber containing skeletal parts of both man and dog" (Nyegaard 1995: 100). It is the only Thule dog burial that I know of from Greenland, but given the apparently testy association between Thule peoples and their dogs under certain circumstances, it is probably not surprising that the broader relationship between them was expressed in this way only rarely, whereas in other regions of the world dog burials are much more common.[11] But it seems fitting to draw this chapter to a close with a simple thought: Because even the single known Thule dog burial draws attention to the close relationship between dogs and people, it is intriguing to explore more fully just what the act of burying a dog signifies. The value of exploring that issue stems in part from the fact, as highlighted fairly early in Chapter 3, that a good deal of what we know about dogs of the past comes from burial contexts, in different parts of the world.

---

[11] It may be significant to note that the one known Thule dog burial is located in extreme southwest Greenland, south of where dog-sledding demonstrably took place (see Figure 3.8). Perhaps dogs there tended to relate to their people in a manner that was more conducive to such treatment at death, given the demonstrably testy, even violent, treatment of some sled dogs farther north.

7

THE BURIAL OF DOGS, AND WHAT DOG BURIALS MEAN

> Hominids, including some early anatomical forms of modern *Homo sapiens*, seem to be unique in the animal kingdom for ritualistic disposal of conspecifics. Often, such disposal included burial.
>
> Lyman 1994: 411

Given that observation, it is legitimate to suggest that "Nothing signifies the social importance that people have attached to dogs more conspicuously than their deliberate interment upon death" (Morey 2006: 159). It may seem awfully self-serving to quote oneself at the beginning of a chapter, but in light of Lyman's observation, the answer to the basic question of what dog burials mean is captured by that statement. That is, people are treating dogs much like other people. The main objective in this chapter is to document and elaborate on that point, but first, it is well worth pointing out that others have also recognized the basic importance of animal burial. For example, G. Clark (1996), dealing with both prehistoric dog and pig burials from several Polynesian sites, wrote as follows: "Deliberate animal burials are the most obvious sign of a close prehistoric human-animal relationship" (G. Clark 1996: 34). Clark's work directly draws attention to the fact that dogs were not the only animals to be ritually interred at death. But other animals, with one conspicuous (localized) exception covered later, were not dealt with this way nearly as consistently as dogs. Even G. Clark's (1996: 32–33) work highlights that point, albeit not dramatically. Specifically, he reported a total of thirteen dog burials from the Polynesian sites, but only seven pig burials.

A previous presentation devoted to dog burials (Morey 2006: Table 1) inventoried many of them, or provided additional sources to consult, but there are many more to report now.[1] There definitely are still more, but an extensive series of documented non-modern dog burials is inventoried in Table B.1. The dogs represented in this inventory were buried under a wide variety of circumstances, sometimes ambiguous as to broader implications. Given that situation, it is worth pointing out that the act of burying an animal can, at face value, be largely devoid of symbolic significance and mostly concern hygienic corpse disposal. While that may be largely true in some cases, the care with which dogs were routinely buried suggests otherwise for most of them, and that aspect of dog burials warrants consideration in some detail. As a beginning generalization, the care with which dogs were often buried has been pointed out before. For example, James B. Griffin, in writing about the Middle Archaic Period of eastern North American prehistory (some six to eight thousand years ago), commented some years ago that "Some of the dogs were buried as though they were someone's best friend" (Griffin 1967: 178).

CARE IN BURIAL

In considering the kinds of situations that might lead to an assessment such as Griffin's, it is useful to recognize two major modes of dog burial. In one mode, people buried a dog individually, by itself, and in another mode they buried one or more dogs in direct proximity to a person. It stands to reason that a dog buried with a person would be buried with considerable care. At the same time, though, in many such cases it is likely that the dog was dispatched in order that it could be buried

---

[1] In order to minimize geographic redundancies in the previous tabular presentation (Morey 2006: 160, Table 1), and have that table fit onto a single journal page, a lengthy footnote identified sources for some additional burials. In addition, that previously published table lacked some burials that were specified in the text. No such tabular restrictions apply here, and that is one reason the table that assembles them here is much lengthier, and appears as Table B.1, in Appendix B, at the end of this volume. Moreover, there are many additional burials, learned about since that original compilation, including a separate section on Polynesia. The Old World series in that original tabulation did not distinguish how many dogs were buried with people, as opposed to the total number. The larger New World series did, and all series are treated that way here. One component of that original compilation is, however, abbreviated. Specifically, several Archaic Period sites from the Green River region in Kentucky are now collapsed into a single summary entry, including sites that were not in the original compilation. The reason is that at a later point in this chapter all those sites appear in a single table, to be dealt with separately.

with the person in question. As such, those situations often suggest something about the nature of the association with that person when they were both alive. The rationale was likely that the two should be together in an Afterlife. For that reason, the two burial modes are distinguished in this account, and individually buried dogs may, under some circumstances, be a more direct indication of a dog as "someone's best friend" (Griffin 1967: 178). At frequent junctures, use of the words of primary investigators themselves serves to describe the burials in question. The primary goal in using that approach is to establish that dogs have routinely been dealt with reverently at death, on a worldwide basis. The coverage of New World examples here is the most extensive, due to my own primary background. In general, the inventory of dog burials in Table B.1 moves from earliest to most recent, within three categories (Old World, Polynesia, and New World). In going over cases that reflect the care that dogs commonly received when buried, it seems useful take a geographic approach again, beginning with the Old World.

Care in Burial: The Old World

It is fitting to begin with simultaneous dog and human burials. As Chapter 3 made clear, the oldest securely identified dog in the world is from a burial at Bonn-Oberkassel in Germany, and dates to about 14,000 years ago. This is an instance where one cannot use any original investigators' words, since this dog was discovered in a quarry in 1914, and most of the bones have long been lost. Here, then, is a case where little more can be said, except to note that the single jaw fragment that has survived "was unearthed from a double grave of a 50-year-old man and a 20–25-year-old woman" (Benecke 1987: 31). Beyond that case, the country of Israel provides other noteworthy early cases. First, from Ein Mallaha, some 11,000–12,000 years ago a dog or wolf puppy, its immature bones too underdeveloped to allow a secure taxonomic determination, was buried with an elderly person. According to the investigators who reported this find,

> The human skeleton, whose sex cannot be determined due to the damage to the pelvis, lay flexed on its right side, and judging from the state of its dentition, was an old individual. Its left wrist was partially under its forehead, and hand upon the thorax of a puppy, which has evidently been buried complete with the human. (Davis & Valla 1978: 608)

The Ein Mallaha burial is actually rather well-storied, and the image of that burial, as provided by Davis & Valla (1978: 608) has appeared in several subsequent sources (e.g., Davis 1987: 147; Morey 1994a: 337; 1995: 141; Clutton-Brock 1995: 11). As to what that situation suggests, the original investigators offered their appraisal of that as well: "The puppy, unique among Natufian burials, offers proof that an affectionate rather than gastronomic relationship existed between it and the buried person" (Davis & Valla 1978: 609). Another compelling case, also from Israel, is at Hayonim Terrace. There, two dogs were buried with people sometime between about 11,000 and 10,500 years ago, as reported by Tchernov & Valla (1997).

Elsewhere in the Old World, at the Mesolithic site of Skateholm, Sweden, dating to about 6500–5500 B.P., there were some fourteen dog burials at a cemetery, at least four of which were with people. At this site, reported and illustrated by Lars Larsson (1990, 1995), there was a case in which a woman and dog were buried together, with the body of the dog situated above the woman's legs. The dog's neck may have been broken to dispatch the animal for the burial. Like the Ein Mallaha case, this burial image also appeared in sources other than Larsson's (e.g., Aaris-Sørensen 1988: 203; Morey 1996: 72). Even among the individually buried dogs there (see discussion that follows), "The most striking spatial pattern, however, is the apparent close relationship between *individually buried dogs and children* under the age of eight" (Fahlander 2008: 36, original emphasis). Fahlander notes that these kinds of burials were situated at the boundaries of the site. Elsewhere, though not involving conventionally conceived care in burial, some ca. 3,200 year-old tombs from the urban Yin Complex in China yielded numerous noteworthy burials:

> One hundred five of the [939] excavated tombs produced dogs which were recovered from undifferentiated fill and 197 tombs (such as M9, M294, M326, and M703) were found to contain dogs carefully placed in the *yaokeng*, or "waist pit," a trench located below the waist of the human interment that was a specially prepared locus for the disposition of sacrificial animals. (J. W. Olsen 1985: 61, original italics)

In this case, what is ambiguous is the number of dogs that were directly associated with the interment of people. According to J. W. Olsen (1985: 61), "there were 439 sacrificial dogs in 339 of the excavated tombs." The correct number of people is unclear, and that is why Table A.2 lists "439?" as the number with people. That is, some of the dogs may have been interred individually. Regardless of that uncertainty, this context highlights the extent to which dogs were sometimes used as sacrificial animals and interred alongside people, some of whom were themselves sacrificed: "Seventeen of the tombs yielded a total of 38 human sacrificial victims" (J. W. Olsen 1985: 61). This situation represents a dramatic example of human and dog sacrifice resulting in simultaneous burial, one not covered in my own earlier study (Morey 2006), for that presentation included a relevant but less dramatic example involving the ancient Greeks (L. P. Day 1984).

A poignant Old World example of simultaneous dog and human burial comes from Unar 2, in the United Arab Emirates. There, a tomb was found containing the skeletal remains of a woman and a dog, dating to just over 4,000 years ago:

> The dog was mostly complete with the fore and hind limbs still being visibly in articulation (Fig. 3). The hind limbs appeared to have been pulled around to accommodate the dog within the southern end of the chamber. (Blau & Beech 1999: 34–35)

Moreover, "Clearly not a later intrusive burial, the dog and the human were interred in the same stratigraphic level" (Blau & Beech 1999: 39). This situation brings to mind the Natufian puppy burial much more than the Yin Complex tombs just covered. MacKinnon & Belanger (2006) have also presented a noteworthy case involving an elderly dog that was buried with a 10- to 15-year-old child some 1,500–1,800 years ago in Tunisia. In particular, from studying the skeleton they inferred that

> This dog was not healthy, but lived with a variety of dental and skeletal ailments for quite some time...[and] Practically all limb articulations in the Yasmina dog are disfigured due to osteoarthritis. (MacKinnon & Belanger 2006: 40)

They make the point that this dog had to have received special care while still alive, to have lived as long as it did (perhaps 15 years or older when it died).

In a more synthetic study, Bodson (2000) has presented a comprehensive account of the motivations for pet-keeping in ancient Greece and Rome. As part of that account, she emphasizes that a variety of animals were sometimes kept as pets, and were also afforded funerary treatment upon death. In surveying the epitaphs that were sometimes written, she notes the following: "As for the epitaphs, they highlight dogs more than any other species and belong mostly to the Hellenistic and Greco-Roman periods (third/second centuries BC to fourth/fifth centuries AD)" (Bodson 2000: 30). As for the owners themselves, "Whatever their age, dog owners openly wept for their animals, and expressed affliction and bereavement" (Bodson 2000: 32). So, the ancient Greeks and Romans clearly valued their dogs, according them ritualistic treatments at death more often than they did other animals.

Even more recent than the settings considered by Bodson is a series of dog burials (sometimes with people) from Scandinavia, some dating to the Viking period, from about 800–1050 A.D. (Prummel 1992: 135). As described by Gräslund (2004), some of those represent what are known as boat-graves. In such cases, a boat was used to house the dead, or represented a grave good. Gräslund (2004) describes several from Sweden and Norway, and examples of those are included here in Table B.1. These are situations that suggest acts that go beyond personal reverence, as suggested by Gräslund (2004: 167) herself: "the dog may well have been a beloved companion, but it may also have had a deeper, symbolic meaning." In a similar vein, Prummel (1992) presented an account of medieval dog burials not only in Scandinavia, but also among peoples in mainland Europe, as well as England. For Scandinavia, Prummel (1992: 135, Table 1) indicates a total of 185 dog graves, containing 246 dogs. Some of the Scandinavian cases indicated by Prummel (1992) are represented in Gräslund's (2004) work as well. Turning to elsewhere in Europe, including England, for the time frame of about 1600–1200 B.P., Prummel (1992: 135, Table 1) indicates a total of 86 dog graves from 55 different cemeteries, representing a total of 114 dogs (Prummel 1992: 139). For both Prummel's mainland European (including England) and Scandinavian cases, Table B.1 includes simple summary entries rather than an attempt to inventory all of them. Burial modes varied in these settings, ranging from dogs buried individually, to others buried with people, and yet others buried with horses, or with horses and people (Prummel 1992: 138, Table 3).

Although technically not part of the Old World, in terms of the three part geographic division employed in Table B.1, it is appropriate to touch base with Polynesia. Polynesia shows, in fact, the kind of modal variation whereby it seems warranted to consider individual burials separately from simultaneous human–dog burials: "In Hawaii, dog burials occur in association with human remains and ceremonial sites, while in New Zealand no clear burial context is apparent" (G. Clark 1996: 32–34). Thus dogs might be buried by themselves in Polynesia, or they might be buried with people, or at least in clear ceremonial settings. Moreover, apparently there is some geographic variation within Polynesia. Regardless, variation in burial modes is a pattern that is true throughout the world.

In the Old World, the circumstances surrounding individual dog burials varied as well. At Botai, in northern Kazakhstan, over 5,000 years ago more than forty dogs, or parts of dogs, were deposited in settings "that show dogs to have been repeatedly interred in small pits, either in houses or just outside them" (S. L. Olsen 2000: 86). Overall, those dogs may have been especially of sacred symbolic value:

> The occurrence of dog deposits or burials on the west side of houses at Botai is provocative. According to Indo-European mythology, the gate to the Otherworld was to the west and was guarded by two dogs. (S. L. Olsen 2000: 87)

Nothing definitive can be established, but this pattern certainly is intriguing. In light of that pattern, it is worth noting that about 2,000 years ago at Côte-d'Or (Vertault), France, among dogs that were apparently sacrificed, "In 70 of the pits, the dogs appear to have been sacrificed and deposited on their left side, head facing west" (Horard-Herbin 2000: 115). Again, the westward orientation may be meaningful.

Simultaneous human and dog burials from the Mesolithic site of Skateholm, in Sweden, were highlighted previously, but one might recall that Skateholm had even more examples of individual dog burials. Lars Larsson, directly involved in the work there, marveled that "an individually-buried dog has been accorded as many grave goods as any human and, indeed, more so than most" (Larsson 1990: 156). And beyond the grave goods, often enough "red ochre was scattered over the dogs' corpses" (Aaris-Sørensen 2001: 32). These observations about dogs being given grave goods are intriguing, and such a practice can be documented in at least one setting in nearby Denmark. That setting is a slightly later Neolithic site, Esbjerg, dating between about 4,800 and 4,400 years ago. As reported by Lauenborg (1982), there were four graves in close proximity, three human and one dog. This dog was buried with an amber bead and a battle axe head. Such a situation certainly suggests symbolic significance, the dog perhaps belonging to one of the deceased people.

Rather clearly, the circumstances that led to individual dog burials were variable. They ranged from ritual sacrifice, which might also lead to burial with a person, sometimes with sacred symbolic overtones, to a personally symbolic burial. The Esbjerg case, just covered, appears to be of the latter sort. But rather than explore this topic further now, it is worth pointing out that these summary statements about Old World dog burials have excluded one particularly intriguing case. That case is Ashkelon, in present-day Israel, surely the most dramatic localized case of dogs being buried that is known, not only in the Old World, but in the entire world. As such, it warrants separate treatment, after a full consideration of New World dog burials.

Care in Burial: The New World

The earliest known dog burials in North America are about 8,500 years old and are from the Koster site, in Illinois (Figure 7.1). Morey & Wiant (1992: 225) described them as follows:

> The three canids occurred within a horizontal area of 24 m². A fourth canid of problematical stratigraphic association and dating has been reported elsewhere (Hill 1972). The three skeletons were in generally well-demarcated pits capped by and extending below the Horizon 11 midden. These pits were basin-shaped, shallow (15 cm maximum depth), and only large enough to accommodate the animals. They are considered graves even though they are not marked, for example, with stone cairns ([R. B.] McMillan 1970) or by direct association with a human interment. (Davis and Valla 1978)

These Early Archaic burials were clearly prepared with care, and, similar to the cited Davis & Valla study (1978), dealt with earlier, "The evidence from the Koster site hints that an affectionate relationship between humans and dogs may have existed over 8,000 years ago in the North American Midwest" (Morey & Wiant 1992: 228). The Koster dogs are apparently not, however, the first known case of ritualistically treated dog remains in North America. That distinction currently goes to a dog skull and mandible associated with a human skeleton from the Rancho La Brea asphalt deposits in California, originally discovered shortly after the beginning of the twentieth century. As reported in 1985, these remains are about 9,000 years old, and the overall assessment is that "this occurrence is a reburial rather than an accidental entrapment or the disposal of a homicide victim" (R. L. Reynolds 1985: 83). Though apparently not a true primary burial, this situation is a logical point of departure for considering simultaneous dog and human burials in the New World.

The oldest simultaneous dog and human burials that are securely documented are evidently from the Braden site, in Idaho, and date to about 6600 B.P.: "at this time the [two] Braden specimens appear to represent the earliest known example of intentional dog burial in direct association with human interments in North America" (Yohe & Pavesic 2000: 103). Braden was excavated in the 1960s, and the circumstances surrounding this situation include the following:

> More than thirty years ago two dog burials were exhumed from what was reported as a 'mass grave' consisting of seven human burials surrounded by at least five additional human interments. (Yohe & Pavesic 2000: 93)

At any rate, this situation presages a considerable number of simultaneous dog and human burials in a variety of places.

From within a broader span of time that encompasses the date from Braden, there is a series of Archaic Period dog burials from the Green River Valley in Kentucky. These will be considered in a separate section below, in order to highlight just how common dog burials can be in a single locality. It is one area with which I have great familiarity, having lived there for several years in the mid-1990s, actively pursuing research interests focused on subjects other than dogs (e.g., Morey & Crothers 1998; Morey et al. 2002). Now, however, it is appropriate to consider a different intriguing case that anticipates some of the points to be covered with the Green River series. That case is the mostly Archaic Period Perry site in Alabama, from which three basic excavation units were reported in two different places (Webb & DeJarnette 1942: 58–91; 1948a). From the combined accounts, fifty-five dog burials can be documented, at least thirteen of which were with people.[2] Of nineteen dog burials known from the later round of work, "There were four dog

---

[2] It should be noted that temporal resolution at some of the Alabama sites is compromised by the fact that there are post-Archaic components, and the Archaic affiliation for absolutely all of the dog burials cannot be verified, though with little question the vast majority of them are Archaic in age. In addition, the number of dog burials indicated here for Perry differs from what is reported by Morey (2006: 160), that number being twenty. Morey (2006) did not study Webb & DeJarnette to the extent that he should have, and did not consult their later report on Perry at all (Webb & DeJarnette 1948a). In fact, the total number of dog burials from Perry that can be verified from the original reports is fifty-five. As for the number of dogs buried with people, because of an uncertain quantity in one section of their combined reports, I can only indicate more than thirteen in Table B.1. That is, there were thirteen reported numerically, but an unspecified number of additional ones. Regardless of those uncertainties, DeJarnette himself once offered his summary evaluation regarding dog burials at Perry and other sites in the present state of Alabama: "The Shell Mound Archaic people showed either

FIGURE 7.1. One of several individually buried dogs at the Koster site, in Illinois, just prior to being excavated. All three dog burials discovered there at this time are about 8,500 years old, making them the earliest known dog burials in North America. The blade of a standard-sized trowel, at the upper right, provides an approximate scale. A separate presentation was devoted to these three dog burials, and this image previously appeared in it, in *Current Anthropology* (Morey & Wiant 1992: 226, Figure 1). At about the same time, a similar image of it, in color, was featured in the popular press, specifically in a column with no separate author in *Discover* magazine (September 1992, page 14), and soon after that in the French Science magazine *La Recherche* as well (Morey 1996: 73). Subsequently, the present image also appeared in Marion Schwartz's book on dogs in the early Americas (Schwartz 1997: 104), and in a college-level introductory archaeology textbook (Feder 2007: 326). Photograph by D. R. Baston.

burials reported as in the sleeping position, i.e. body as 'curled up' in grave with bones in anatomical order. Two of these were in association with human graves" (Webb & DeJarnette 1948a: 22). The disposition of all the others is not entirely clear, though Webb & DeJarnette (1948a: 21) provided a photograph of two, by themselves, for which "the limbs

ceremonial or personal regard for the dog; dogs seem to have been buried with the same care accorded people" (DeJarnette 1952: 274).

of the skeletons have been folded. In a human burial they would be classified as 'fully flexed'." Situations such as this one obviously complicate efforts to treat dogs buried by themselves separately from those buried with people. That is, both modes could obviously involve comparable care and positioning of the remains, at the same basic location. Another case worth calling attention to is the comparably aged Cherry site, in Tennessee. There, Magennis (1977) was explicitly focused on human mortuary practices, and reported on four dogs buried with people.[3]

From much more northerly climes two sites are worth noting. Port au Choix, Newfoundland, is nearly 4,000 years old, and included a substantial human cemetery where remains of four dogs were buried with people. Two of the dog skeletons, buried with one person, were complete, and one had been killed by a club blow (Tuck 1976: 77). Ipiutak, located at Point Hope on the northwestern coast of Alaska, dates between about 1500 and 1150 B.P. and also includes numerous human burials, four of which were associated with dogs. There, describing Burial 137, Larsen & Rainey (1948: 250) indicated that "The articulated [human] skeleton rested on its back, with the hands over the pubic region and the head towards the east . . . and an almost complete dog skeleton near the left leg." Another dog there, buried distinctively by itself is noted again below.

But for the present, we move back south, all the way to South America. There, at Sipán, in Peru, two dogs figured into pre-Inca Moche royal burials well over 1,000 years ago. At one setting, Tomb 1, pertaining to a man who was interred near the principal figure, "He was wearing a beaded pectoral, and had several unidentified copper objects on top of his body. Inside his coffin was a dog, stretched out with its head near the man's feet and its tail by the man's waist" (Alva & Donnan 1993: 123). It is logical to suppose that the dog was that man's. A perhaps

---

[3] From the Cherry site in Tennessee, Walker et al. (2005: 85) reported, citing Magennis (1977), that there were two dog burials from that site. There were in fact at least four, as revealed by Magennis's own words:

> The practice of burying dogs with human interments continues at the Cherry site. There were four occurrences of dogs buried in direct association with humans. One was placed with an adult male, two adult females were buried with dogs, and one indeterminate sex adult was afforded the same treatment. The importance of this practice in relation to the entire mortuary ritual is unknown. (Magennis 1977: 79–80)

even more intriguing case concerned a small cane coffin near the feet of a principal coffin in Tomb 2 that contained a young man. In that smaller coffin was a child with the following animal accompaniments:

> The dog lay on its right side on top of the child, with its head over the child's pelvis and its tail over the child's feet. The snake lay extended in the northwest corner of the coffin near the child's feet. The dog was reminiscent of the dog inside the cane coffin of the adult male buried next to the principal figure in Tomb 1. However, the snake remains an enigma; no other Moche burial previously excavated contained a snake. (Alva & Donnan 1993: 159)

So the snake is an enigma, while the dogs were considered suitable accompaniment for royal burials. How the dogs died is not clear, just as how they died in other cases where they accompanied human burials is not always clear. Similarly, though the time frame is likely a bit later than at Sipán, more than 100 years ago Nehring (1887) dealt with the reality of dogs (and other animals) buried with people at the necropolis at Ancon, also in Peru (see Table B.1).

Yet another setting in South America further illustrates the complications that can arise in trying to treat simultaneous human and dog interments separately from dogs buried by themselves. The site in question, Chiribaya Baja, is in southern Peru. There, at a human cemetery of the Chiribaya peoples, dating within the span of about 1,100 to 650 years ago, archaeologists have found the remains of more than 40 naturally mummified dogs, buried in separate plots alongside their owners. Each dog had its own grave and some were buried with blankets and food. Overall, "it seems that ancient Peruvians also treated their dogs like members of the family" (de Pastino 2006). These remains have been rather widely publicized (e.g., de Pastino 2006; Lange 2007), and some of them have previously been the target of some professional investigations (e.g., Dittmar et al. 2003). So these dogs seem to have been treated with a sense of personal reverence, beyond broader spiritual implications, though such implications may well have obtained also. Moreover, these dogs apparently were not sacrificed. Schwartz (1997: Table 4.1) provides some additional South American cases and the text of the relevant chapter (4) describes some of them as well.

The following case, at a somewhat later and more northerly site, the Norris Farms Cemetery in Illinois, dating to at least several hundred years ago, suggests the perception by people of dogs that seems to have obtained at Sipán in Peru, as covered earlier:

> A complete skeleton of a four-month-old male dog (*Canis* cf. *familiaris*) was found at the feet of a crippled adolescent male (Burial 56). No evidence of how this animal died was observed. Its proximity to a pottery vessel and spoon may imply that the dog was a food offering, but the young age of the boy might also suggest that the dog was a pet. (Santure & Esarey 1990: 93)

Another case indicates the special role of dogs in human burial rites, but without the complete skeleton of the animal. Barbara Lawrence (1944: 73) offered the following observation concerning a late prehistoric setting in the North American Desert Southwest (New Mexico): "The association of dog skulls with all burials indicates that a special significance was attached to the dogs, more so than has heretofore been recorded for the Southwest." This would seem to be a localized expression of the association between dogs and people. Special significance is also suggested by what was found at the Ausmus Farm Mounds, in east Tennessee, where the relevant burials are in the vicinity of 1,000 years old. There, with Burial 3, a person, William Webb (1938: 109) noted that "A dog had been buried across the feet of this skeleton." Though not as clearly associated, Webb (1938: 110) also wrote of Burial 5 that "Northwest of this burial at a distance of 23 inches was a burial of a small dog."

There are other cases of simultaneous dog and human burials, but rather than dwell further on that reality, it is worth considering more fully the care with which dogs were often buried in the New World when interred by themselves. Highlighting the three Early Archaic Period Koster Site dog burials from Illinois at the beginning of this section anticipated this effort, and it is now fitting to expand from there. One might recall the quoted passage at the beginning of this section, from Morey & Wiant's (1992: 225) work, in which reference was made to dog graves being marked by stone cairns. The following passage represents the key example, from the cited source, describing,

> the discovery of a small, adult dog, buried in a prepared grave, from the Archaic deposits of Rodgers Shelter in Benton County, Missouri. The skeleton, found during the summer of 1966, was covered by a small tumulus of dolomite stones beneath 2.7 m of later cultural deposits. The position of the rocks near the base, which were tilted down and inward,

suggests that a shallow basin-shaped pit had been prepared for the burial before the rock covering was added. (R. B. McMillan 1970: 1246)

The Rodgers Shelter dog was buried about 7,500 years ago, not as old as Koster, but still quite early. It is significant of course that it was in a clearly marked grave. An earlier study of dog burials (Morey 2006: 161) unfortunately reported the Rodgers Shelter burial as an example of an individual dog burial in an otherwise unmarked grave. Clearly, though, the grave was marked, and it is a pleasure to correct that oversight.

Meanwhile, from within a span of time that encompasses that date several dogs were buried at Dust Cave, Alabama. One was reported in the 1990s (Morey 1994b), and subsequent work revealed more, bringing the total to four. Regarding the disposition of the Dust Cave dogs, "All the dog remains were in shallow pits on the west side of the site. The bones of the canid skeletons were in correct anatomical order, lying on their sides, with their tails curled between their legs" (Walker et al. 2005: 85). Clearly, care is reflected in those burials. One of the most compelling cases of care for a dog concerns an unusually old dog from the Middle Archaic (ca. 7000 B.P.) Anderson site in Tennessee. This dog had experienced multiple maladies in life, ranging from various injuries to arthritic development to a persistent infection stemming from unhealed rib fractures. A reasonable assessment of this dog's lot in life is that

> The pathological condition of this individual suggests that the owner insured the safety and well-being of the individual throughout life since it is doubtful that, given all of the traumatic and age degenerative manifestations, the dog could have survived in the absence of care. Further evidence supporting the latter contention is the fact that the dog was ritually interred. (Breitburg n.d., as quoted in Dowd 1989: 122)

Like the Tunisian case covered earlier (MacKinnon & Belanger 2006), this is a situation where one does not need to be concerned about sacrificial implications, or any other ambiguous circumstance. This was a case of a previously traumatized elderly dog being treated with great consideration, and then being buried with care when dead.

From the same general time frame, the Archaic Period Eva site in Tennessee (Table B.1) had at least 18 dog burials, 14 of which were by themselves. Lewis & Lewis (1961: 144) provided several photographs and some noteworthy descriptions, of which a burial designated as number 122 is a useful example: "This was the burial of a very large dog lying in a curled up position. Two splinter bone awls and a large stone pick were definitely associated. This is one of the 15 dog burials in the Three Mile component" (Lewis & Lewis 1961: 131). Also in Tennessee, at a multi-component site in the Tuscumbia River drainage, in apparently Archaic Period deposits, "Feature 3, a dog burial, also emerged in level 2 and continued into level 3. The majority of the animal was contained in the unit; it had been buried in a curled up, sleeping position" (S. D. Moore 1991: 43).

Quite clearly, Archaic Period dogs were buried rather often in what is now the Tennessee region. In fact, when they are absent, the situation can be considered unusual. For example, as one of several unusual characteristics of a particular shell midden site, the following trait was duly noted: "An additional dissimilarity was the absence of dog burials accompanying human interments or as separate interments at Penitentiary Branch; *Canis familiaris* elements were recovered, however" (Cridlebaugh et al. 1986: 65). So no dog burials were found at that site, though there were dog remains. Not surprisingly, the authors were clearly familiar with other Archaic Period sites in the same vicinity, such as Cherry and Eva, where dog burials are well documented.

Moving beyond Tennessee, a series of Archaic Period shell midden sites in Alabama includes both separate dog burials and simultaneous dog and human burials. In addition to the Perry site, described previously, there are numerous other sites of interest. For example, there were nineteen dog burials from Flint River, but unlike some cases at Perry, "They were not placed in the grave with human burials here, as they have been found elsewhere" (Webb & DeJarnette 1948b: 37). A comparable situation, among nine dog burials, was the case at Whitesburg Bridge: "None were in direct association with human graves" (Webb & DeJarnette 1948c: 18). Unfortunately, the numerical situation with another site, Little Bear Creek, is ambiguous.[4] But in general terms, after dealing with some of the recorded features, Webb & DeJarnette

---

4 The numbers reported here for Flint River, Whitesburg Bridge, and Little Bear Creek, in Alabama, differ from those reported by Walker et al. (2005: 85). The numbers reported here are taken directly from the original sources, but as a coauthor of that article, I bear partial responsibility for the disparities. As one example, I note that Walker et al. (2005: 85) reported two from Little Bear Creek, whereas I indicate an unspecified number. The reason is that Webb & DeJarnette (1948d) did not specify a number due to substantial difficulties maintaining proper field records, a problem exacerbated by genuinely bad weather conditions, including flooding. Rather, they simply reported that "the burial of dogs in the midden was quite common" (Webb & DeJarnette 1948d: 21).

(1948d: 21) offered the following statement: "Besides these features, the burial of dogs in the midden was quite common.... Seemingly as much care was given to placing the dog in position as was ordinarily given to a human burial." It should be clear that the care with which dogs were often buried in Archaic times is highlighted rather frequently in the literature, and several later cases (see Table B.1) warrant coverage as well.

As one example, Elizabeth Wing shared a personal communication from one of the excavators of a site dating to some 2,000 to 1,500 years ago, at Sorcé, Puerto Rico, regarding some of the seven dogs found buried there:

> They were buried complete with their legs pulled together, as if the four legs were tied with a rope, the tail was between the legs, and the heads of both specimens YTA3 (YTA 3 J 6 and YTA 3 1 5) were twisted and resting on top of the body. (Wing 1991: 380)

The next case that warrants attention is one that was directly anticipated in closing the previous chapter. At Ipiutak in Alaska, dating between about 1,500 and 1,150 years ago, five of six dogs, or parts of dogs, were buried with people. Regarding the one buried individually, Burial 109, "The fragments of a log tomb like those found in the group including Burials 42–64 were found at a depth of 50 centimeters. It contained a complete articulated dog skeleton, which was extended with the head towards the west" (Larsen & Rainey 1948: 248). Lacking an indication of anything else in this tomb, one can surmise that it was likely this dog's tomb. A later site that merits special attention is Broad Reach, from coastal North Carolina. At this site there were thirteen dog burials, all by themselves, though apparently two of them were in a basic cemetery area. The deposits yielding these dogs are apparently Middle or Late Woodland in age, in this area dating sometime between about 2,400 and 400 years ago. All of these dog burials were encountered and excavated in the summer of 2006 (Millis 2010). The special attention indicated above is represented by Figure 7.2, a photograph of one of the thirteen dog burials. One can surely appreciate from this image that the burial appears to be generally consistent with those several, already highlighted, that were described as being in a curled up, sleeping position. So while Archaic Period dog burials in North America are, overall, most noteworthy for their frequency and the care that is reflected, especially in eastern and midwestern North America, it did happen sometimes after the Archaic Period.

In fact, one of those post-Archaic settings is apparently a dramatic exception to a generalization, dealt with more in the next section, about dog burials being less common in post-Archaic times (e.g., Haag 1948: 253). The exception in question is the Hatch site in Virginia, from deposits dating within the range of about 800–400 B.P. This setting was originally described by Gregory (1979), who in the 1970s reported on 41 dog burials. Since that original work, "over 100 [105] intentional and single and multiple dog burials have been excavated" (Boyd & Boyd 1992: 263). These likely were of broader spiritual/symbolic significance, something like Ashkelon (see below), rather than signifying the kind of personal reverence strongly implied by so many Archaic and even some later burials.[5]

One of those later contexts suggesting personal reverence concerns three dogs buried at Lambert Farm, Rhode Island, dating within the span of about 1,150 to 500 years ago. The first one was buried beneath a stone slab, and the excavators found that

> The dog was laid to rest in an extended position, stretched out, lying on its right side with its head pointing northeast and its face toward west. Someone had even placed in the grave two shells, a knobbed whelk and a valve of softshell clam, the only shells found under the stone slab. (Kerber 1997: 67–68)

The second dog was buried on top of a stone slab. As for the third one,

> Surrounding the dog's skull were several complete softshell clams, perhaps left by the owner as grave offerings. The dog was placed in a flexed position, curled up, lying on its left side with its front left paw carefully

---

5 That personal reverence, reflecting friendship, may not have been the primary objective with the Hatch site burials is suggested by the fact the "Two such dog burials were recovered in association with human arms severed at the elbow" (R. Dent 1995: 255). Like the Ashkelon burials in Israel (covered later in this chapter), albeit on a smaller scale, a broader sense of spirituality may have been operative at Hatch. What seems especially surprising is that this Hatch site phenomenon has not been widely publicized, though it has been noted in relevant regional literature beyond Gregory's preliminary report (e.g., Boyd & Boyd 1992; R. Dent 1995). The Hatch site is included on the National Register of Historic Places, and the nomination form for its inclusion, from 1989, notes the unusual number of dog burials, suggesting that at least some of them "have some form of ritual significance." At any rate, in a review of the book containing Boyd & Boyd's (1992) chapter, Seeman (1994: 582) noted that the more than 100 dog burials at the Hatch site "might be an eastern North American record!" Again, this particular phenomenon brings to mind Ashkelon in several important ways.

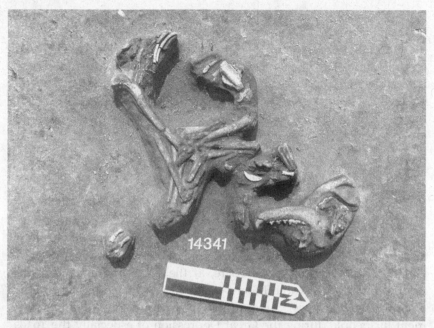

FIGURE 7.2. One of thirteen dog burials found in 2006 at Broad Reach, a Middle/Late Woodland Period (ca. 2400–400 B.P.) site in coastal North Carolina. Photograph by Mike Fisher.

tucked under its head, which was pointing east and its face toward south. (Kerber 1997: 74)

Nearby, in Maryland, about 700 years ago the following happened to a dog that had earned an affectionate name among the excavators: "Seneca was intentionally buried in a shallow pit just northeast of the structure discovered last year. The terrier-sized dog was laid to rest in a curled position, with its four paws gathered together and placed on top of each other" (J. Dent 2003: 3). An even more recent noteworthy case concerns eight dogs buried in east Tennessee, in late prehistoric or early historic times. The following account of a dog buried at the Chota-Tanasee historic Cherokee site is highly reminiscent of the old Archaic dog that was buried at the Anderson site in Tennessee, covered earlier:

> It had lived to be of considerable age, and in view of the loss of several teeth, its arthritic condition, and the damaged right hind foot which very possibly produced a limp, without some special care it probably could not have survived as long as it did. (Parmalee & Bogan 1978: 105)

Clearly, several thousand years ago, as well as in historic times, traumatized elderly dogs were cared for by people in North America, and then buried affectionately when they died.[6] What is emerging from a consideration of dog burials, separately or with people, is that great care is often reflected in their burial treatment. Archaic Period sites in the Green River Valley of Kentucky reinforce that perception, and also highlight just how common this phenomenon can be in a single temporally and geographically restricted setting.

ARCHAIC DOG BURIALS IN THE GREEN RIVER VALLEY, KENTUCKY

The Green River Valley in west-central Kentucky is a genuinely storied region in the annals of North American professional archaeology, apparently beginning with early investigations by the archaeological explorer C. B. Moore (1916). There, at the most famous of the sites in question, Indian Knoll, Moore removed nearly 300 human burials and associated artifacts. Subsequently, as a part of providing employment for people during the Great Depression of especially the 1930s, William Webb supervised extensive excavations at a series of Green River sites, including Indian Knoll, under the auspices of the Works Progress Administration, or WPA. Overall, with Indian Knoll as something of the "signature" site (Morey et al. 2002: 523), the WPA-era work along the Green River was instrumental in developing the very concept of an Archaic Period in eastern North America (see Jeffries 1988a, 1988b; Chapman & Watson 1993). The fundamental importance of the work there is underscored by the fact that William Webb's original lengthy report on Indian Knoll (Webb 1946) was republished in 1974 as a book, in its entirety, by the University of Tennessee Press.

Figure 7.3 is a map that includes the relevant stretch of the Green River, with the approximate locations of several Archaic Period sites that have yielded dog burials indicated. In conjunction with that map, Table 7.1 provides a breakdown of dog burial information, as documented for those and other sites within that stretch. In several cases

---

[6] Writing specifically about dogs among the Cherokee people, Carrie McLachlan has well captured the way in which the burial of a dog can reflect personal reverence, as well a broader order of spirituality: "Burial may suggest special affection towards a particular animal, but it may also reflect the cosmological significance accorded dogs in human thought worldwide" (McLachlan 2002: 4).

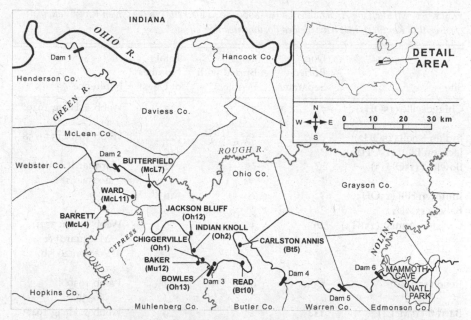

FIGURE 7.3. Map of a stretch of the Green River Valley in Kentucky, showing the locations of several Middle/Late Archaic Period archaeological sites that have yielded dog burials, as inventoried on Table 7.1. Adapted from Morey et al. (2002: Figure 1, page 522).

no citation to a reference is offered, and only totals are given.[7] One should also be aware that the University of Tennessee maintains a digital archive of photographic images from the WPA-era work in several states, including a separate compilation of images from the work with Green River Archaic sites. One way in which that archive is incomplete is that, for reasons surrounding modern issues, it includes no images of buried people, with or without dogs. For images of

---

[7] Totals listed for all sites are information provided to me by George Crothers, director of the William S. Webb Museum of Anthropology in Lexington, Kentucky. That is where the dog remains in question are curated, along with the original information. Sometimes there are discrepancies between the numbers indicated in the original reports, and those that can now be verified. That is why the numbers on Table 7.1 do not always add up precisely. Sites with total number of dog burials only, are information from the Webb Museum collections and records, usually in the absence of published documentation. In those cases, they are curated at the Webb Museum as burials, and many of those dogs were analyzed by Haag (1948: 122–23) Where a source is cited for such a site, it was uninformative regarding the different modes of dog burial at that site. In all, the explanation for many discrepancies is simply not clear.

TABLE 7.1. *Middle/Late Archaic Period (ca. 8000–3000 B.P.) sites in the Green River Valley of west-central Kentucky that have yielded documented dog burials*[7]

Site	Dogs Buried Separately	Dogs Buried with People	Total Number of Dogs	Reference
Chiggerville (15OH1)	7	5	12	Webb & Haag 1939: 11, 17
Indian Knoll (15OH2)	8?	13?	24	Webb 1946: 155–158[8]
Jackson Bluff (15OH12)	?	?	10	
Bowles* (15OH13)	2?	?	12	Marquardt & Watson 2005a: 61[9]
Jimtown Hill (15OH19)	?	?	18	
Baker (15MU12)	?	?	10	
Carlston Annis* (15BT5)	26	3	29	Webb 1950a: 272; Marquardt & Watson 2005b: 111[10]
Read (15BT10)	23	0	23	Webb 1950b: 360–362[11]
Barrett (15McL4)	11?	2?	9	Webb & Haag 1947: 14[12]
Butterfield (15McL7)	7?	1	7	Webb & Haag 1947: 33[13]
Ward (15McL11)	16?	7?	25	Webb & Haag 1940: 81–82[14]
Kirkland (15McL12)	2	?	9	Webb & Haag 1940: 72[15]
Morris (15HK49)	?	?	4	Rolingson & Schwartz 1966
TOTALS	102+ (?)	31+ (?)	**192**	

* *Note:* Bowles apparently had two different spans of occupation, one within the 8000–3000 B.P. time range, but also a later one, dating between about 2900–1200 B.P. Carlston Annis also shows a similar pattern, but with the later date range more restricted (ca. 2700–2400 B.P.). Marquardt & Watson (2005a: 64) have provided an extensive compilation of radiocarbon determinations from the Green River sites, both uncalibrated and calibrated. See also graphically depicted summary of Green River site date ranges in Morey et al. (2002: 540, Figure 8).

[8] Webb (1946: 155) reported twenty-one dog burials from Indian Knoll, eight buried by themselves, the remaining thirteen with people. The Webb Museum, however, contains twenty-four.

[9] As explained by Marquardt & Watson (2005a: 61) the two dog burials from Bowles were originally found in the 1930s, but there is merely a "7-page summary of work at the site" (Marquardt & Watson 2005a: 61), unpublished, by David P. Stout, his notes being on file at the William S. Webb Museum of Anthropology, University of Kentucky, Lexington, Kentucky. The Webb Museum has a total of twelve dog burials from Bowles.

simultaneous dog and human interments, one has to go to the original reports, usually published in the 1940s or 1950s. But a substantial number of buried dogs, enumerated on Table 7.1, can be viewed in this archive.

Given the centrality of Indian Knoll to work in this area, that is a logical place to begin with more detailed coverage. To start that coverage, it is worth calling on the words of one of William Webb's frequent collaborators, William Haag, who was writing in a broader sense, of Archaic Period sites in the Southeast, especially Kentucky and Alabama: "... it is of importance that dogs often accompanied human burials and received as much care in their disposition as did the humans. In some instances more than one dog was placed with a single human skeleton" (Haag 1948: 121). Indeed, at Indian Knoll itself there were twenty-four dog burials in all, thirteen of which were with people. Commenting on this circumstance, Webb wrote that

> It appears that many dogs were buried with the same degree of attention to grave pits, and placement of body as was accorded to their human contemporaries. In some cases, the dog was buried in human graves, in such close association as to indicate simultaneous interment. (Webb 1946: 155)

(See Citation of footnotes 10, 11, 12, 13, 14, 15 on previous page 170)

---

[10] Webb (1950a: 272) reported twenty-eight dog burials from Carlston Annis more than fifty years ago, though the Webb Museum can account for only twenty-seven from that time. But more recent work, beginning in the 1970s, yielded two more: "In our excavations at Bt5, we encountered a total of 12 to 15 human (and 2 dog) burials." (Marquardt & Watson 2005b: 111). That is why the total is twenty-nine.

[11] Webb (1950b: 360–62) reported sixty-three dog burials from Read, but that reported number is apparently an error, as the collections at the Webb Museum only have 23. Following the original report, I also erroneously indicated sixty-three from Read in my earlier study (Morey 2006: 160, Table 1).

[12] Webb & Haag (1947: 14) reported the following about Barrett: "In each of two graves, the body of a dog had been placed at the time of interment at the foot of the human body. An additional 11 dogs had been buried separately, seemingly with as much care as was accorded their human contemporaries." This account would bring the total to thirteen, but only nine can be accounted for presently at the Webb Museum.

[13] Webb & Haag (1947: 33) report that at Butterfield "Eight dog burials were found in the midden." The Webb Museum can presently account for only seven.

[14] From Ward, Webb & Haag (1940: 81) report that "In the midden were found 23 dog burials." The Webb Museum, however, presently accounts for twenty-five.

[15] From Kirkland, Webb & Haag (1940: 72) reported that there were ten dogs. The Webb Museum can presently account for only nine.

Webb went on to provide photographic documentation of several examples, along with verbal descriptions. But before offering those descriptions, he gave his general assessment of what such situations signified:

> From such definite evidence of intentional association as is presented in Figure 16, described in detail below, one must conclude that dogs were often killed at the time of burial of their owner, and buried with them perhaps as a symbol of continued association in the spirit world. (Webb 1946: 156)

One of his described (and photographed) examples was Burial 232:

> Burial 232 was in a pit 3.1 feet deep below the surface of the midden. The body was fully flexed on the right side. The body of a dog fully flexed was laid on its right side just at the back of Burial 232. It appears that the dog's body was intentionally arranged to take a position similar to the human skeleton. The left foot and nose of the dog skeleton rests upon the left shoulder of the human skeleton. (Webb 1946: 158)

In broader view, Indian Knoll may well be the most famous of the Green River sites where there were dog burials, one that was buried separately being shown in an earlier major work on dog burials (Morey 2006: 163, Figure 3). In fact, a photograph of that burial, dog burial No. 1, appears at the beginning of this book. But there were many other such sites, and Webb himself dealt with most of them. For example, what was found at the Carlston Annis site prompted this comment:

> As is usual in shell middens, dogs were often buried with the same care as that given to human burials. A total of twenty-eight dog burials were found, at this site, some in round grave pits in the subsoil, see Figure 3-D, and some were laid on the shell midden and covered with shell. (Webb 1950a: 272)

The Barrett site provides yet another example:

> In each of two graves, the body of a dog had been placed at the time of interment at the foot of the human body. An additional 11 dogs had been buried separately, seemingly with as much care as was accorded their human contemporaries. See Figure 4C. Of these, nine were placed in the midden and two were buried in well-formed, round-grave pits extending into the hardpan below the midden. (Webb & Haag 1947: 14)

FIGURE 7.4. One of twenty-five dog burials found at the Archaic Period Ward site in Kentucky, dating between about 7,500 and 4,000 years ago, probably the latter portion of that span (see Marquardt & Watson 2005a: 64). Original photograph from more than 50 years ago (Webb & Haag 1940: 82, Figure 9) enhanced and made available by George M. Crothers, William S. Webb Museum of Anthropology, University of Kentucky, Lexington, Kentucky.

Like the Broad Reach site in North Carolina, covered earlier, a Green River site that merits special attention is Ward. Again collaborating with William Haag, Webb described the situation there as follows:

> In the midden were found 23 dog burials. Of these, 7 were in human-grave association. This would seem to definitely establish that dogs were domesticated, and so highly regarded as to warrant burial in manner and place generally acceptable for human burials. This regard for the dog even permitted it to share the same grave pit with its human contemporary. Scattered dog bones were even more rare than scattered human bones and all dog burials were as carefully placed as the human burials. (Webb & Haag 1940: 81–82)

Webb and Haag photographically illustrated one such burial, by itself, and described it as "A typical dog burial at the Ward site" (Webb & Haag 1940: 82). Figure 7.4 is their picture of that burial, enhanced, as made from the original negative of that photograph, taken more than half a century ago. Some observations about their quoted passage are in order. This is one of their first treatments of dog burials from Green River sites, though not absolutely the first one. As an early one, the comment about the joint human–dog burials establishing that the dogs were domesticated gets at the fundamental nature of that domestic relationship. They produced an earlier report on the Chiggerville site (see Figure 7.3, Table 7.1), where twelve dog burials were found, five in direct association with human burials (Webb & Haag 1939: 11, 17). Notably, while they included in that report a photograph of Burial 31, a joint human–dog grave, they offered no descriptive or synthetic comments on the overall phenomenon. With one interesting exception, Read (see later), it seems that fieldwork at Chiggerville and Ward, in the spring of 1938, was likely their first exposure to such situations at Green River Archaic Period sites. Overall, one can discern a growing familiarity with and appreciation of this distinctive phenomenon over the years. The growing interest in this phenomenon is reflected in several ways, starting with the initiation of a separate section on dog burials in many subsequent Green River reports. Those sections routinely contain descriptive and evaluative statements. An example of the latter is Webb's comment in the Indian Knoll report about dogs being killed for burial with people as perhaps a "symbol of continued association in the spirit world" (Webb 1946: 156). Another indication of the growing interest in this phenomenon is in the Indian Knoll report as well. That report included an entirely separate brief section near the end, devoted to a study of the Indian Knoll dogs (Skaggs 1946). That is where Opal Skaggs erroneously suggested that the Coyote, or an animal very much like one, was the progenitor of the dogs, as covered in Chapter 2. Yet one more such indication is the fact that William Haag later produced a comprehensive osteometric analysis of dogs that included, but was by no means restricted to, the Alabama and Kentucky Archaic Period series (Haag 1948).

One of the intriguing aspects of dog burials is to "wonder why, at some sites, dogs were sometimes buried with people, but not at others" (Morey 2006: 170). Thus far, coverage of the Green River sites has featured only cases where both burial modes were evident. But this region provides at least one compelling exception as well. The example in question, the Read site (see Figure 7.3 and Table 7.1), was reported to have sixty-three dog burials (Webb 1950b: 360), which would constitute the greatest number of documented dog burials at any of the Green River sites. As indicated in commentary stemming from Table 7.1, though, that reported number is apparently a clerical error, as only twenty-three skeletons can be located now, and Haag (1948: 122) also reported on twenty-three. But apparently none of those dogs were buried directly with people:

> Dogs seem to have been buried in the shell debris in much the same manner as their human contemporaries. Sometimes the body was laid in the shell and covered with shell and debris, but sometimes the burial was made in a pit, after the manner of human graves, see Figure 3 C. At this site no dog skeleton was found in a human grave, although many were in the midden not far from human skeleton [sic]. There was no demonstrated attempt to place a dog in burial association with a human body. (Webb 1950b: 360–362)

In short, it continues to be quite puzzling that dogs were buried there only by themselves.

Bearing in mind that this WPA-era work was extensive, there is little chance that sampling deficiencies are responsible. That factor is underscored by the fact that Webb (1950b: 368–376) individually inventoried 247 human burials from Read. Thus there were surely many occasions when a dog could have been buried with a person, but as a localized expression of mortuary practices, the people responsible for the Read site declined to do so. It is noteworthy that fieldwork at Read commenced in December of 1937, before the other sites that have been covered here. But the report on Read was not forthcoming until 1950, and in the intervening years Webb clearly gained considerable familiarity with the occurrence of Green River Archaic dog burials. Accordingly, the Read report warranted a separate section on dog burials when the report was finally written (Webb 1950b: 360–362).

At any rate, it seems as appropriate to draw this Green River section to a close by pointing out how work in that setting contributed directly to some broader empirically based generalizations. An example occurs in William Haag's (1948) comprehensive osteometric analysis of archaeological dogs. Haag's work covered a broad range of settings, including Green River Archaic sites, Alabama Archaic sites, sites in the desert southwest of the United States, Alaska, including St. Lawrence Island and Kodiak Island, and the northwest coast. For comparative purposes, he included series of recent dogs from Siberia, Alaska, and Greenland. Finally, Haag also dealt with post-Archaic archaeological sites in North

America, including Kentucky and Alabama, dating within the Woodland and Mississippian time frames (post-3000 B.P.). One might recall from summary comments in Chapter 3, that my own initial foray into the world of dog-related research (Morey 1986) relied on some of Haag's raw data, and others have as well. At any rate, based on analysis of this extensive series of several hundred archaeological dogs, Haag offered some generalizations. One was the prevalence of dogs in Archaic times, as known especially from burial contexts, throughout much of North America: "The dogs from the Woodland and Mississippi sites are relatively few in number" (Haag 1948: 243). A few pages after that statement, he provided his general assessment as to the reason:

> The apparent fact that the dog lost much of its importance to the aborigines with the advent of agricultural practices may indicate that the dog was primarily used by hunters. It is in the Hunter-Fisher-Gatherer horizons all over the world that the dog is prominent in their [sic] cultural context. (Haag 1948: 253)

That sentiment has been echoed later by others. For example, struck by the presence of cut marks on dog bones from the post-Archaic Fisher site in Illinois (see Appendix B), Paul Parmalee (1962: 406) commented that "the dog was not a revered animal as was often the case in Archaic and other early cultures." Similarly, working in the Southeast, Bruce Smith (1975:110) suggested that the "almost total lack of evidence for intentional interment of dogs at Middle Mississippi sites suggests dogs were not held in much esteem in the villages" (Smith 1975: 110).[16]

These consistent observations, made up to more than half a century ago, seem especially astute in light of some of the ground covered in Chapter 3. One might recall the much more recent suggestions, stemming from genetics-based inferences centered on work with modern canids, that it was only with the advent of agriculturally based

---

[16] Diane Warren, also working in southeastern North America, has emphasized this point in more recent times: "Dog burials decrease in frequency after the Archaic and are rare at Mississippian (1,100–400 B.P.) and historic period sites in the region" (Warren 2000: 105). Moreover, the same basic pattern is evident elsewhere in the world, sometimes in even more dramatic terms. For example, in writing introductory comments to Kotondo Hasebe's (2008) analysis of skulls and jaws from Jomon period (ca. 12,500–2,000 years ago) sites in Japan, Holger Funk (2008: 17) noted that "In the agricultural Yayoi period (rice cultivation) following the Jomon period the descendants of the Jomon dogs lost their meaning as hunting companions as well. There were no more careful burials, presumably dogs for the first time served as food." In North America, of course, there were post-Archaic careful burials, just reduced in per capita frequency.

economies that dogs could even be recognized. But dogs could be recognized before that, as firmly demonstrated by the Bonn-Oberkassel dog from Germany, about 14,000 years old, and pre-dating agricultural economy there. In fact, the archaeological visibility of dogs diminishes somewhat with the advent of agricultural economies, and Haag and others recognized that clearly. Equally clearly, there were some dogs then and this chapter has drawn attention to some examples, but using burial data as a guide, there were fewer. And that is true even though, generally speaking, human populations were larger in post-Archaic times, and one might reasonably expect that dog populations were larger too. To be sure, it's hard to know whether people simply buried dogs less frequently in post-Archaic times, if there really were fewer dogs, or whether maybe the reduced frequency of burials is partly an artifact of the shorter time scale of the post-Archaic span. Perhaps, as Haag suggested, dogs tended to assume a less conspicuous role with the advent of agricultural economies. Whatever the truth of that situation is, one setting in which dogs played a role of dramatic proportions was post-agricultural in general, but that role surely had little to do with economic considerations of that kind.

THE ASHKELON PHENOMENON

It is a standard old cliché to suggest that the truth can be stranger than fiction. If you are already familiar with the situation at Ashkelon, you will know what inspired the use of that cliché to begin this section. If you aren't, the initial portion of this section is for you: "Ancient Ashkelon, now quietly nestled beside the Mediterranean in the south of Israel, is shaped like a giant 150-acre bowl, with the sea wearing away at much of the western half" (Stager 1991: 27). Thus, describing its location, Stager begins one of the accounts of Ashkelon, an archaeological site where excavations dealt with remains that date to the Persian era, in the time frame of about 2,550–2,300 years ago. Excavations there in 1985 yielded the first evidence of what turned out to be a major dog cemetery. In the years since, more than 1,200 individually buried dogs have been found. Indeed, the truth can sometimes be stranger than fiction. At a later point in the same article Stager (1991: 30) divulged that "This is by far the largest animal [dog] cemetery of any kind known in the ancient world." To impart an initial sense of the magnitude of this site, Figure 7.5 provides a plan map of the site area, indicating its proximity to the Mediterranean Sea. It also has two areas that have yielded hundreds

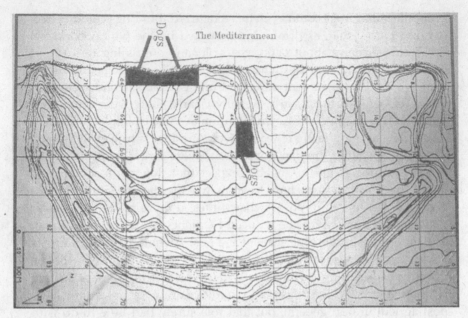

FIGURE 7.5. General plan of the Ashkelon site in Israel, showing the two main areas of Persian time frame dog remains. The ones nearest the sea are in Grids 50 and 57, while the others, slightly inland, are in Grid 38. Image made available by Paula Wapnish and Brian Hesse, part of the Leon Levy Expedition to Ashkelon, directed by Lawrence E. Stager, Harvard University.

of dog burials shaded in and labeled. Using the scale at the bottom left of the map one can appreciate, by virtue of the grid overlay, just how large these areas are. Figure 7.6 makes possible further appreciation for the nature of the work there, being a photograph of work in progress a number of years ago. This picture is from grid 50, representing part of one of the dog burial areas illustrated and labeled in Figure 7.5.

Wapnish & Hesse (1993) documented in substantial detail the basic characteristics of the Ashkelon sample. From a detailed analysis of the skeletal remains, they reached several inferences regarding the structure of the living population. For example, the mortality profile did not suggest that the animals had died some kind of short-term catastrophic deaths, but that they died from natural causes over time. Moreover, there was no evidence suggesting human mistreatment or outright brutality as a contributing factor. There also was no osteological evidence suggesting that the animals might have been eaten, as dogs sometimes were in other settings (see Chapter 5). Nor was there any indication that they were selectively bred for any specific appearance or purpose.

FIGURE 7.6. Ongoing excavation of part of the dog cemetery at Ashkelon, in Israel. This area is Grid 50 (see Figure 7.5). The site is a vast area, requiring that workers get from place to place there by motor vehicle. Photograph made available by Paula Wapnish and Brian Hesse, part of the Leon Levy Expedition to Ashkelon, directed by Lawrence E. Stager, Harvard University.

Halpern (2000: 134), assessing the available information, put the matter this way: "The dogs were not bred; they were not pets; they were not sacrificed, eaten, or even killed. Despite this, they were so carefully and individually interred that many of the skeletons, tails included, could be recovered without disarticulation." Halpern (2000: 133) consequently referred to them as feral animals, and "standard near Eastern mutts," meaning, in this case, that they were free-ranging animals that were attended to by nearly anybody at any given point in time. In considering just what he meant by careful interment, it is best to turn to Wapnish & Hesse (1993: 58):

> Each corpse was carefully placed in its grave.... The dogs were buried on their sides with tails carefully arranged to curl toward the feet, sometimes reaching between the lower hindlimbs. In a few cases the feet were so entwined that they may have been bound at the ankles before burial.

Wapnish and Hesse's description is compelling, but an even more effective means of imparting this reality is to show a picture of one.

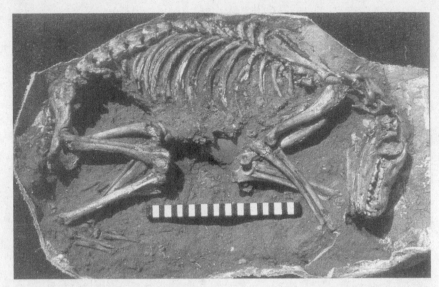

FIGURE 7.7. The partially cleaned skeleton of a dog buried at Ashkelon, in a plaster jacket shell, the manner in which some of the dogs there were removed. A different small picture, apparently of the same skeleton, appears in Stager (1991: 27). Photograph made available by Paula Wapnish and Brian Hesse, part of the Leon Levy Expedition to Ashkelon, directed by Lawrence E. Stager, Harvard University.

Accordingly, Figure 7.7 is a photograph of one of the adult dog burials at Ashkelon just after removal, in a plaster cast, an occasional means of dealing with the challenges of safely documenting and removing them.

In considering the care with which these dogs were buried, it is no accident that Halpern (2000), in assessing the overall situation, referred to "The Canine Conundrum at Ashkelon" as part of the title of his work. One part of that 'conundrum' is that well over half of the animals were puppies, an unusually high frequency. In keeping with that pattern, Figure 7.8 provides a picture of two of the Ashkelon dogs, an adult and a puppy. One possible explanation for the pattern of predominantly puppies, suggested by Wapnish & Hesse (1993: 61), is that puppies would be especially likely to die from diseases. A provocative alternative explanation can be found in work by Billie Jean Collins (1990), who was dealing with the ritual treatment of dogs among Hittite peoples in Anatolia, pre-dating Ashkelon by a few hundred years. Collins enumerated some of the uses of dogs in Hittite society, but also indicated that "It seems very likely as well that some symbolic or religious significance was attached to puppies that was not attached to fully grown animals.

FIGURE 7.8. Two buried dogs from Ashkelon, an adult and a puppy, in keeping with the overall prevalence of puppies there. This illustration appeared previously in the *Journal of Archaeological Science* (Morey 2006, Figure 2, page 163). Photograph made available by Paula Wapnish and Brian Hesse, part of the Leon Levy Expedition to Ashkelon, directed by Lawrence E. Stager, Harvard University.

Puppies had two primary uses in Hittite ritual, namely, prevention and purification" (B. Collins 1990: 211). Prevention could involve sacrifice of the animals, in a belief that such an action would protect important people from evil. Purification, on the other hand, might involve the ingestion of a puppy portion, in a belief that it should have a healing effect on an ill person. An alternative mode of purification, sometimes called transfer, could involve different procedures: "These procedures are touching the puppy to the body of the patient, applying medicine made with dog excrement, waving the puppy over the patient, spitting into the puppy's mouth, and passing through the divided carcass of a puppy" (B. Collins 1990: 214).

There isn't, of course, any way to know for certain that ancient Hittite peoples undertook such activities. Collins is extrapolating from fragmentary written accounts that have survived the ravages of time, been recovered and then translated into English. In any case, from what she relates, albeit pertaining to a different but related context that Wapnish & Hesse (1993) also called attention to, it seems that puppies held special symbolic significance to peoples of this general era, from this general region. So the Hittite situation at least suggests how puppies might have been imbued with spiritual qualities by the people at Ashkelon, making their prevalence there more understandable. But whichever explanation for the prevalence of puppies is preferred, it is clear is that dogs in general were perceived as having spiritual qualities. In turning to that dimension of what dog burials mean, it is fitting to begin with the overall setting most relevant to Ashkelon. Lawrence Stager, in writing about Ashkelon, made clear the basic nature of the spiritual significance commonly attributed to dogs in that general region:

> In the ancient Near East, dogs are often associated with particular deities and the powers they wield. We cannot yet be sure with which deity the dogs in the cemetery at Ashkelon were associated. There are several possibilities, in several cultural guises, often interrelated as one deity merges into another. But in the end, a common theme emerges – deities with healing powers are often associated with dogs. (Stager 1991: 40)

With that statement in mind, the care with which dogs were buried at Ashkelon likely involved much more than a conventionally understood social bond between dogs and people. To be sure, to attribute spiritual powers to dogs does not preclude that sort of social bond, and the mere act of carefully burying a dog signifies spiritual relevance in the sense of a concept of an Afterlife. This chapter has covered many examples of dogs being buried carefully, in different settings in the world, and clearly those other cases represent events that took place on a lesser scale of magnitude. Ashkelon continues to pose a "Canine Conundrum" (Halpern 2000), but instead of trying to solve that conundrum, it seems appropriate to consider the broader perception in the ancient world of dogs as having spiritual qualities. In the immediate region of the Near East, Wapnish & Hesse (1993: 67–70) identify several other known cases of dog burials, in some ways reminiscent of Ashkelon, but on a much smaller scale. Beyond Ashkelon and immediate environs, another intriguing situation warrants consideration.

Dog Mummies in Ancient Egypt

Ancient Egyptians, well-known for mummifying humans, also commonly mummified dogs. Wapnish & Hesse (1993: 70) get across a sense of the scale of this practice: "Dog mummies have been found in large numbers at Roda in Upper Egypt, at Thebes, at Suares near Maghagha, and at Abydos." They also indicate one probable reason why so few of them have been thoroughly studied:

> The Roda burials (of indeterminate date) were also summarily pre-pared corpses. This may explain why so few dog skeletons of the many

thousands interred were carefully studied or collected, although some dog mummies, such as those at Thebes, were considerably more elaborate. (Wapnish & Hesse 1993: 70)

Mary Thurston indicates another probable reason: "In their search for treasure, nineteenth-century British excavators found cemeteries and underground vaults filled with the mummies of dogs, cats, ibexes, falcons, and other creatures" (Thurston 1996: 34). In short, many probably didn't survive the ravages of time. But for present purposes, what is most noteworthy is the number of different kinds of animals that ancient Egyptians ceremoniously mummified and placed in repositories that can legitimately be called cemeteries. This commitment to mummification immediately sets Ashkelon apart: "the dogs of Ashkelon were not mummified, as were most canines ritually buried by the Egyptians" (Stager 1991: 35). The Ashkelon dogs were individually buried, over a restricted span of time, one that Stager (1991: 31) originally estimated may have lasted no more than about fifty years. That is a really brief span of time, considering the magnitude of what took place there.

In ancient Egypt, some dogs were buried with royalty, presumably individually valued animals, but there were originally many thousands of dog mummies, the majority of which "probably were a product of institutionalized Anubis worship" (Thurston 1996: 36). Anubis was a symbol of both death and resurrection, and was routinely depicted as a black, dog-headed figure. He was a watchdog of the land of the dead, and recalling the point made earlier about certain dog burials facing west, Anubis was associated with the west, and supervised mummification. This was clearly a concept of spirituality that superseded a sense of personal reverence, though it can't, as a matter of definition, preclude such a sense.

DOGS AND SPIRITUALITY: BEYOND THE NEAR EAST

The perception of dogs as possessing spiritual qualities comes in several different guises, and Ashkelon is the most dramatic localized example that is known. The guise there, however, involved recognized deities, going well beyond the sense of personal reverence reflected in so many dog burials elsewhere. In some cases, where dogs were apparently put to death to be buried with a person, it seems to have been anticipated that the dog would accompany the person to the spirit world, as suggested by archaeologist William Webb (1946: 156). In other cases, where dogs were buried alone, such a connotation is more elusive. But the care with which they were buried suggests rather clearly that their burial was an act of personal reverence. Dogs were being treated much like people, and as David White (1991: 14) has pointed out, such treatment signifies that the role of the dog was being extended into the perceived world of the dead. Such a perception typifies how modern dogs are often perceived as well, signified in part by the existence of modern dog cemeteries, covered in Chapter 10. At the same time, such concepts are also sometimes reflected in prominent literature. An excellent example can be found in a classic work by the great American writer Ambrose Bierce, in *The Devil's Dictionary*. There, roughly a century ago, he offered a formal definition of the word "dog," replicated in full here:

> Dog, *n*. A kind of additional or subsidiary Deity designed to catch the overflow and surplus of the world's worship. This Divine Being in some of his smaller and silkier incarnations takes, in the affection of Woman, the place to which there is no human male aspirant. The Dog is a survival – an anachronism. He toils not, neither does he spin, yet Solomon in all his glory never lay upon a door-mat all day long, sun-soaked and fly-fed and fat, while his master worked for the means where-with to purchase the idle wag of the Solomonic tail, seasoned with a look of tolerant recognition. (Bierce 2000: 58)

Bierce's definition, always aiming at a humorous result by way of his sardonic wit, well captures the logic underlying the careful burial of deceased dogs. But in at least one conspicuous case, the dog was not the only canid accorded a ritualistic burial.

WOLVES AND SPIRITUALITY

Given the basic similarities between dogs and their immediate ances-tors, wolves, it should come as no surprise to find that wolves were often perceived as possessing spiritual qualities as well. For example, Fritz et al. (2003: 291) have observed that

> Most of North America's indigenous people were familiar with wolves and often regarded them as spiritually powerful and intelligent ani-mals.... Some tribes believed that wearing the skin of the wolf brought about a supernatural union of human and wolf.

To be sure, though, Native Americans perceived spiritual quali-ties in a variety of animals, and did not elevate wolves above other animals (Fritz et al. 2003: 291). But in keeping with Fritz et al.'s (2003: 291) point about wolves, it is worth identifying a particular North American example, known from tangible archaeological remains. Keyser (2007) has called attention to a series of pictographs and petroglyphs of shield-bearing warriors at the late prehistoric Bear Gulch site in Central Montana. In this art, the wolf plays a conspicuous role, with more than 100 headdresses identified at Bear Gulch and at nearby Atherton Canyon. The pertinent ethnographic literature suggests that the analogous historically known item is what is known as the wolf hat. Keyser relates the association of wolf hat headdresses and shield-bearing warriors to the symbolic association of the wolf with war, prevalent among North American plains groups. The association was due to such factors as the wolf's quickness and stealth, traits highly valued in warriors. As Keyser (2007: 66) also noted, "It also has been suggested that wolves were revered because their pack behavior was similar to the organization and actions of a plains war party (Thompson 2000: 12)." So once again, the wolf's pack behavior comes into play. Clearly, the live wolf was a highly regarded animal. But be that as it may, the primary focus of this chapter is burials.

A Prehistoric Wolf Burial

The site dealt with here is in Siberia, and was found in the 1880s by workers on the Trans-Siberian Railway. Accordingly, this site, a cemetery area, became known as "The Lokomotiv" (Bazaliiskiy & Savelyev 2003). The site dates within the span of about 7000–6000 B.P., and it featured a burial of what is known as the Tundra wolf, *Canis lupus albus*. The specific circumstances of this burial suggest that it is, to put matters in the vernacular, loaded with symbolic spiritual significance: "The wolf lay with its head slightly raised in relation to the trunk, with its paws pressed to its body, in a dynamic posture" (Bazaliiskiy & Savelyev 2003: 27). That description, augmented by the published photograph (Bazaliiskiy & Savelyev 2003: 27, Figure 9), makes it apparent that this wolf was on the move. To the west of its front paws lay an oval area, surrounded by ochre. Significantly, the skull of a mature man was buried with it:

> The skull was found in full anatomical order in a small pit especially dug for it under the wolf's skeleton. Its jaws were pressed together, with atlas and axis under the ribs of wolf's skeleton, which suggests that the human head was buried with the wolf. (Bazaliiskiy & Savelyev 2003: 28)

It also seems highly noteworthy that this large, cold-adapted wolf was found buried a long way from its natural habitat. On this front, the authors suggest that "It seems likely that its appearance is owed to human agency" (Bazaliiskiy & Savelyev 2003: 28). These factors combined suggest that the burial of this wolf was a deeply symbolic act, entailing spiritual significance. The authors also report, after noting the common occurrence of dog burials, that there is no previous evidence for an individually buried wolf (Bazaliiskiy & Savelyev 2003: 28). At any rate, this burial, along with the North American plains rock art just highlighted, and coupled with ethnographic accounts, indicate that the wolf was highly regarded in a spiritual sense in different parts of the world. But as anticipated at the outset, a different sort of animal also needs to be considered.

CATS: THE ANCIENT EGYPT PHENOMENON

Cats, *Felis catus*, are the (localized) exception to the generalization offered earlier about the routine occurrence of ritually treated deceased dogs compared to other animals. Given the current popularity of both dogs and cats as household pets, this basic disparity and its localized exception warrant consideration. In ancient Egypt, the treatment of cats in a ritualistic fashion did carry spiritual connotations: "The cat was one of the most sacred of all animals to the Ancient Egyptians and they were mummified in enormous numbers" (Clutton-Brock 1981: 110). The cats' remains have been found in contexts that can be regarded as cemeteries, and such mummified cats have also been found associated with tombs in ancient Egypt (Clutton-Brock 1981: 110). To be sure, there are a few other cat burials known, for example one in ancient Cyprus (wild? domestic?) that is several thousand years earlier than the classic Egyptian contexts (Vigne et al. 2004). Notably, as highlighted earlier, ancient Egyptians also maintained cemeteries for a variety of other animals, including non-domestic animals such as crocodiles, birds, and gazelles (Budiansky 2002: 30; Thurston 1996: 34). What emerges from a consideration of the ancient Egypt phenomenon is that spirituality was likely less a matter of personal reverence to those people, and more a matter of appealing to broader spiritual deities. Given this pattern, it seems that ancient Egypt can be regarded as Ashkelon-like, only for cats and a variety of other animals, including dogs. That is, unlike so many dog burials, where a strong social bond is suggested by the animals' burial treatment, both Ashkelon and ancient Egypt signify spirituality of a broader order.

But whereas ritualistically treated cats, which may or may not signify personal reverence, occur only sporadically beyond ancient Egypt, ritualistically treated dogs, "buried as though they were someone's best friend" (Griffin 1967: 178), occur throughout the world. Time after time, as this chapter has covered, those dogs were buried with great care, indicative of personal reverence.[17] The reasons that dogs, as opposed to cats (or other animals), have so routinely been treated in ways that signify personal reverence at death is the topic of the following chapter. In anticipating that chapter, and closing this one, it is fitting to turn once again the Ambrose Bierce for his formal definition of "cat" in *The Devil's Dictionary*:

> Cat, *n.* A soft, indestructible automaton provided by nature to be kicked when things go wrong in the domestic circle. (Bierce 2000: 33)

One should contrast that definition with his definition of "dog," quoted earlier. Certainly there is nothing there that implies spirituality, either as personal reverence, or what might be called spirituality writ large. Bearing in mind that Bierce was essentially a modern writer, the different perceptions of dogs versus cats that are reflected in his classic work call for an explanation, sardonic wit aside. That effort follows.

---

[17] Recalling some of the prehistoric instances described many years ago, by Webb, De-Jarnette, and others, Costello (2004) reported on recent (1996) archaeological work in the city of Los Angeles, the part of the city now known as Chinatown. There, in deposits dating to the A.D. 1800s, the crew encountered a dog burial:

> The 9-month-old dog was laid to rest coiled in a sleeping position, wearing its collar, and protected with a metal capping. (Costello 2004: 17)

Allowing for the modern accoutrements, the terms used to describe this burial are much like many prehistoric cases. Put simply, the perception of dogs by people is a long-standing phenomenon.

8

WHY THE SOCIAL BOND BETWEEN DOGS AND PEOPLE?

> ...dogs are not just a silly version of the wolf.... Instead, dogs have adapted specifically to their new and unique habitat – namely, human societies.
>
> Kaminski 2008: 211

Indeed they aren't and indeed they have. Given those points, the in-depth consideration of the routine burial of dogs (Chapter 7), by people who apparently thought of dogs as something other than just silly wolves, seems to lead directly to the question posed by this chapter's title. And in addressing that question, the focus shifts well beyond the archaeological realm, and squarely into the province of biology and physiological psychology, including neuroscience. By way of initial background, some notable scholars have offered relevant guidance and insights. For example, some half a century ago, Konrad Lorenz (1954: 85; 1975: xii) suggested that among animals, dogs are most like people in their capacity for true friendship. And well before that, Charles Darwin's own cousin, Francis Galton, recognized this capacity of dogs: "The animal which above all others is a companion to man is the dog, and we observe how readily their proceedings are intelligible to each other" (Galton 1973: 187). In more recent times, Wolfgang Schleidt (1998: 2; Schleidt & Shalter 2003: 59) has noted that they exceed even our close phylogenetic relatives, chimpanzees, in this capacity. And in writing of social bonding behavior, Feddersen-Petersen (2007: 108) has noted that "in wolves and dogs, this can be much stronger and longer lasting than in species closely related to ourselves, such as chimpanzees."

In keeping with that point, I assembled and presented a body of data on the worldwide practice of ritually interring dogs upon their deaths (Morey 2006), a study elaborated on in the previous chapter. There, a central point was that the deliberate burial of dogs has taken place for more than ten thousand years, in widely separated geographic areas. Moreover, although individual circumstances vary widely, burials signal the emergence of a domestic relationship between dogs and people. As pointed out previously in this volume, there are also modern dog cemeteries in different parts of the world (Thurston 1996: Chapter 11), to be covered more fully in Chapter 10. In addition, people sometimes single out dogs that have died recently for special, sentimental musings that are shared with others (e.g., Hoeflich 2005). The source just cited is a local newspaper column, but Kuzniar (2006: Chapter 4) has provided several examples of peoples' sense of loss at the death of a dog, taken mostly from recognized works of literature.

It is the primary purpose in this chapter to propose a theoretical basis for the evolution of this relationship, by developing a framework that surely can be productively modified and refined over time. Foremost, it is important to consider altered behavioral patterns conditioned by neurological changes that have occurred as a result of domestication, demonstrated by studies of dogs and experimental work with foxes. Related hormonal factors are relevant, too, and receive some attention. In addition to dogs, this chapter also revisits a topic introduced in the previous chapter: how the common household cat fits into the picture. This step is taken for comparative purposes, given the widespread popularity and cosmopolitan distribution of cats in modern times (e.g., Necker 1977: 3). As with dogs, the emphasis is on behavioral patterns conditioned by neurological factors. Finally, it is important to consider the evolutionary consequences of the patterns outlined, with special reference to how people and dogs have converged with respect to some of their behavioral capacities.

Clearly, dogs have long occupied a unique niche within human society and "Considering the attention commonly lavished on dogs in mortuary contexts... it seems clear that they are about as close to being considered a person as a non-human animal can be" (Morey 2006: 165). In order to examine the development of this relationship, it is useful first to review some behavioral attributes of dogs' immediate ancestors, wolves.

THE RELEVANCE OF WOLF PACKS

Friendship, however one might wish to define it, is commonly understood as a social phenomenon, involving interactions between members of a group. Dogs and their ancestors, wolves, are both quite social in nature, as covered in Chapter 4. Sociality is likely the primary basis for the initial development of the domestic association between wolves / dogs and people, and the hierarchical structure common to both kinds of societies presumably underlies this behavioral compatibility. As noted in Chapter 4, for many dogs the human household is truly an ecological arena in which they pursue their interests (Morey 1997). Indeed, J. P. Scott (1950: 1019) once noted that dogs behave in human society much as wolves behave in wolf society, as characterized by its familiar pack social dynamics. Finnish zoologist E. Pulliainen (1967) later suggested, based on experiences with wolves raised from a young age by people, that it was also possible to say that wolves behave in human society much like they behave in wolf society.

Yet, despite that fundamental compatibility, wolf pups raised by humans lack the proficiency displayed by dogs when faced with certain human social signals (Hare et al. 2002; Miklósi et al. 2003; Hare & Tomasello 2005; Topál et al. 2005; Hare 2007; Miklósi 2007b). Michael Fox notes that wolves and coyotes reared by humans may learn to interact sociably with their handler, yet they are "notoriously difficult to train" (Fox 1978: 253). In fact, contrasting dogs with wolves, as well as nonhuman primates, Hare and colleagues (Hare et al. 2002) indicate that dogs, including puppies, display greater proficiency with human social cues than do either wolves or non-human primates (see also Bräuer et al. 2006). As such, they suggest that dogs' social-communicative repertoire was honed over the course of domestication (Hare et al. 2002: 1636). Clearly, the social compatibility between dogs and people cannot be explained entirely through analogies to wolf packs. Thus, in addition to considering the particular mechanisms that brought those changes about, it is important to consider what selective factors favored dogs' increased social proficiency with people. In simplistic terms, it is reasonable to suggest that, given the basic social structure of wolves as a prerequisite, some earliest domestic dogs / wolves were favored in a selective sense to the extent that they fit successfully within the structure of a human "pack". The premium on doing that was facilitated by the basic structural similarity between a wolf pack and a conventional human family, a point highlighted in a passage by Mech & Boitani (2003: 6), quoted in Chapter 4. More recent work by Feddersen-Petersen (2007: 105) emphasizes the same basic point:

Pack-living wolves are social canids par excellence. They develop a very high degree of sociality, a fact that may be judged as a kind of pre-adaptation for domestication: many capacities fit with the very high degree of social contacts and interactions of humans, living in reproductive units (families).

Indeed, the wolf pack is a fitting analogy for the human family. Given that similarity, it is predictable that dogs should have evolved by undergoing heritable physiological and related behavioral changes that maximized their compatibility with people. While theoretical considerations predict changes along certain lines, it is important to establish what those changes were, the degree to which they occurred, and, as clearly as possible, their behavioral correlates.

DOGS' BEHAVIORAL AND BRAIN CHANGES UNDER DOMESTICATION

In beginning these considerations, it is useful to point out that Hare & Tomasello (2005: 441) and Hare (2007: 61) have characterized much of what has happened with dogs as "emotional evolution." It would be a mistake, however, to infer that emotional evolution precludes the existence of socially undesirable behavior among dogs. For example, is worth noting that Fox (1978), in specifying comparable emotional (neurological) structures in people and dogs, included those "which are linked with psychosomatic and emotional disorders" (Fox 1978: 258). Jensen (2007b: 64) has made a comparable point more recently. Moreover, in the modern era, people have at times purposely produced dogs with heightened capacity for aggression in order that they "provide protection through inter-specific aggression (e.g. most guarding breeds) or for 'entertainment,' in the form of heightened intra-specific aggression of fighting breeds, including 'pit-bull' type dogs" (Lockwood 1995: 132). Accordingly, it is probably more realistic to frame dogs' domestication in terms of an increased capability for emotional regulation, consistent with changes in serotonin levels (see below). But even beyond the examples just noted, aggressive behavior is not entirely absent among domesticated dogs. Certainly, people are also capable of genuinely aggressive intraspecific (as well as interspecific) behaviors, in some cases with lethal consequences. Yet, within the niche of the human household, successful dogs are able to regulate aggressive, dominance-seeking behaviors. Consistent with that suggestion, dogs may be subjected to therapeutic intervention (Mugford 1995, 2007) when they exhibit behaviors, such as aggression, found undesirable by humans.[1]

Considering matters more broadly, Björnerfeldt et al. (2006) have suggested that there has been a general relaxation of selective constraints on the mitochondrial genome of dogs, compared to wolves. The implications of this factor are seen to lie with the major phenotypic variation in dogs, both structural and behavioral. While various interrelated selective factors are likely at work, dogs' compatibility with people, especially within a group, has apparently long been valued. Accordingly, the domestic dog's capacity to engage people socially is a vital component of the bond between dogs and humans. Notably, certain aspects of dogs' sociability have been linked with enhanced attention to visual cues, in comparison to wolves.

Attention Capabilities

Attention capabilities clearly have import for a variety of animals (Dukas 2004), and it seems that flexibly augmented attention skills in dogs have substantially enhanced their compatibility with people. When faced with an insoluble task, "dogs initialized communicative face/eye contact with the human earlier and maintained it for longer periods of time compared to the socialized wolves" (Miklósi et al. 2003: 764). Further, Call et al. (2003: 258) indicate that dogs raised both with and without human interaction follow human cues from an early age, but wolves, again raised both with and without human interaction, do not. Call et al. (2003) and Schwab & Huber (2006) also highlight domestic dogs' capacity to comprehend a presence or absence of attention in humans. For example, dogs were capable of deciding to steal or avoid a piece of forbidden food depending upon whether a human witness watched or feigned distraction (Call et al. 2003). Moreover, "In contrast [to a variety of other animals], the majority of dogs effectively use many different visual cues given by humans" (Schwab & Huber 2006: 169).[2] In

---

[1] Some recent work, focused on understanding the genetic factors underlying behavioral patterns are keyed especially on isolating, if possible, the genetic basis for aggressive behaviors. As pointed out by Houpt (2007: 442) when considering what to do about the public health/safety problem of excessive canine aggression in some situations, "That is a question all of us interested in public safety, the human animal bond, and in veterinary animal behavior should ponder."

[2] Moreover, the importance of people in such situations is suggested by a recent study by Udell et al. (2008) in which six dogs, admittedly a small sample, more readily

that same vein, Gácsi et al. (2004) emphasized dogs' remarkable adeptness at gauging the context of different situations, using not only direct facial cues, but also the orientation of the human body, including the head (see also Virányi et al. 2004). Specifically, based on experimental data involving begging behavior in dogs, they (Gácsi et al. 2004: 151) reported that

> dogs chose [who to beg from] more definitely if there was a clear difference in the head orientation of the experimenters (facing vs back). Nevertheless, their performance seem [sic] to be better than shown by chimpanzees tested in a similar situation (Theall and Povinelli 1999), and we should add that while the chimpanzees were rewarded only for correct choice, no such differential rewarding was used here.

Overall, they found that dogs perceive the role of the human face orientation in social interaction, but they do not pay as much attention to the visibility of the eyes themselves.

In keeping with that point, a study by McGreevy et al. (2004) found a strong correlation between nose length and the distribution of retinal ganglion cells in the dog. They suggested that the documented distribution likely reflects the enhanced value for dogs of being able to focus optimally on human faces. More specifically, they called attention to the probability that breeds with shorter faces, and more forward-oriented eyes, have retinal ganglion cells distributed in a manner that should enhance that capacity. In that light, it is worth emphasizing that, from an archaeological vantage point, the proportional reduction of snout length relative to wolves is one of the most consistent traits that characterizes early dogs (e.g., S. J. Olsen 1985: 19; Morey 1992). In fact, in presenting information on certain North American prehistoric dog specimens from Idaho, Barbara Lawrence (1967: 47) stated that

> since a short, broad nose in dogs is correlated with a steep forehead, one can further say that the animals in question had this most doglike of all characters well developed.

It does seem as though this trait, a major component of paedomorphic, or juvenilized, morphology in dogs (see Chapter 3), served to enhance

understood and learned cues given by people than by comparably sized cues offered by nonhuman objects. An example of the latter circumstance is that the dogs were about twice as successful at following a point made by a person than one made by a mechanical arm of similar size.

their overall compatibility with people.[3] And commenting on dogs' proficiency with human social cues, Kubinyi et al. (2007: 42) noted that "Dogs are inclined to look at our faces, and this inclination provides them with a broadened opportunity for learning about human gestures."

To be sure, the difference between dogs and chimpanzees, as reported by Gácsi et al. (2004: 151), and noted earlier, is not dramatic (see Kaminski, Call & Tomasello 2004). And certainly chimpanzees also have markedly short faces, much like peoples', a factor that surely facilitates their capacity to focus on faces, among other sorts of visual information. Given their other anthropoid characteristics and established close phylogenetic affinity to people, it is reasonable to expect them to be much like people in some important behavioral and emotional capacities. And clearly they are. Equally clearly, however, dogs exceed chimpanzees in some of these capacities, to the extent that Schleidt (1998) has suggested that humaneness is a canine trait. Thus, in the absence of other anthropoid characteristics or genetic affinity, dogs have undergone distinctive physiological changes that render them more receptive to certain forms of human communication than even our closest relatives, the chimps. Indeed, the details of brain structure in dogs changed markedly during domestication, while the animals remained, genetically, all but indistinguishable from wolves.

Changes in Cerebellum Structure

Two researchers, D. Atkins and L. Dillon, have compared the cerebella of dogs, wolves, and some other canids, subsequently grouping their brain samples by structural affinities (Atkins & Dillon 1971). By convention, the cerebellum is considered to be concerned primarily with motor coordination, in people (Churchland 1986: 104; Nolte 1993: 343;

---

[3] One can surely appreciate that when Clutton-Brock (1984: 205) suggested that the "large eyes and appealing form of the puppy" (see Chapter 3) endeared the animals to people, this might be correct, but in different terms than she envisioned. That is, 'large eyes' likely reflects, in part, their more forward orientation, by which they focused on human faces more successfully. In other words, the paedomorphic cranial morphology of dogs, predictable for broader theoretical reasons covered earlier in this volume, goes hand in hand with the enhanced capability of dogs to focus optimally on human faces. In a manner of speaking, those broader theoretical causes underlying paedomorphic morphology and the more specific capability to focus on human faces seem to be different sides of the same coin.

TABLE 8.1. *Primary differences between the cerebellum in brains of wolves* (Canis lupus) *and of dogs* (Canis lupus familiaris), *as documented by Atkins & Dillon (1971: 97–101)*

Cerebellum Part	Wolf	Dog
Lobule 1	Typically three folia	Usually one folium
Lobule 2	Supplied by a single common tract	Supplied by two tracts
Lobule 3	Small, weakly developed	Substantial, strongly developed
Lobule 4	Fissura prima directed anteriorly	Fissura prima directed posteriorly
Lobule 5	Lobule small "in some"	Lobule larger in some
Lobule 6	Tract to it branches from the tract to lobule 5	Lobules 6 and 7 receive branches from a short common tract
Lobule 7	Little difference between dogs and wolves or other canids	Little difference between dogs and wolves or other canids
Lobule 8	Little difference between dogs and wolves or other canids	Little difference between dogs and wolves or other canids

Note: Beitz & Fletcher (1993: 932) have conceived of the dog cerebellum as having two major lobes, and Atkins & Dillon (1971: 97) conceived of those two lobes as being comprised of four lobules each, for a total of eight lobules. Beitz & Fletcher did not offer that scheme, but they do note that it is possible to subdivide the main lobes into lobules.

Kirschen et al. 2005: 462) and in dogs, as well (Redding 1971; Jenkins 1978: Chapter 14; Beitz & Fletcher 1993: 932). However, functional neuro-imaging studies have subsequently implicated the cerebellum in cognitive and affective occupations, including allocation of attention and verbal working-memory (Dolan 1998; Kim et al. 1999; Chen & Desmond 2005; Kirschen et al. 2005). The cerebellum has also been linked to the orientation of visual attention (Barrett et al. 2003; Gottwald et al. 2003) as well as emotionality and fear-learning (Sacchetti et al. 2005). The results of the Atkins & Dillon (1971) comparative study are summarized in Table 8.1. Notably, they placed dogs and wolves in entirely separate groups. The nearly identical genetic makeup of wolves and dogs had not yet been established, and they suggested that dogs and jackals likely shared an immediate non-wolf ancestor (Atkins & Dillon 1971: 104–105). To be sure, the idea of direct jackal ancestry for dogs was current at the time, likely a consequence of Lorenz's (1954) work, an idea that he later rescinded (Lorenz 1975), as noted in Chapter 2. At any rate, the noted disparity between the cerebellar structures of wolves versus domesticated canids may indicate, at least in part, selection for flexible attention and augmented memory as well as certain emotional attributes, particularly the conditioning of fearful responses.[4]

In general, of course, genetic affinities are especially important in establishing taxonomic relationships. Atkins and Dillon thought that cerebellar structure should reflect such relationships, but in the same era, Will (1973) conducted a similar comparison of canid cerebella. She concluded, based on the extent of observed variation, that "The cerebellum in the genus *Canis* is not of taxonomic significance" (Will 1973: 72). Some three decades later, Lyras & Van der Geer (2003: 507) made much the same point. For present purposes, it is useful to emphasize that brain structures can change, accompanied by little or no discernable genetic change. The case of dogs versus wolves exemplifies that point well. But given the importance of behavioral patterns in canid domestication, it would be useful to compare other brain regions, such as the frontal lobe of the cerebrum. In people, the frontal lobe is associated with what are sometimes known as mental states (Churchland 1986: 351). Specifically, beyond being involved with certain motor functions, the frontal lobes guide, control, and execute "the expression of emotions, speech, or internal mental states" (Koch 2004: 129). In other words, the cerebellum may not be the optimal brain area to focus on for these purposes, but it is relevant, and is the only one for which I could find direct comparisons of dog brains with other canid brains. Much more recently than the Atkins & Dillon (1971) or Will (1973) studies, Lyras & Van der Geer (2003) compared the external brain anatomy, using fossil cranial endocasts, of a variety of canids, taxonomic comparisons being made at the generic level. One of the discriminating features in their work was "found in the frontal pole" (Lyras & Van der Geer 2003: 509). Significantly, the frontal pole involves especially the prefrontal cortex, a part of the brain that is conspicuously important, a point developed toward the end of this chapter. At any rate, Radinsky (1973) and Lyras &

---

[4] Janice Koler-Matznick (2002) has called upon the Atkins and Dillon study in modern times, to make the point that dogs and wolves differ in brain structure, therefore, in her view, casting doubt on the wolf as the ancestor of the dog. But what happened was that the brain structure of the wolf changed as a consequence of domestication, and that brain structure now characterizes the dog. She reports that "They [Atkins and Dillon] conclude that the cerebellum of the DD [domestic dog] most closely approximates the coyote, which is closely allied with the jackals, and that wolves show numerous brain traits distinct from other species" (Koler-Matznick 2002: 104). Again, dogs differ from wolves in brain structure, even as the two animals are genetically nearly identical. Quite simply, in canids at least, the details of cerebellar structure do not speak to ancestor-descendant relationships.

Van der Geer (2003) divided the bulk of the Subfamily Caninae into two basic groups, doglike and foxlike, a distinction that is neurologically and behaviorally important (see discussion that follows).

Aside from comparisons between animals at different taxonomic levels, Beitz & Fletcher (1993) have presented a comprehensive assessment of the brain of the dog. These authors treat the dog brain as consisting of three basic regions, the cerebrum, the cerebellum, and the brainstem. The frontal lobe, already noted, can be considered as part of the cerebrum. These authors also make the point that lobe terminology in dogs has limited usefulness, given the somewhat inconsistent analytical divisions that get drawn.[5] That factor should be born in mind when evaluating the summary of Atkins & Dillon's (1971) analytical results on Table 8.1.

AUDITORY COMMUNICATION

Auditory communication capabilities also factor into the social bond between dogs and people, and further set dogs apart from other domestic animals. Based on his experiences with his own dogs, Stanley Coren inferred their receptive vocabulary to be comprised of some twenty-five signals or gestures and about sixty-five words or phrases, for a total receptive vocabulary of about ninety items:

> If they were human children, they would be demonstrating the level of language customary at around eighteen to twenty-two months of age. Chimpanzees that have learned sign language can obtain scores equivalent to a child of around thirty months of age. (Coren 1994: 115)

As Coren also anticipated in the same discussion, more recent work reveals that one dog was capable of learning human linguistic cues for over 200 different items (Kaminski, Call & Fischer 2004). Moreover, the dog, named Rico, showed the ability to "fast map," meaning the ability to infer the identity of novel objects by exclusion. In tests of this ability, Rico could correctly fetch novel objects with greater than

[5] Uncertainties about terminology in this matter are hardly new. For example, more than a century ago, in a systematic review of what was known at the time about the structure of the dog brain, J. N. Langley (1883: 250) wrote as follows: "In any case the division of the cortex into areas is only approximate for it is impossible to say to which of the boundary convolutions the cortex at the bottom of the fissure belongs, if indeed it does belong to one more than to the other. . . ." If nothing else, continuing difficulties with such matters serve to highlight just how complex the structure and operation of the brain is.

random probability by being cued with their unfamiliar name, when confronted simultaneously with many objects that were familiar. Further, Rico could still do this with greater than random probability ten minutes after the initial task. And as Bloom (2004) points out, in a commentary on the original study, four weeks later Rico showed some retention of the names he had learned. In this capacity, Rico's abilities were akin to those of a human 3-year-old, and his overall word-learning abilities exceeded those of chimpanzees (Bloom 2004: 1605). With a focus on the human-like social skills of dogs, Hare & Tomasello (2005: 440) have stressed this finding as well. Experimental evidence also indicates that dogs may perceive subtle phonetic variation in human commands (Fukuzawa et al. 2005), suggesting sensitivity to the acoustics of human speech.

Receptive vocabulary is not, however, the only medium of vocal communication worth noting. It is no surprise to be told that dogs bark; wolves bark also, though not as frequently (Coppinger & Feinstein 1991: 120; Yin 2002: 189). In wolves, barking is a minor component of their overall vocal repertoire (Schassburger 1987), and occurs most frequently in two contexts, alerting other wolves to some circumstance, and territorial demonstrations (Yin 2002: 189). In dogs, barking serves a variety of purposes and the contexts range widely, from defense or warning, all the way to play (Bradshaw & Nott 1995: 117). Wolves usually bark in specific contexts, and in general, "the bark probably evolved to draw visual attention to the vocalizer" (Harrington & Asa 2003: 73). In both wolves and dogs, barks emitted in different situations have different sound qualities that are documented (Yin 2002; Feddersen-Petersen 2004; Yin & McCowan 2004). The documentation of this pattern is surely no surprise to a great many owners of household dogs. For example, Elizabeth Marshall Thomas (2000: 88–90) described in anecdotal fashion some barking patterns of her own household dogs. After noting that her dog Pearl "barked in several different voices" (Thomas 2000: 88), depending on what Pearl wanted, Thomas went on to contrast those barking behaviors with those directed toward intruders:

> However, when it came to barking at intruders, Pearl's voice was entirely different. What's more, she barked in different tones for different creatures. Amazingly, she used one tone for dogs, another, slightly different tone for large nondogs, and a third very different tone for small nondogs such as skunks, porcupines, and ravens. (Thomas 2000: 88)

Notably, and again surely not surprisingly to many dog enthusiasts, recent test data indicate that people are often able to classify (recorded) dog barks in terms of their emotional content. And they can do so without knowing the context in which the barks were produced, and with no previous experience with a given dog breed or in owning a dog (Pongrácz et al. 2005). That is, they can identify emotional information by sound alone. That factor surely fosters enhanced communication between people and dogs, and serves, in general, to reinforce their social bond. In fact, dogs in general have a remarkable capacity to communicate with people: "In so many respects, modern dogs seem to be better adapted to communication with humans than with other dogs" (Feddersen-Petersen 2007: 117). Accordingly, it seems that dogs belong with people. That reality is highlighted by the fact that truly feral dogs, "dogs that do not have owners and which have gone wild" (Hubrecht 1995: 182), as opposed the Ashkelon dogs (see Chapter 7), fending completely for themselves in human-dominated settings, generally lead short, difficult lives, with population maintenance depending on successful recruitment of other individuals (e.g., Boitani et al. 1995, 2007). Dogs simply do not thrive when deprived of regular human care and interaction.

These, then, are some of the documented characteristics of dogs, especially as they pertain to how the animals have come to relate to people. Some unique experimental work bears directly on these issues.

THE FARM FOX EXPERIMENT

In recent years, Lyudmila Trut (1999, 2001; Trut et al. 2004) has described the ongoing results of an experiment in Russia that was initiated some half a century ago, in 1959, by the late D. K. Belyaev. At a farm where Silver Foxes (*Vulpes vulpes*) were being raised for their pelts, Belyaev had noticed that while most individuals were quite mistrustful of people, a small proportion, about 10 percent (more females than males), seemed to be less so. Consequently, he began a strict selective breeding program in which these calmer individuals were allowed to reproduce and raise offspring only among themselves (for details see Belyaev 1979; Belyaev & Trut 1975, 1982). Among other things, foxes in this selected group showed some remarkably doglike traits, both physical and behavioral. One truly remarkable aspect of the farm fox experiment has been documented change in physiological operation of the animals' brains.

In particular, Belyaev (1979: 306) found that "Tame females exhibit statistically significant changes in certain neuro-chemical characteristics in such regions of the brain as the hypothalamus, midbrain, and hippocampus."

Specifically, in the tame foxes researchers detected increased levels of serotonin, a neurotransmitter linked with lowered impulsivity and reduced aggression in a number of animals, including dogs (Reisner et al. 1996). In conjunction with these documented changes, the foxes acted quite doglike in many respects, showing affection toward, and soliciting attention from, people. Belyaev's experiment was designed to promote the emergence of domesticated behavior, and what makes its results dauntingly complex is the interface between genetic and neurological factors. In dogs, DNA-based changes, so subtle that dogs and wolves can now be regarded as the same species (see Chapter 2), apparently underlie neurological patterns that have substantial implications, some of which are noted above. Furthermore, in the foxes a regime of social/behavioral selection impacted both neurochemical and neurohormonal mechanisms (Trut 1999: 166), and there is every reason to expect that the same was true in early dogs. Several dozen generations have passed in the foxes, and experimental results indicate that the traits are inherited (Trut 1999: 164–165), involving especially early development. Inheritance has been verified by, among other things, backcrosses between foxes in the line selected for tame behaviors and others selected for aggressive behaviors (Kukekova at al. 2004, 2008). In any case, other recent work with the foxes also highlights the complex interplay between genetic and neurological factors (Lindberg et al. 2005; Kukekova et al. 2006: 520–521). Similarly, recent work with wolves and/or dogs highlights the importance of subtle DNA-based alterations. that have substantial effects on neurological expression (Saetre et al. 2004; Lindblad-Toh et al. 2005: 804, 816), as molded by environmental influences.

In terms of direct relevance, Hare et al. (2005) have advanced an intriguing possibility, based especially on the fox work. Specifically, they suggested that the enhanced proficiency of dogs in reading human communication signals may be less a matter of direct selection on communicative abilities themselves, and more of a correlated byproduct of selection against fear and aggression. At the same time, they also indicated the likelihood that simultaneously "selection acted directly on dogs' ability to read human communicative cues" (Hare et al. 2005: 229). Surely it can be appreciated that in this model, the two factors likely go hand in hand. That is, reduced fear and aggression facilitated opportunities for enhanced communication skills, and such skills were at a selective advantage.

Recalling the primacy of neurochemical and neurohormonal mechanisms (Trut 1999: 166) in the foxes, the neurohormonal component should not be ignored. In those foxes Trut was referring especially to a delayed fear response, linked to changes in plasma levels of corticosteroid hormones. For dogs, without repeating details here (see Chapter 4), Crockford (2000b, 2002, 2006) has suggested that different levels of the thyroid hormone thyroxine posed a factor that conditioned certain earliest wolves to domestication, by acclimating them to human-dominated habitats. Hers is in effect a neurologically based model, given that "thyroxine manufacture begins in the brain" (Crockford 2000b: 15). The thyroid gland is a major component of the overall endocrine system, and "Hormone synthesis is the one principal function shared by all components of the system" (Hullinger 1993: 559). Crockford invokes the Russian fox work, noting that her model accommodates both distinctive morphological and physiological changes that are documented. Moreover, this thyroid hormone model, still in need of empirical backing (e.g., Richardson 2007), is seen to be of broader relevance to a wide range of animals. As indicated in Chapter 4, Crockford (2006: 50) wrote that "all of the physical and behavioural characteristics that change when wild animals become domestic are controlled by thyroid hormone." Ironically, even if the test of time should reveal its basic validity, it is that very broader relevance that relegates this model to a subsidiary role here. The reason is that a factor that accounts for patterns seen in a variety of domestic animals cannot account for the distinctive relationship between people and dogs. Quite simply, these other animals have not been routinely treated as dogs have, for example by being ritually interred upon their deaths. Chapter 7 called attention to a (localized) exception to that generalization, cats, but there is a more immediate issue.

WHY NOT FOXES?

This particular heading reflects the fact that fox domestication occurred only under experimental conditions, but not otherwise. Accordingly, it is legitimate to wonder why that is the case, given how amenable to domestication they were under experimental conditions. In broaching this topic, it is useful to be familiar with the basic distinction drawn between doglike and foxlike canids in terms of overall brain structure. The basic neuromorphology of doglike canids was apparently established by the Middle Pleistocene, as noted by Lyras & Van der Geer (2003: 511), though some specific aspects of their brains underwent additional modification. But after noting that pattern, Lyras & Van der Geer (2003: 511) went on to suggest that: "In the *Vulpes*-lineage, on the contrary, it appears that the external morphology of the brain was already more or less fixed in the Hemphillian," which would be some 5–10 million years ago (see Radinsky 1973: 171). It thus seems to be the case that doglike canids, including wolves, continued undergoing some modification in basic brain morphology for considerably longer than foxes. And one should realize that the *Vulpes*-lineage includes the foxes of the Farm Fox experiment. As already noted, the frontal region was one of the brain areas that figured into the basic distinction between foxlike and doglike canids. The role of the frontal region crops up again when contrasting dogs and cats, but now it is important to address more fully the question posed in this section's heading.

In the first place, foxes of the genus *Vulpes* can be described as solitary species (Gittleman 1986: 33), in contrast to most members of the genus *Canis*. In fact, writing about the Red Foxes (*Vulpes vulpes*) that he studied at Kent's Green, in England, Roger Burrows (1968: 124) stated the following: "The adult fox is a solitary animal except during the breeding season, and as I have never seen two adults together, I cannot speak from first-hand experience about what happens when they meet."[6] Historically, the New World Red Fox has been known as *Vulpes fulva*, not *Vulpes vulpes*, though as Burt & Grossenheider (1976: 72) noted, some authorities in that era regarded both as the same species, *V. vulpes*. At this point, *V. fulva* has apparently largely been dropped from usage, and both Old and New World Red Foxes are considered *V. vulpes*.

But regardless of shifts in specific taxonomic nomenclature, animals of the genus *Vulpes* are not social in the same sense as most members of *Canis*. Writing in the same era as Roger Burrows, J. P. Scott put the matter this way: "Foxes are even more solitary than coyotes and jackals, and they have apparently never evolved patterns of agonistic behavior which would permit them to live in larger groups than the mated pair" (Scott 1967: 373). Accordingly, Scott held the wolf as the mostly likely ancestor of dogs. Another factor that is important in considering the

---

[6] Regarding the exception noted by Burrows, at certain times "groups with up to six members (usually one adult male and 2–5, probably related, vixens) may share a territory, depending on habitat" (Macdonald & Reynolds 2004: 133).

differences between foxes and wolves is their differential body size. Put simply, foxes are smaller than wolves. One clear consequence of that fact involves their dietary practices. Wolves, of course, are group hunters, often dispatching prey of larger body size, a feat they accomplish by virtue of operating in social groups, often led by a mature breeding male (see Sand et al. 2006). The ebb and flow of individual relationships within a group is mediated by the kinds of agonistic behaviors that Scott had in mind, as quoted above. Though wolves incorporate other smaller items into their diet as opportunity permits, as Patricia Moehlman (1989: 144) has observed, "Among most large canids cooperative hunting is an important if not critical method of obtaining food."

But in keeping with their smaller size, foxes are different. In England, Burrows (1968: 102–118) found that rabbits and small rodents were the most common animal food items, along with an occasional small bird, and some fruit. In southeastern North America, Ashby (1974: 15–18) found that rabbits were the most common prey throughout his study range, while foxes also ate mice and voles, and some plant foods in late summer and early fall. More recently, Macdonald & Reynolds (2004: 132) have made a comparable point about dietary practices among Red foxes in particular. As far as larger prey animals, such as wolves might hunt, Ashby (1974: 17) noted in his study that in part of Michigan White-tailed Deer (*Odocoileus virginianus*) "shot and wasted by hunters during the November deer season made up the bulk of the winter food of foxes." How they subsisted in winter prior to modern human hunting seasons is a matter of conjecture.

When putting all these factors together, it really is no surprise that canids of the genus *Vulpes* never entered into a domestic association with people, except for recent experimental conditions. They are essentially solitary animals. As maintained throughout this volume, sociality is, first and foremost, what underlies the domestic association whereby wolves became dogs. The foxes could relate to people much like individual dogs, but only under those experimental conditions. Given their inherently solitary nature, along with the fact that they are neither as clean nor as relatively self-sufficient as cats (see below), that pattern is not surprising. Indeed, dogs are social animals, requiring care and social interaction. So, earliest dogs/wolves changed in ways that facilitated the strong social bond that has developed between dogs and people. As just anticipated, though, sociality of that kind is not essential for a domestic association between people and a fundamentally different sort of animal.

DOGS AND CATS: A GENUINE CONTRAST

[The situation with cats] is even more interesting if one considers that the social structure of their ancestors was most probably more similar to that of foxes than dogs. (Miklósi & Topál 2005: 464)

One of the points to emerge from the previous chapter is that cats have not been ritualistically treated at death nearly as consistently as dogs, with the conspicuous exception of ancient Egypt (see Clutton-Brock 1981: 110–112; Morey 2006: 168–169). Since ancient Egyptians also maintained cemeteries for a variety of wild animal species (Budiansky 2002: 30), this seems to have been a practice that, among them, was commonly applied to animals for reasons other than their social compatibility with people.

It is noteworthy that a telephone survey of several hundred Rhode Island residents, reported by Albert & Bulcroft (1987, 1988), found that "Owners who selected dogs as their favorite pets reported feeling more attached to their pets than did people whose favorite pets were cats or other animals" (Hart 1995: 162). The reasons for this kind of difference in perception probably stem from the fact that "cats are solitary and mostly nocturnal hunters" (Morey 2006: 169).[7] In contrast, both dogs and people are socially gregarious and, in general, active by day, though wolves do sometimes hunt at night (e.g., A. Murie 1944: Chapter 2). Those are behavioral factors that surely underlie the differences in perception, and it is especially useful to focus more closely on their different patterns of sociality. In that light, it is relevant to note that in a study emphasizing visual cues, Miklósi et al. (2005) found that cats seem to lack some components of human attention-getting behavior that are present in dogs.

Moreover, Christof Koch (2004: 130) has made the observation that "While accounting for only 3.5% of the volume of the feline cortex, PFC [prefrontal cortex] occupies 7% of canine cortex (dog lovers note)." Based on fossil cranial endocast data, Radinsky (1973: 191–197) has considered the expansion of this region in canids during Miocene and Pliocene times (up to about two million years ago). In the doglike canids

---

7 The common cat is indeed a solitary hunter, but as Grandin & Johnson (2009: Chapter 3) have pointed out, cats are more social otherwise. In fact, "Free-ranging cats naturally form cat colonies" (Grandin & Johnson 2009: 90), anywhere from two to more than 50 cats, though cats sometimes live alone, too. The key point here is that such aggregations of cats, or cat colonies, differ from wolf packs, for which parents and their offspring (conventional families) form the basic unit underlying pack social structure.

it was of modern proportions by latest Pleistocene times. The expansion of this region probably indicated increasingly flexible and complex behavior patterns, and it is associated with those canids having pack social structure (Radinsky 1973: 195). In people, the prefrontal cortex is activated (as documented with imaging studies) when people deal with the kinds of situations that make demands on one's mental acuity, or intelligence (Passingham 2002: 191). From the perspective of cognitive evolution, Povinelli (2004: 30) has referred to "the seat of higher cognitive function, the prefrontal cortex." Additionally, "For these high-level, executive functions, the prefrontal cortex comes into its own" (Koch 2004: 129). But one doesn't have to appeal simply to what is documented in humans to suggest the relevance of this pattern to dogs. Dealing with this region in their comprehensive assessment of the dog brain, Beitz & Fletcher (1993: 927) note that "Association cortex of the frontal pole (prefrontal cortex) is necessary for intelligent behavior."

In more specific terms than Radinsky, noted above, it is worth calling attention to the fact that the prefrontal cortex portion of the brain is wired to several parts, including the hypothalamus. According to Koch (2004: 130), "PFC is the only neocortical region that talks directly to the hypothalamus, responsible for the release of hormones." In addition, in people (and presumably canids as well), emotional responses to internally driven scenarios "are generated, via mediation of the brainstem structures, amygdala, and hypothalamus" (Churchland 2002: 228). The hypothalamus has widespread connections within the brain (Beitz & Fletcher 1993: 917), and in controlling the endocrine system, the dog hypothalamus "plays a major role in affective behavior, containing sites that elicit rage, escape, or pleasure responses" (Beitz & Fletcher 1993: 916). It should be recalled that the hypothalamus was one of the brain areas where Belyaev (1979: 306) found statistically significant changes in certain neurochemical characteristics of fox brains, in the farm fox experiment described earlier. Specifically, there were increased levels of serotonin, associated with lowered impulsivity and reduced aggression. Another affected brain area was the hippocampus, where signals engage memory functions (Churchland 2002: 94). In fact, the hippocampus is involved in declarative (explicit) memory (Churchland 2002: 351; Koch 2004: 195), and the role of dogs' augmented attention and memory skills have already been pointed out here. Additionally, the hippocampus can be considered a part of the limbic system (Beitz & Fletcher 1993: 919), a system consisting of "emotional structures capable of generating specific feelings or affects reflected in overt emotional reactions"

(Fox 1978: 258). It seems clear that some important brain functions have been modified in canids as a consequence of domestication. As one conspicuous example, hippocampal changes documented in foxes undergoing domestication seem to relate to the need for the animals to evaluate their socially modified surroundings.

Regarding communication, previous discussion already pointed out how people can commonly classify the emotional content of a dog's bark, by the sound alone. The same is not true of cats. A study by Nicastro & Owren (2003) of human responses to cat meows revealed that context and familiarity with a given cat, or long term familiarity with cats in general, substantially enhanced a participant's ability to classify the meows. In short, to a human listener meows tend to be non-specific in their meaning, unlike dog barks, as noted.

Overall, cats did not enter the domestic association with people with the kinds of physiological prerequisites and behavioral tendencies in place that characterize wolves. As a consequence, cats could not evolve according to the patterns that typify dogs, and accordingly, they have come to relate to people in their own distinctive terms. Cats are basically solitary and mostly nocturnal animals, so many people find them clean, quiet, relatively self-sufficient, and, therefore, pleasing to have around, especially if they are individually affectionate, which they can be. They can also be useful, given their propensity to clear peoples' living quarters of "unwanted scavenging animals such as rats and mice" (Clutton-Brock 1981: 111), and they may have been regarded that way long ago: "Cats were likely used as mousers" (Udell & Wynne 2008: 256). But cats are not equipped to relate to people in the same social manner as dogs, irrespective of circumstances whereby "they [cats] are not typically bred for purposes that require a close partnership with humans, even today" (Udell & Wynne 2008: 256).

DOG "HUMANIZATION"

Even if some of the specific physiological factors invoked in this chapter should not turn out to be of quite the import modeled, in combination, the result is an animal that is remarkably compatible with people. For example, as part of a study previously highlighted here, Brian Hare and colleagues offered the following statement: "Our conclusion is that as a result of the process of domestication, some aspects of the social-cognitive abilities of dogs have converged, within the phylogenetic

constraints of the species, with those of humans through a phylogenetic process of enculturation" (Hare et al. 2002: 1636). With that point
in mind, Hare (2007: 61) has more recently stressed the probability that
"dogs' social skills in cooperative- communicative contexts are a case of
convergent evolution with humans." More than three decades earlier,
Konrad Lorenz put the matter in his own distinctive style:

> the domestic dog really has been "humanized" in regard to certain of
> its faculties. To love one's brother as one does oneself is one of the
> most beautiful commands of Christianity, though there are few men and
> women able to live up to it. A faithful dog, however, loves its master
> much more than it loves itself and certainly more than its master ever
> can be able to love it back. There certainly is no creature in the world in
> which "bond behavior," in other words personal friendship, has become
> an equally powerful motivation as it has in dogs. (Lorenz 1975: x)

As Lorenz's comment highlights, what is so noteworthy is the manner
in which dogs have evolved as a consequence of the domestic relationship. They have evolved in ways that are documented morphologically
and neurologically, and manifested behaviorally. To offer a summary
answer to the question posed by this chapter's title, an especially strong
social bond has come to characterize the association between people
and dogs because they relate well socially to each other, and they do
so because of some intriguing neurological similarities that have developed. The consequence of those similarities is that, in several important
ways, dogs are much like people, and people are much like dogs. In
the following chapter, I seek to highlight some additional dimensions
to the issue of just what dogs can do, and just how strongly they relate
to people.

9

OTHER HUMAN-LIKE CAPABILITIES OF DOGS

IN THE PREVIOUS CHAPTER, ONE OF SEVERAL CHARACTERISTICS OF dogs that posed an important topic was the development of social skills that are remarkably parallel to those of people. For example, work by Hare & Tomasello (2005), among others, focused on this very characteristic of dogs. This is truly a remarkable feature of the evolution of dogs, but in this chapter the principal concern is some other remarkably human-like characteristics that are found in dogs. These characteristics are not, strictly speaking, social skills themselves, but they do concern the execution of such skills. A good place to begin is with some of the emotional reactions, toward people, that have been attributed to dogs under particular circumstances. At one level, it will come as no surprise to many people to hear that a particular dog seems to like them. Here, though, a much more consequential capability, one that reflects nothing less than life or death importance, is the initial topic.

SEARCH-AND-RESCUE DOGS

As particularly good examples of the close association between people and dogs, dogs are commonly used to search for and rescue people and to assist handicapped people. Dogs that perform these functions are highly selected and extensively trained for a particular task. (Macpherson & Roberts 2006: 113)

Following that passage, I deal here with the first of the examples identified as an example of the close association between people and dogs. Most readers are surely at least vaguely familiar with the phenomenon of search-and-rescue dogs. Common enough contexts for these trained dogs include situations where, for example, a hiker has disappeared in a wilderness area. It need not involve the wilderness, however, since, as another relevant example, children sometimes seem to vanish into thin air in a more populated setting. They may have wandered off and gotten lost or, more problematical, a child might have been abducted. In many such cases trained dogs seek to find missing people, using a variety of cues, but especially stemming from their keen olfactory sense. Those special skills are, of course, also put to use in efforts to locate criminal suspects, and/or illegal contraband, such as narcotic drugs or dangerous explosives. For example, as Coppinger & Coppinger (2001: 253) noted, "The dogs of the customs services that detect drugs or explosives in huge container shipments save agents a lot of tedious searching." These are kinds of situations, of course, that call for the use of dogs' capabilities that people lack, or at least don't have developed to the degree of a dog.[1] Here, however, the objective is to emphasize some characteristics of dogs that are quite similar to human characteristics, and in certain situations the use of their previously mentioned distinctive capabilities draws attention to certain provocative similarities.

A particularly compelling case took place in Mexico City in 1985, in the wake of a major earthquake there. This disaster prompted an international search-and-rescue mission by Caroline Hebard and her trained dogs (Whittemore & Hebard 1996) and an account of this situation is in the cited book, devoting multiple chapters to this event. The book also gives an account of other search-and-rescue efforts by Hebard and her trained dogs, both in the vicinity of her home in New Jersey and in other countries, including but not limited to, Mexico. The entire book is written in an informal mode, intended for public consumption. That feature is highlighted by the fact that the initial passages, prior to the beginning of the text itself, include quoted words from a former American president, the late Ronald Reagan, uttered in 1988. His words were especially about search-and-rescue efforts after a major earthquake in Armenia in 1988.

But in Mexico City, as a consequence of the aforementioned earthquake, a large hospital collapsed, and at least 800 people had died within the debris. But using especially their keen sense of smell, the dogs also detected signs of life in the wreckage. The dogs, including Hebard's own dog Aly, "gave strong live alerts" (Whittemore & Hebard 1996:

---

[1] Indeed, the great Dutch ethologist and Nobel Prize laureate Nikolaas ("Niko") Tinbergen singled out dogs' superior olfactory capabilities more than half a century ago: "In distinguishing between scents and in their ability to recognize a particular scent among a mixture of others, dogs are far superior to man" (Tinbergen 1951: 21).

125), displaying different signals if they detected no signs of life. And a good many live people were rescued. But one pattern was apparent as the hours and then days wore on, and the dogs found fewer and fewer live victims. Quite simply, they became less motivated, continuing their work only with reluctance, and seemed to lose their appetites for food. Then, to remotivate them, the handlers devised an impromptu strategy. A live, uninjured man, a veterinarian, was sent to the wreckage, where he concealed himself amidst the rubble. And repeatedly, the seemingly despondent search dogs approached the area, detected the live person, and happily gave a live alert. Clearly, living people served as a reward, and the animals that experienced this fabricated rescue situation "remained always eager" (Whittemore & Hebard 1996: 137). Hebard apparently developed a keener appreciation of the emotional strain on a dog of finding dead body after dead body. These dogs wanted to find live people. That they reacted as they did when they didn't find live people points to a particularly human-like quality of dogs.

Dogs Get Depressed

The assertion represented by this heading will surely come as no surprise to a majority of people with dogs as pets, and this reality is recognized by specialists as well (e.g., F. D. McMillan 2004; Hedhammar & Hultin-Jäderlund 2007: 245; Landsberg et al. 2007: 471–480). Certainly anecdotal evidence for its accuracy abounds. For example, a basic internet search for the phrase "depression in dogs," using the standard google.com search facility is variable in outcome, but results in at least several hundred hits. To be sure, some of the links use the term "depression" in a direct physiological sense, as in cardiac depression, or a reduced heart rate. Though having perused only a tiny fraction of the available links, it is clear that many of them refer to depression in the more commonly used sense. As an example, one posted by Rebecca Ash indicates that "canine depression is very real" (Ash 2005). Table 9.1 assembles some of the causes indicated by Ash that commonly lead to depression in dogs, as well as the commonly recognized symptoms.[2] And as informal as Ash's compilation is, it should be noted that pharmacological remedies are sometimes actively sought

[2] This informal compilation makes no pretense at methodological rigor. Despite that limitation, what seems most compelling about it is the remarkable parallel between commonly recognized causes and symptoms of depression in dogs, and those in people.

TABLE 9.1. *Commonly recognized causes and symptoms of depression in dogs (adapted from Ash 2005)*[2]

Common Causes of Depression in Dogs	Recognized Symptoms of Depression in Dogs
Loss of a playmate or companion	Loss of appetite (possibly including major weight loss)
Separation anxiety	
Change in routine or scenery, such as a residential move	Minimal water intake
Loneliness or isolation	Lethargic demeanor, excessive sleeping, lack of interest in customary sources of happiness
Chemical imbalances	Changes in normal behavior patterns
Unpleasant weather	
Grief (e.g., death of a close individual)	
Empathetic response (e.g., to a depressed person)	

for dogs experiencing such problems (e.g., Hsu 1983; F. D. McMillan 2004: 89). Other kinds of remedies are sometimes sought as well. As a good recent example, Takabatake (2007) has explored the use of music therapy for the treatment of some cases of depression in dogs. Tellingly, he concludes that music can indeed have a positive effect on a dog's well-being. And there is little question that the use of music is also considered beneficial to some people experiencing such problems.

To be sure, the informal and anecdotal nature of much of the widely accessed information on depression in dogs, such as that by Ash (2005), might, understandably, lead one to infer that the very topic represents little more than the long-standing tendency of people to anthropomorphize dogs, as well as many other animals. To do so, however, would be a mistake. A conspicuous example to illustrate this is contained in a book written a few years ago by the noted primatologist Frans De Waal (2001). De Waal is the author of several books, and was, at the time he wrote that book, the C. H. Candler Professor of Primate Behavior at Emory University, in Atlanta, Georgia. His book emphasized, not surprisingly, some of the apes and, among other issues, dealt with the reality of cultural expression among some of those animals. In one of the last chapters in his book, De Waal (2001: Chapter 10) also deals explicitly with the topic of depression in dogs. Among other things, he directly calls attention to Caroline Hebard's work with her own dog Aly, and the other dogs of the search-and-rescue team that worked in the wake of the 1985 Mexico City earthquake. In that chapter, De Waal has an entire section devoted to depressed rescue dogs (De Waal 2001: 330), a heading under which the account of Hebard's work occurs. In that section, he describes the situation when the dogs were locating little other than dead human bodies as follows:

> They [the dogs] required longer and longer resting periods, and their eagerness for the job dropped off dramatically. After a couple of days, Aly clearly had had enough. His big brown eyes were mournful, and he hid behind the bed when Hebard wanted to take him out again. He also refused to eat. All other dogs on the team had lost their appetites as well. (De Waal 2001: 332–333)

That passage recounts the situation in very much the same terms that Hebard does, a consequence of having drawn directly from her work. More importantly, for present purposes, are the implications that De Waal ascribes to this situation:

> Instead of performing a cheap circus trick, they are emotionally invested. They relish the opportunity to find and save a live person. Doing so also constitutes some sort of reward, but one more in line with what Adam Smith, the Scottish philosopher and father of economics, thought to underlie human sympathy: all that we derive from sympathy, he said, is the pleasure of seeing someone else's fortune. Perhaps this doesn't seem like much, but it means a lot to many people, and apparently also to some big-hearted canines. (De Waal 2001: 333)

So, the dogs were emotionally invested in finding live people, and got depressed when they didn't. Moreover, the manner in which their state of depression was expressed was remarkably like that of many people known to suffer from depression. This is indeed an intriguing parallel between dogs and people, one that speaks to an important similarity. It is worth emphasizing again that the original account of this situation (Whittemore & Hebard 1996) appeared in an informal book, intended for a wide public audience. Despite that venue, its implications were taken quite seriously by a prominent primatologist.

A fascinating and illuminating contrast is provided by a more recent account of search-and-rescue dogs associated with the terrorist attacks on the United States on September 11, 2001 ("9/11" in customary parlance), centered on the World Trade Center building in New York City. Nona Kilgore Bauer (2006) has provided what amounts to a lengthy though informal dog-by-dog account of activities, featuring their handlers as well. This was a situation where the word "rescue" turned out to be a misnomer because there were no survivors to be found where they searched. For some of the dogs, the toll of finding dead body (or body part) after dead body (part) was similar to Whittemore & Hebard's accounts, as just covered. For example, pertaining to a dog that was part of one of the first teams to arrive at the World Trade Center, Bauer (2006: 91) writes that "Polly [the dog] was eager and ready to work for the first several days, but as the days wore on with no survivors, she lost her normal playful attitude and became more serious and withdrawn." Also at the World Trade Center, according to her handler Sarah, "After the first day, Anna [the dog] became very quiet and her attitude went down because she was finding only human remains" (Bauer: 2006: 27). Likewise, at the Pentagon, also attacked, officer Alice Hanan was quoted as follows:

'Locating human remains is a gruesome task that can deeply affect the emotions of the search dog,' Alice said. 'Stryker [her dog] was always focused and effective whenever he was working, but over our 11-day deployment he lost his normal playful attitude.' (Bauer 2006: 160)

But here is where the fascinating and illuminating contrast surfaces. For the majority of the dogs, and dozens were highlighted, the handlers specified that the dogs' spirits remained good, and the search was like a game, with the reward at the end in the form of a toy and/or praise. On the one hand, it is relevant to note the informality of these accounts, perhaps underlying a real possibility that a lack of objectivity among the handlers is in play with such assessments. On the other hand, they knew their dogs well, and were surely attuned to their moods. Accepting their informal assessments, what emerges from a consideration of this pattern is the fact that there were no "live finds" to be made among dogs that were trained mostly in that capacity. That is, in their deployment in this tragic situation, the dogs had no basis for developing a working contrast between live finds and just bodies. Consequently, many, though obviously not all, maintained their working spirits. This is indeed a real contrast to the dogs featured in Whittemore & Hebard's (1996) account of dogs who worked in the aftermath of the Mexico City earthquake. Those dogs were keyed on finding live people, of which there were some, and got depressed when they didn't.

Similarly, though not on nearly as monumental a scale, 1990 flooding in eastern Ohio, at Shadyside on the Ohio River, prompted the use of some trained dogs (Graham & Graham 1990). The final human death count from this disaster reached twenty-six, of which the dogs located nine. Several dog teams were involved, and the leader of one of the rescue teams was quoted as saying that there was "quite a bit of depression among the dogs. They were acting very leery, but they also had a drive like I've never seen them have before" (Graham & Graham 1990: 2). Graham & Graham's informal account is reminiscent of some of the accounts shared by Bauer (2006) involving the World Trade Center disaster. That is, some of the dogs were clearly impacted by the absence of "live finds," but they worked hard nonetheless to accomplish their objectives.

To sum up this section, dogs thrive when they find live people, but often are emotionally impacted when they don't. Be that as it may, another quite compelling similarity between dogs/wolves and people does not have the same life versus death importance, but should evoke genuine pleasure rather than the discomfort associated with accounts of depression.

MUSICAL EXPRESSION

It is surely not customary for people to think of dogs when they think of musical expression in general. Though a variety of animals are known sometimes to make sounds that are regarded as melodic to the human ear, true music is often considered to be the province of people alone. One clear example of that perception is offered by Ian Cross, a member of the faculty of music at Cambridge University in England:

> *Musics are those temporally patterned **human** activities, individual and social, that involve the production and perception of sound and have no evident and immediate efficacy or fixed consensual reference.* (I. Cross 2001: 98, italics in original, bold emphasis added)

In arriving at that definition, Cross did note certain animal sounds that are often labeled in a way that connotes music, such as bird songs, a term that is surely familiar to most everybody. In fact, as is well known, certain types of birds are sometimes referred to as "song birds." But in commenting on bird song, as well as a phylogenetically more plausible possibility for real music in the animal world, Cross wrote as follows:

> But though there is complex sound pattern in birdsong and in human music, in traversing the evolutionary space that separates birds and

TABLE 9.2. *Specific points made by acclaimed music critic Harold Schonberg (1971) in his review of a recording that features wolves howling (American Museum of Natural History 1971). Immediately preceding those points, he wrote as follows: "From it [the recording] the following can be easily recognized:" (Schonberg 1971: 1)*

Wolves sing.

There are lupine vocal registers – soprano wolves, contralto wolves, tenor wolves, even a bass wolf or two.

Wolves have a characteristic call in which the interval of the major sixth (C to A) predominates.

humans the incidence of the use of complexly patterned sound diminishes vastly. When we reach our closest evolutionary neighbours, the primates, we find a very limited stock of complex sound patterns being produced and employed. Most researchers into birdsong acknowledge that the relation between human music and birdsong is one of analogy, not homology, a surface resemblance. (I. Cross 2001: 97)

From these considerations, Cross proceeded from the premise that genuine musical expression is a uniquely human trait, and sought to model how it came into being. Unfortunately, in considering only birds and primates, Cross missed a splendid and well documented example of musical expression among animals. Given the focus of this book, it is surely no surprise that I am about to highlight music in certain canids, including dogs. It is further no surprise that this treatment begins with the immediate ancestors of dogs, wolves.

Wolves Sing

In 1971, a recording was released that prominently featured the howling of North American wolves, both wild and captive, singly and in groups (American Museum of Natural History 1971). In the same year, that effort was reviewed in print by a Pulitzer Prize winning music critic for the New York Times newspaper, the late Harold Schonberg (1971). In that review, he wrote, point by point, as quoted word for word on Table 9.2. Following that point by point summary, Schonberg continued as follows:

Judging from the record, some wolves have much better voices than other wolves. The best of these virtuosos are capable of six- or seven-second phrases on one breath. Each phrase is a glissando swoop, up and down, like a very lonesome, sentimental, fire siren with a soul. (Schonberg 1971: 1, 34)

Surely it can be appreciated that this acclaimed and veteran music critic expressed absolutely no reservations about the matter: that is, wolves sing (see Table 9.2). Among other modern people who in general are a part of industrialized societies, this kind of perception stems beyond the ranks of professional music critics. For example, Olaus Murie, who did some work for the U.S. Biological Survey, frequently wrote about the American wilderness and its wildlife. He departed this earth in 1963, and James Glover summarized his legacy forty years later (Glover 2003). One of Murie's published works was a field guide to animal tracks, one of the first in the well-known Peterson field guide series (O. Murie 1954). In that book, he related an account regarding how he and his family were camping in northeastern Alaska. At dawn one morning,

> Soon we realized that we were being serenaded by two wolves, one upstream, the other below our camp. First one, then the other, raised its muzzle and howled. Apparently we were intruding on their home ground. At any rate, we lay there in the crisp autumn morning, comfortable in our sleeping bags, and listened to this song of the Arctic wilderness with a feeling of awe. (O. Murie 1954: 93)

Just earlier than that, Adolph Murie, a naturalist and wildlife biologist, also wrote about wolves he encountered in Alaska:

> Considerable ceremony often precedes the departure for the hunt... There they all howled, and while they howled the gray female galloped up from the den 100 yards and joined them. She was greeted with energetic tail wagging and general good feeling. Then the vigorous actions came to an end, and five muzzles pointed skyward. Their howling floated softly across the tundra. (A. Murie 1944: 31–32)

Evidently some professional wildlife specialists find the howling of wolves quite compelling. Another example is zoologist John Theberge, who has been involved in the technical documentation of the properties of wolf howls (e.g., Theberge & Falls 1967), as covered shortly. Later, though, working in Ontario, Theberge (1971) dealt more informally with the howling of wolves in a nontechnical column intended for a wide readership. At one juncture in that column, he related the case of a wolf pup that had seriously injured its foot, and had to be destroyed.

Stemming from that event, "All through that night, over and over again, an adult wolf housed in an adjacent pen gave a very melancholy but beautiful howl" (Theberge 1971: 41). It seems that the wolf expressed an emotion familiar to people, using howling to convey it. One might recall the reality of dogs getting depressed, and the melancholy howl of that wolf seems entirely consistent with that pattern. On a more positive note, writing about the wolves of Isle Royale National Park, in the Great Lakes region of North America, Scot Stewart ended his column by expressing his satisfaction that "our children, and our children's children may have the opportunity to experience the thrill of hearing the music of the wolf" (Stewart 1982: 49). And in more recent times, Ian McAllister (2007: 174, 177), related his own comparable account from Canada's North Pacific coast:

> It was one of those winter days when the air is so still and so quiet that sound seems to amplify with distance. The howls picked up an ecstatic momentum. As they cried, some of the wolves stood on their hind legs, pawing at the air with heads held back, as if to loft their voices higher into the sky.

In light of these various passages, it seems fitting to return to professional music critic Harold Schonberg once again. Noting how hunters in the wilderness do not necessarily feel comforted by the howl of a wolf, he touched on what were then some of the presumed purposes of howling, and suggested that:

> the lonely hunter should merely relax and enjoy the beautiful music. The wolf only wants to sing him to sleep. oo-oo-oo-oo-OO-OO-OHHHHHH – oo-oo-oo-oo. (Schonberg 1971: 34)

A more technical assessment of the qualities of a wolf howl has been offered in recent years by Harrington & Asa (2003: 74):

> Howls are harmonic sounds with a fundamental frequency of 150 Hz to more than 1,000 Hz for adults (Theberge and Falls 1967; Harrington and Mech 1978; Harrington 1989; Tooze et al. 1990), The lowest frequencies usually occur briefly at the beginning or end of a howl, and the highest frequencies typically occur during the first half of the howl (Theberge and Falls 1967; Tooze et al. 1990). The mean pitch of adult howls varies between 300 and 670 Hz (Harrington and Mech 1978; Tooze et al. 1990). Both pups and adults respond differently to playback of adult and pup

howls, replying to and approaching adult howls more readily. (Harrington 1986)

These assessments, whether offered in technical terms or not, often focus on the qualities of individual wolf howls. But since wolves often howl in groups, for a non-technical assessment of the musical quality of a group howl, one can turn to the words of Richard Ballantine, who in the 1990s monitored a particular group of wolves in the state of Idaho: "Each wolf has its own distinctive howl. When wolves howl together it sounds like a chorus" (Ballantine 1996: 148). At least two aspects of Ballantine's account are worth emphasizing. First, each individual wolf howls distinctively and, second, when they howl as a group, which they often do, it sounds to the human ear "like a chorus." An implication of individually distinctive wolf howls is important shortly, but for now, it is surely no surprise to be told that their dog descendents will howl as well, as many people well know.

Dogs Sing (Howl), Too

Regarding this reality, Mark Derr once related the following account of dogs howling in an Alaska town, when encouraged to do so:

> On a warm, overcast August evening in Big Lake, Alaska – a town so small it does not appear in my atlas – Martin Buser looks at me after feeding the last of his eighty huskies and asks, "Do you want to hear me howl with them?" Almost before hearing my answer, he tilts his head back and, starting low, raises his pitch in a long howl. The dogs join in in twos and threes until the entire yard is harmonized. After about a minute, they quit abruptly, a full orchestral stop. (Derr 1996: 34)

Derr continued by commenting on situations that more generally prompted those dogs to howl:

> Buser's huskies howl after meals, at dusk, when excited, and whenever barking alone will not suffice. On the trail, Buser – record-setting winner of the 1,159-mile-long Iditarod Trail Sled Dog Race in 1994, his second win in three years, and three-time recipient of its award for humanitarianism – huddles and howls with his dogs to build morale and team spirit. (Derr 1996: 34)

As for what individual dog howls sound like, it is initially worth sharing a personal anecdote about a particular dog. A few years ago, before she went mostly deaf, I twice played a recording, the one reviewed by Schonberg, at home where our late dog Jingo (featured at the end of Chapter 4) could hear it. The first time, she quickly began to howl in response, sounding much like a wolf. But by the second time, she apparently had realized that the wolves weren't interactive, or at least they weren't interactive members of her pack, so she howled at them only once, and then stopped to play with a toy. So when deprived of the expected sociality of howling, Jingo tried to demonstrate her behavioral compatibility with people in another way, through play. How similar is dog and wolf howling in general? Beyond the personal account just shared, it is worth pointing out again (see Chapter 2) that the striking similarity between the howl of a dog and that of a wolf played a major role in prompting Konrad Lorenz to rescind his 1950s view that the golden jackal was the primary ancestor of the dog, and infer instead that it was the wolf. Specifically, in the 1970s he commented that

> The golden jackal, like the coyote, howls with a falling pitch, beginning with a high sharp "yip, yip, yip" and then dropping into a melancholy-sounding decrescendo while, conversely, practically all breeds of domestic dogs howl exactly like wolves, beginning softly at a low pitch and rising gradually in pitch as well as in loudness. (Lorenz 1975: viii)

So, while I knew it to be true in Jingo's case, a good many years earlier Lorenz had conveniently expanded that point to include most dogs in general. In any case, the reality of musical expression in dogs and wolves holds some noteworthy implications that, at least in this case, go beyond the realm of music itself. Accordingly, attention now shifts to a particularly major implication.

CULTURE AND MUSICAL EXPRESSION

Naturally, it is important to begin this section by considering just what the term 'culture' means. Given my primary training in the social science of anthropology, it is fitting to point out that among the numerous definitions that have been offered over the years, one of the most frequently cited was provided long ago by Edward Tylor, one of anthropology's leading pioneers:

> Culture... taken in its wide ethnographic sense, is that complex whole which includes knowledge, belief, art, morals, law, custom, and any other

capabilities and habits acquired by man as a member of society. (Tylor 1871, as quoted in O'Brien & Lyman 2000: 386)

Tylor believed that culture was unique to people, and set them well apart from other animals because it enabled them to achieve genuine progress over time, culminating in civilization. Watson (1995) has emphasized the substantial role of this definition in contemporary anthropology. However one might wish to define it, the social learning environment is the key, and it seems useful to follow archaeologist Robert C. Dunnell's (1971: 121) lead in regarding it as basically "*shared ideas* – and nothing more" (original emphasis). A great deal has been learned since Tylor's time, of course, and one of the important lessons has been that culture, as a basic phenomenon, is not the exclusive property of people. It is established, for example, that some of our close phylogenetic relatives, specifically chimpanzees and bonobos, have a simple form of that complex whole that entails customs or habits that supersede genetic control. That point is acknowledged by most contemporary anthropologists (e.g., Boesch & Tomasello 1998; Hohmann & Fruth 2003), and more than a quarter of a century ago biologist John Tyler Bonner (1980) wrote an entire book that was devoted to exploring how culture developed in the nonhuman animal world.

Given the recognition that culture, again as a basic phenomenon, is not the exclusive property of people, it is surely no surprise to learn that a present objective is to develop that point with respect to wolves and/or dogs. Given that objective, a point that was made by a wolf specialist some three decades ago merits notice:

> As an apex predator, the wolf demonstrates a superb adaptation, physiologically, psychologically, and socially. In this chapter I shall discuss four interrelated aspects of that adaptation – group hunting practices, intrapack aggressive behavior, reproduction, and wolf individuality. Though summary statements can be attempted, it is becoming increasingly clear that wolf behavior varies greatly according to place, time, individual personality, and the *social and cultural traditions of individual packs*. (Sullivan 1978: 31, emphasis added)

If shorn of its taxonomic specificity, Sullivan's statement pinpoints several characteristics that apply to people as well, such as behavioral variability and individual personalities. All are part of the package of customs and habits, among other things, that are part of Edward Tylor's definition of culture from well over a century ago, as repeated above. Dogs, of course, do not live as wolves, and people are the central feature of the environment that shapes their habits and customs. As indicated by the heading of this section, musical expression is the aspect of behavior to be considered here. The point that dogs will howl musically like wolves has already been emphasized, and now it is time to carry that point a step further. Specifically, it is worthwhile to consider the role of music as a mode of cultural expression. And because the phenomenon of wolf howling has been investigated in some depth, the strongest position lies in considering musical expression among wild wolves, though analogous principles likely apply to dogs as well.

Music vs. Musics

For present purposes, it is useful to distinguish between a general capability for musical expression, and the existence of different varieties of that expression within one species of animal. With reference to people, if the former can be designated as "music," the latter, how music is understood and used in each specific society, could be designated as "musics," a point made clear by Bohlman (1999).[3] Surely there can be little doubt at this point that wolves and dogs are capable of music, in the sense just used. The really interesting question at present is whether they are capable of exhibiting musics, again in the sense just used. On logical grounds, recalling Sullivan's (1978: 31) reference to "social and cultural traditions of individual [wolf] packs," it is logical to suspect that they are. Phrased in the vernacular, the question becomes whether or not different wolves exhibit different styles of howling. Certainly people exhibit different styles of singing, and of course for the most part, people can distinguish between the voices of different individuals. That reality would seem to go hand in hand with the development of recognizably different singing styles. Such differences can have genuinely important pragmatic ramifications, as, for example, when people need to distinguish between group members and outsiders. The point

[3] One might have noticed that Ian Cross's (2001: 98) definition of musical expression concerned what he referred to as "musics." There is no indication, though, that he intended to use this term in the same way that I am. Rather, he was seeking to draw a distinction between human music and music-like sounds among non-human animals, which he regarded as not really music. And at the very least, music must exist for there to be musics in the sense that I use the term.

worth calling attention to here is that wolves almost certainly can do this: "The likelihood of individual recognition (Theberge & Falls 1967; Tooze et al. 1990) would allow wolves to distinguish between packmates and strangers" (Harrington & Asa 2003: 76). One might recall from earlier that Richard Ballantine (1996: 148) reported in his assessment of a particular wolf pack in Idaho that "Each wolf has its own distinctive howl." So, a person could distinguish one wolf voice from another, and from a more rigorous vantage point, Harrington and Asa also suggested the wolves likely can do that. John Theberge, in his informal column, summed up part of his piece by writing: "In short, the properties of the howls of individual wolves differ, as do the singing voices of humans" (Theberge 1971: 41).

Harrington & Asa (2003: 76), as well as Theberge (1971), earlier, made the comments quoted above when considering the different functions of howling. They covered several proposed functions, and were especially struck that there was "little doubt that howling helps coordinate movements among separated packmates" (Harrington & Asa 2003: 76). That statement was the context in which they suggested the reality of individual voice recognition. It is certainly worth calling attention to what is surely a noncontroversial point, that however much people often simply enjoy the music they make, human music can also serve functions that are analogous to the functions, reliably inferred or merely suggested, of wolf howling.

Another frequently proposed function of wolf howling is "the strengthening of social bonds" (Harrington & Asa 2003: 76). Harrington and Asa indicate that the evidence on this front is more equivocal, and likely stems in part from more intuitive reasons than is the case in assessing how it likely helps coordinate movements. Such reasons include the pattern whereby "chorus howls are highly contagious events within the pack" (Harrington & Asa 2003: 76). For present purposes, it is relevant to point out that the strengthening of social bonds is surely one function of music among people. And, certainly singing among people can also be highly contagious. To be sure, it is empirically problematical to document a social bonding function of chorus howls, just as it is difficult to document empirically an analogous function for social bonding through singing in people, though most people surely would not balk at that suggestion. The difficulty, of course, is that people grasp the effects of human singing in human terms, but are trying to infer the effects of wolf singing in wolf terms, an effort that can only be problematical.

Because of that factor, Harrington and Asa conclude their consideration of the possible social bonding function of howling in wolves in what seems to be an entirely reasonable way:

> Until evidence of a relationship between chorus howling and some objective measure of social bonding is produced, we must consider this role a hypothesis to be tested. However, whatever form of arousal wolves experience during chorus howling may serve to ensure that wolves howl when appropriate to fulfill other important functions, such as reunion or spacing. (Harrington & Asa 2003: 76)

In short, the suggested social bonding function of wolf howling remains equivocal, but even so, other more tangible functions are served simultaneously.

Be that as it may, considerations thus far have not yet adequately addressed the question of music vs. musics. To do that, a useful beginning point is to identify an organization called Raised by Wolves (RBW). RBW is a licensed nonprofit research center in New Mexico that emphasizes the scientific study of wolves. Their efforts are interdisciplinary, and include both qualitative and quantitative approaches to the study of behavior and social dynamics among wolves. The research center itself is home to some twenty wolves and dog–wolf hybrids that live in different enclosures there. RBW has put out, online, a multipage document, entitled "Essential Wolves," that deals with wolves generally, beyond the specific ones at the research center (Raised by Wolves, Inc. 2006). In that document the author makes the point that cultural characteristics have been documented among several different animal species. An intriguing suggestion follows when turning to wolves:

> It is not so far fetched to propose that similar evidence will be discovered when wolf songs are studied in the same fashion, revealing that different wolf packs sing different songs, and that there is an element of wolf songs that is culturally transmitted. (Raised by Wolves, Inc. 2006: 2)

Given that suggestion, one should recall Sullivan's (1978: 31) point, quoted earlier, that there are social and cultural traditions among wolf packs. There is certainly no logical reason why wolf songs shouldn't be part of that package, beyond the basic behavioral tendencies that Sullivan likely had in mind. It is also worth noting that the above RBW passage indicates that wolf songs hadn't been studied in such a fashion up to that point. And given that this statement was made by an organization devoted to the interdisciplinary study of wolves, there is no reason to suspect that RBW simply overlooked the key study/studies. The intriguing possibility here is that wolves exhibit recognizably different styles of group singing and not just as individuals, in much the same way that is true for people (e.g., a standard church choir versus a commercial pop band).

Wolves, Dogs, Music, and Spirituality

Much has been made in this volume about the sense of spirituality that people often have about both wolves and dogs. That was especially the case in Chapter 7, when dealing with burials, but also elsewhere at different points. To be sure, the variety of spirituality in play seems to vary across different settings. That is, Chapter 7 incorporated example after example in which dogs were buried in a manner that apparently reflects a sense of personal reverence. At the same time, other cases rather clearly reflected a broader kind of reverence. Ashkelon is a prime example of the latter, of spirituality writ large, as it were, but other cases involving dogs, including some in North America (e.g., the Hatch site in Virginia), seem to reflect that as well. The point is that, depending on circumstances, dogs could evidently be perceived in either light. Wolves, though, seem to have been perceived mostly in the broader sense. Even the one known wolf burial, covered in Chapter 7, seems indicative of a broader kind of symbolism that doesn't reflect, though it could well entail, a sense of personal reverence. Given their close domestic association with people, as members of human family units, it is not surprising that the kind of spirituality perceived in dogs often reflects personal reverence. That said, it is logical to suspect that the broader spiritual relevance routinely associated with wolves is conditioned, if only in part, by their musical output, as in howling. One might recall acclaimed music critic Harold Schonberg's assessment of a wolf howl as being "like a very lonesome, sentimental, fire siren with a soul" (Schonberg 1971: 34). That of course is a contemporary person's perspective, but it seems likely that earlier people, including the Native Americans referred to by Fritz et al. (2003), highlighted in Chapter 7, had an analogous assessment, corresponding to their particular view of such issues.

And having touched on spirituality one more time, it seems fitting to end here by once again calling on Ambrose Bierce for his take on the matter of personal reverence, the sense of reverence indicated by so many dog burials in different parts of the world and over such a long period of time.

Reverence, *n.* The spiritual attitude of a man to a god and a dog to a man. (Bierce 2000: 200)

Overall, the nature of the attachment between dogs and people is often so deep that as Alice Kuzniar (2006: 137) put the matter, "Indeed, it often seems as if the nature of the attachment to the dog can make its death more unbearable than that of a beloved human." This recognition of hers stands in curious contrast to her perception of just how modern people do or do not commemorate the passing of a dog, as highlighted in the next chapter. And certainly in this book's last chapter it is time to consider selected roles of dogs in modern times, one of which directly suggests continued personal reverence of people toward dogs.

10

ROLES OF DOGS IN RECENT TIMES

> When we think of dogs, we tend to think of animals that were selected
> for behavior performed in the service of people. Dogs pull sleds, guard
> property, herd sheep, guide the blind, track and retrieve game, and
> so on.
>
> Coppinger & Schneider 1995: 22

With that statement, Raymond Coppinger and Richard Schneider began
their substantial treatment of the many ways in which different dogs
have become specialized for the working tasks that they carry out.
Among the different roles that dogs play in recent times, Coppinger
and Schneider focused especially on certain ones. Specifically they dealt
mostly with sled dogs, sheep dogs, and other herding or livestock-
management dogs. The "and so on" from their passage captures the reality that these kinds of specialization are but a mere subset of the roles
carried out by "working dogs," the explicit target of their contribution
to a broader collection of pieces about dogs. As with their contribution,
there would be little point in even trying here to offer a comprehensive survey of the roles that dogs play in modern times. Of those that
they highlighted, one, sled dogs, is clearly a role played in the past,
emphasized here in Chapter 6. Another one, involving sheep dogs and
other herding or livestock-management dogs, is largely restricted to
recent times, since it involves livestock, and there is no way to know
with certainty if prehistoric peoples with livestock used dogs with their
stock, though it is logical to suppose that they may have at times. As
covered in earlier chapters, it is also probable that dogs were used for
assistance in hunting, just as they are in modern times. The simple fact is
that "Dogs play an astonishing range of roles in human society" (Udell
& Wynne 2008: 248).

Of the vast variety of roles that dogs play in society, the examples covered in this chapter emphasize, first and foremost, the distinctively strong social link between dogs and people. Such a tack is directly rooted in a theme that has been emphasized throughout this volume and is the basic rationale for the approach at this point. This effort does, however, bypass their participation in search-and-rescue missions, since that role was covered in Chapter 9, especially by means of Whittemore & Hebard's (1996) work. The coverage here starts instead with what seems at first to be counterintuitive, given the overall topical emphasis in this volume.

WAR DOGS

To be sure, it may initially seem odd to bring attention to the use of dogs in modern or recent warfare in a volume that stresses the social bond between dogs and people. In a curious way, though, it works. That is, dogs used in human warfare are committed, through whatever cause, to advance the interests of one group of people, at the direct expense of some other group. "War Dogs" is in fact the very title to a useful book by Michael Lemish (1996). He notes early on, in introducing the development of the use of dogs in modern military operations, that this development was not uniform. For example, he reports that "By far, the European countries showed a keener interest in developing and expanding upon the dog's role in warfare" (Lemish 1996: 5). Lemish's book is an account of how dogs were eventually used in the American military, and he contrasts the later use of dogs in the American military with the situation in European countries. Accordingly, he proceeds to outline the roles that dogs are known to have played in World War I. Within the span of that war, he starts with the "Red Cross dogs" (Lemish 1996: 12), dogs that had been trained to supply wounded soldiers with canteens of water or other spirits. First, of course, the dogs located these soldiers, and sometimes called attention to their presence by returning a helmet or cap, or some other item from the body of the wounded soldier. Different European countries' forces used such dogs, in slightly different but analogous ways. Interestingly, "Except for the United States, every country embroiled in the war considered dogs a valuable commodity" (Lemish 1996: 17). In addition, they were used as messengers, especially in the event of disrupted telephone line service. Messages were typically put in a small container that was attached to their collar, and sometimes they even carried messenger pigeons in a saddlebag. The dogs were especially adept at utilizing the elaborate trench systems that were in use during World War I. Late in this war, in 1918, American officials recommended that dogs be trained for a variety of military duties, a step that would have created the first official canine unit for the American army. The plan was dropped, however, and for the rest of the war the American military was dependent on the services of dogs maintained by the British and the French.

World War II and Later

It wasn't until World War II that the American military formally sought to use dogs in military operations. Eventually, what became known as the K-9 Corps was established. The results included some efforts, with mixed results, to involve dogs directly in combat activities. At any rate, Lemish charts the use of dogs in military operations after World War II, through the Korean War, and the Vietnam War as well. As the official historian for the Vietnam Dog Handler Association, Lemish became primarily involved in search-and-rescue operations involving the use of dogs. One aspect he consistently highlights is their role in boosting the morale of military personnel, by providing a source of companionship. During World War II operations, for example, "They provided a subtle impact on the war effort, as they comforted the human spirit in the highly charged atmosphere of combat" (Lemish 1996: 39). In keeping with the productive role of dogs in warfare, Figure 10.1 is a picture of a modern war dog memorial, a statue located outside of the building housing the Veterinary School clinic and hospital at the University of Tennessee, in Knoxville. It represents a Doberman pinscher and commemorates the service of those dogs in the Pacific theatre of World War II. That was one of several breeds that were used, and for a time it was widely believed, erroneously, that the Doberman was the official dog of the Marines. Although as a breed Dobermans were eventually regarded as not optimal for active military duty, some Dobermans stood out for their exceptional contributions. For example, in one engagement, some dogs were falling short, but "Other dogs picked up the slack – like Jack, a Doberman who scented an enemy machine-gun placement one hundred yards distant" (Lemish 1996: 101). In another example, "Carl (441), a Doberman pinscher, alerted his handler, Pvt. Raymond N. Moquin, a full thirty minutes before a Japanese attack. Fully prepared, the Marines wiped out the attacking Japanese" (Lemish 1996: 137).

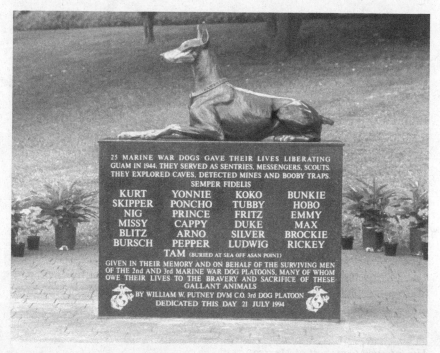

FIGURE 10.1. A World War II war dog memorial outside of the building housing the veterinary clinic and hospital at the University of Tennessee, in Knoxville, Tennessee. Photograph by Phillip D. Snow, College of Veterinary Medicine, University of Tennessee, Knoxville, Tennessee. This is a replica of the original, at the War Dog Cemetery in Guam.

Given Lemish's observation that European countries were receptive to such a role for dogs substantially earlier than the American military, it is useful to call attention to an aspect of European military dogs that highlights how social interaction positively influences the performance of modern military dogs. Specifically, a study by Lefebvre et al. (2007), based on an extensive questionnaire of dog handlers in the Belgian army, found that handlers who took their animal home and/or practiced sports with them, experienced dogs that were more obedient and less aggressive toward human family members compared to dogs that were not handled in such ways.

War Dogs in Iraq

It is also worth drawing attention to the fact that dogs serve in recent conflicts, in the twenty-first century. As one example, the conflict in Iraq has raged for several years, and in that war, dogs are sometimes victims of violence. One reporter, relating some information about this circumstance, put the matter this way: "Their services are so valued, though, that wounded dogs are treated much like wounded troops. 'They are cared for as well as any soldier,' says Senior Airman Ronald Harden, a dog handler in Iraq" (Donn 2007: A10). The larger context of that observation was an account of how regulations now allow for the adoption of war dogs by people, under some circumstances. One such circumstance related in Donn's report concerned the adoption of a German shepherd named Rex, by Air Force Technician Sergeant Jamie Dana. The two had been riding together in a military vehicle when a bomb blew its door off, barely injuring the dog, but nearly killing Dana. As of the writing of that report, Dana and her dog live together in Pennsylvania, and are quite happy with the arrangement. According to Dana, "He loves everybody. He sleeps beside my bed" (Donn 2007: A10). Those are modern dogs, but it is also useful to highlight a particular example of the European use of dogs in warfare that predates the modern era by some five centuries, one with some frankly troublesome implications.

The Spanish Conquest of Mexico

Most people are probably at least vaguely aware that in the 1500s, the Spaniards landed in present-day Mexico and set about making the indigenous peoples subject to the Spanish crown. At that time, the Aztecs had primary power throughout much of Mexico, operating out of their major city Tenochtitlan, at essentially the location of present-day Mexico City. Leon-Portilla (1992) has assembled a variety of accounts, as handed down by the indigenous peoples, some of which are represented in what are known as the Aztec Codices. The Aztec Codices, many of which are preserved in prominent libraries, constitute written accounts by observers themselves, or as handed down orally and later put in writing. Leon-Portilla (1992), who assembled and edited these accounts, explains more about their particular origins, as well as their historical dispositions. For purposes of this section, much of what follows is derived from Leon-Portilla's work. First, though, it is beneficial to set the stage. The year after Columbus's famous 1492 voyage he returned to the New World, and the Spaniards began by settling islands in the Caribbean. The indigenous people were to be regarded as Spanish subjects, not slaves. But in the Caribbean, the gold, much desired by the

Spaniards, played out quickly, and the local population suffered from introduced epidemic diseases, as well as the trauma of conquest. In the meantime, Mexico itself remained "undiscovered until 1517" (Berdan 1982: 163) when a Spanish ship, driven off course, ended up landing on Yucatan, and some forty men returned in 1519 with Hernán Cortéz. Through their far-ranging contacts, the Aztecs got word of the arrival of the Spaniards, heard accounts of them, and suspected that one of their (Aztec) gods had returned, as their own legends foretold.

The current Aztec leader, Motecuhzoma (more commonly known as Montezuma), sent some messengers to return with word of what and who arrived, some account of what they wanted, and what they had brought with them. Quite obviously, they brought ships, unknown to the indigenous peoples, strange clothing, and an appearance unlike people that the Aztecs had previously encountered, including pale skin. They also arrived with European weapons of that era, such as guns and cannons, the latter of which thoroughly terrified the Aztec messengers. But beyond these kinds of goods, they also brought horses and dogs. According to Leon-Portilla's transcribed account, the messengers reported to Motecuhzoma:

> Their dogs are enormous, with flat ears and long, dangling tongues. The color of their eyes is a burning yellow; their eyes flash fire and shoot off sparks. Their bellies are hollow, their flanks long and narrow. They are tireless and very powerful. They bound here and there, panting, with their tongues hanging out. And they are spotted like an ocelot. (Leon-Portilla 1992: 31)

Apparently they found the dogs, mastiff-types, rather ominous. Though Motecuhzoma's envoys wished to keep the Spaniards away from Tenochtitlan, Cortéz and his men marched inland. Before getting as far as Tenochtitlan, the Spaniards encountered some other notable places. One was the city of Cholula, not of direct Aztec affiliation, but part of the Aztec empire. At the time of Cortéz, it was second only to Tenochtitlan in size, with up to 100,000 people. As far as the linkage to Tenochtitlan, Aztec princes were formally anointed by a Cholulan priest. At any rate, Cholula became the site of a massacre at the hands of the Spanish. One of the original native sources, in the Codex Florentino, indicates regarding this massacre that "Their dogs came with them, running ahead of the column. They raised their muzzles high; they lifted their muzzles to the wind. They raced on before with saliva dripping from their jaws"

(Leon-Portilla 1992: 41). Cholula was destroyed, with quite a few of the immediate survivors of the massacre taking their own lives. Among those who didn't, it became widely believed that the God of these newer people must be more powerful than their own. At that point, some native peoples began operating in the service of the Spaniards.

Consequently the Spaniards, assisted by certain native peoples such as some from the city of Tlaxcala, eventually took over Tenochtitlan in 1521, after besieging the city for seventy-five days. Even after the Aztecs surrendered, the Spaniards set dogs upon some of them, and fed the dismembered bodies of some of the Aztecs to those dogs. Leon-Portilla (1992: 143–144) shares this part of some indigenous accounts of these happenings, including a picture from a historical graphic rendition known as the Proceso de Alvarado. Rather clearly, the use of dogs in the Spanish Conquest of Mexico evokes grisly rather than soothing images of how dogs relate to people. At the same time, it should be borne in mind that the dogs were those of the Spaniards, and basically did the bidding of Spaniards, or at least acted unencumbered by them. As suggested earlier, in a curious way, this use of dogs reflects their relationship with a particular group of people, in this case at the direct expense of another. This account is also consistent with Lemish's (1996) observation, noted earlier, that in more recent times Europeans have been especially receptive to the idea of using dogs for military purposes. At this point, though, it is time to shift gears and highlight a modern role for dogs in which they help save human lives, not end them.

DOGS IN THE HUMAN HEALTH SERVICES

Even though dogs were likely sometimes used in various ways in the past in the service of human medical interests, the role to be highlighted here concerns only the modern era. As we all know, modern medical practice advances regularly, as new and better techniques for preventing, diagnosing, and treating human health maladies are developed. Among those advances one must count the use of dogs, especially their keen olfactory sense, to help medical personnel detect human maladies that can evade detection even by trained medical people in the short run. A good example of this capability involves ways in which dogs have been used to detect human cancer of different kinds in its relatively early stages (e.g., Willis et al. 2004; Balseioro & Correia 2006).

Dogs are also being used as a model for understanding different aspects of human aging and its sometimes associated dementia (e.g., Studzinski et al. 2005). In a related vein, based on interviews with over 900 elderly people in urban settings in southern California, Siegel (1990) reported that dog owners felt more attached to their pets than owners of other pets did. She also found that the frequency of medical doctor contacts was lowest overall among those who owned dogs, as opposed to those with other pets. In reviewing some of the then-current information on this subject more than a decade ago, Beck (1996: 3) found that "While animal ownership generally had value, the most remarkable benefits to health were for those who own dogs."[1] In a genuinely troublesome setting, a study based on twenty-two residents with Alzheimer's disease in a special care unit in a midwestern city, where a dog was resident, resulted in this conclusion: "The presence of the resident dog decreased the occurrence of behavioral disturbances during daytime hours for the 1-month study period" (McCabe et al. 2002: 693).

Yet another dimension to this kind of modern use for dogs is exemplified by the role of these animals as "seizure-alert" dogs for owners who have epilepsy (Strong et al. 1999; Ortiz & Liporace 2005). In the Ortiz & Liporace study just cited, two patients who felt more secure with their dogs were admitted, accompanied by their dogs, to a hospital Epilepsy Care unit in Philadelphia. The results were equivocal, but the authors recognized that the restrictive environment of an in-hospital care unit likely affected both people and the dogs. Given the small sample size (two patients), it is not surprising that the authors recognized the need for larger samples, to investigate the phenomenon adequately. The equivocal nature of the direct evidence for this capability in dogs is apparent from other ongoing work (e.g., Kirton et al. 2004; J. Martin 2004; Litt & Krieger 2007). In fact, it has been suggested by certain people with firsthand experience in this area that "true alerting behavior is the result of the dog and human developing a strong bond, which can only evolve over time" (J. Martin 2004). In short, certain dogs

---

[1] It bears noting that afflicted people in general seem to benefit especially from dogs. For example, Spence & Kaiser (2002) have dealt explicitly with the benefits of companion animals for chronically ill children. Their account is a general endorsement of the role of such animals; however, in keeping with Beck's findings, dogs are the only animals mentioned specifically, either directly or parenthetically. It does seem that, in general, people relate to dogs in ways that set dogs apart from other animals.

may be responsive to subtle cues that operate effectively only when a particular dog and his/her person are fully familiar with each other. David Spencer (2007: 308) recently expressed the importance of the mutual bond this way:

> We do not understand how dogs might be able to do this, and we do not know much about how good they are at predicting seizures correctly. Mostly we know that people with seizure alert dogs report feeling "better" and "more confident" and enjoy the companionship of their seizure dog. (Spencer 2007: 308)

Another study that is consistent with those points involved the use of a dog in treating a patient with aphasia, and who thus had marked communication deficits (LaFrance et al. 2007). The patient was receiving speech and language therapy in an effort at rehabilitation. Overall, in addition to the necessary therapy, LaFrance and colleagues found that the presence of the dog was a positive influence in leading to overt social behavior, in both verbal and nonverbal modes. Dogs have also been used in efforts to accomplish therapeutic results with children and adolescents who have undergone inpatient psychiatric treatment. In a notable example, results of animal-assisted therapy (AAT) using dogs for sixty-one patients were evaluated in comparison to a group of thirty-nine comparable patients, for whom there was no AAT (Prothmann et al. 2006). A test was used on the patients to assess changes along four primary dimensions: vitality, intraemotional balance, social extroversion, and alertness. Patients that received AAT using dogs showed notable improvement in all dimensions of the test, whereas those who did not receive this sort of AAT did not show such changes. Prothmann and colleagues suggested in conclusion that dogs can be productively used in psychotherapeutic work with such young people.

In a particularly interesting study, Friedmann & Thomas (1998) examined the effects of pet ownership, and several other factors, on the one year survival rate for people who had experienced acute myocardial infarction (heart attack). The authors obtained information on 369 people, with a mean age of about 63 years old, of whom 112 owned pets. One key finding was that pet ownership, along with high social support, tended to predict survival, independent of the severity of the heart attack or different demographic and psychosocial factors. What is especially interesting about this study, given points that have been stressed in this volume, is that dog ownership and survival rate were significantly related, whereas cat ownership and survival rate were not.[2] Another appropriate example is the recent availability of extensive information on dog genome sequences which has, according to Björnerfeldt (2007: 26), "increased the value of the dog as a model organism for the study of human diseases."

Given the current topic, it is worth emphasizing that the use of dogs in the modern human health services is not a topic that is entirely restricted to the medical literature. In fact, on two consecutive Sundays (April 22 and 29, 2007), the Public Broadcasting Service (PBS) in the United States aired the two parts of a television show, collectively entitled "Dogs that Changed the World." The first installment, "The Rise of the Dog," dealt with the origins and early development of the dog. But the second installment, "Dogs by Design," dealt with the development of modern breeds, some of the problems associated with that, and then highlighted the use of dogs in modern medical practice. One segment of the show focused on an adolescent boy with diabetes, and a dog that was sensitive, apparently by means of olfactory or other personal cues, to an impending insulin-related diabetic seizure. As aired, the young man in question left no doubt as to his gratitude to his faithful dog. Also in the realm of media coverage, another good example harks back to what was covered above about the use of seizure-alert dogs with epilepsy patients. In the fall of 2007, at least one major newspaper ran a story on a diabetic teen whose dog Garbo was reportedly quite effective at sensing and alerting to dangerously low blood sugar levels in this young person (Mayer 2007). Again, whether olfaction is generally responsible for this success, or whether more subtle cues between person and dog are at work is difficult to say, but the result is the same: successful seizure-alert.

Finally, for this general topic, although it doesn't involve medical practice strictly speaking, another role of dogs is worth covering. In this case their role doesn't entail a specific functional activity, but might better be thought of as a source of inspiration. Specifically, Grandin & Johnson (2005) develop an intriguing case that there are many notable

---

[2] Clearly, some of the positive medical outcomes linked to dogs or other animals do not yet have a securely demonstrated basis. In fact, Kruger & Serpell (2006) have recently drawn attention to this point, stressing especially situations that involve mental health outcomes for patients. The purpose here is not to assert that the positive influence of dogs in different situations has been amply demonstrated in all cases, but to suggest that dogs are used more frequently than other animals in developing a case for the value of animals in medical situations, including the use of animal-assisted therapy for mental health patients.

parallels between how animals perceive the world and process information (visually), and how autistic people do those things. They call
on dogs as a major example supporting their more general case, a role
dogs share with other animals. At face value, such a perspective seems
harsh, even insulting, to people who cope with autism. That is, it seems
to suggest that autistic people are little more than animals. But what
immediately removes this work from that pejorative implication is that
the lead author, Grandin, is herself autistic. And they make a noteworthy case that autistic people tend to understand many animals better
than other people do. In keeping with that implication, Grandin, a successful animal scientist, serves as a consultant to the livestock industry,
seeking more humane ways of dealing with the animals.

Lateralization Studies

In keeping with the neurological implications of Grandin & Johnson's
(2005) work, dogs are also playing a prominent role in studies of lateralization tendencies (e.g., "handedness") in different animals, including
humans. For example, Aydinlioğlu et al. (2000: 129) noted near the
beginning of one study that "as stated by Tan (1987), animal studies
would be essential to understand the origin of human handedness."
The study they cited (Tan 1987) investigated handedness (right versus
left paw preference) in dogs using a relatively small sample of twenty-
eight test subjects. The protocol was to close an individual's eyes by
means of an adhesive plaster, and then monitor the use of left versus
right paw to remove it. Tan inferred a tendency for dogs to be right-
pawed (and therefore similar to people) based on this study, though
an important complication with that assessment turns out to be that
nineteen of the twenty-eight dogs were female, and nine male. Subsequent work has indicated why the predominance of females in Tan's
study renders such a generalization premature. One example is a study
by Poyser et al. (2006) in which seventy-nine dogs were subjected to
three different kinds of tests, including one like Tan's (1987). Of the
two others, one involved manipulation of food, the other of a ball. The
authors found a tendency for males to be left-pawed on a test involving
a ball, but "we also found that this tendency declined with repeated
presentation, indicating that the phenotype is labile" (Poyser et al. 2006:
219). Slightly earlier than that study, and also based on three different
testing procedures, Wells (2003) found that females preferred their right
paws on all three tasks. As Quaranta et al. (2004: 525) summed up the

situation, "dogs show paw preferences at the population level, with males preferring use of the left paw and females preferring the use of the right paw."

In addition to helping provide insights into behavioral tendencies and brain lateralization, studies that feature dog paw preferences also entail genuinely health-related implications. For example, in a study of immune functions in seventy-six dogs, Quaranta et al. (2004) found that immune parameters involving such factors as lymphocytes, leucocytes, and proteins appeared to be similar in right-pawed and ambidextrous dogs, but consistently different in left-pawed dogs. They summed up their findings by noting that "modulation of immune responses strongly depends on the direction (left or right) of lateralisation," (Quaranta et al. 2004: 524). In a more recent study, Branson & Rogers (2006) investigated the degree to which paw preference tendencies in dogs might be related to noise phobias in the animals. They initially explained their rationale for undertaking the study:

> Because human patients with alexithymia and PTSD [post-traumatic stress disorder] have been found to show a higher incidence of ambi-laterality and that this has been linked to impaired interhemispheric communication (Parker et al., 1999), we thought that investigating the relationship between noise phobia and brain lateralization in dogs might extend our understanding of the relationship between brain activity and emotional behavior. (Branson & Rogers 2006: 177)

Working with nearly fifty adult Australian household dogs, the authors used both owner questionnaires and recordings of fireworks and thunderstorms to assess the degree of noise phobia exhibited by the dogs. Overall, they found that dogs with weaker paw preferences showed the greatest reactivity. The relevance to people is clear from one of their conclusions: "dogs exhibit behavioral disorders that may be homologous to some psychiatric disorders in humans" (Branson & Rogers 2006: 182). One might note from that conclusion that they are suggesting homology, not mere analogy. In short, this dimension of work with dogs also underscores the degree to which dogs can be regarded as an appropriate model for people, by that highlighting in yet another way the remarkable affinity between dogs and people.

But to end this particular section, another role fits in, in an odd way, with a point about dogs that was once emphasized by James Serpell. Specifically, he called attention to the apparent incongruity of a situation

in which people commonly dote lovingly on their dogs, while simultaneously often expressing disgust with an animal that "eats shit, sniffs
genitals, and bites people" (Serpell 1995d: 254).[3] The first of these traits
merits attention in light of a study by Wasser et al. (2004). These authors
described how specially trained dogs in Alberta, Canada, were being
used to detect animal scats, for use in wildlife management studies. The
dogs were especially trained to detect scats of grizzly bear (*Ursus arctos*)
and black bear (*Ursus americanus*), over an area of some 5,200 square
kilometers of the Yellowhead Ecosystem in Alberta. The scats provided
information on levels of physiological stress and reproductive activities.
Thus this methodology of using dogs to detect wildlife scats is helping
to address a variety of issues in wildlife research and management. To
be sure, it is but one of the many, many purposes that dogs serve in
modern times. Be that as it may, though, other domestic animals have
played a variety of different roles as well. But one thing about dogs
stands out: "However, unlike other domesticated animals, dogs also
make excellent companions" (Coppinger & Schneider 1995: 22). It is
this reality that leads to the final role to be covered in this chapter.

THE MODERN MORTUARY ROLE OF DOGS

The phenomenon of dogs being ritually buried upon their deaths has
been a recurring focus during the course of this volume. But now we
reach this point in the last chapter, where it is time to deal meaningfully with the topic of modern dog cemeteries. They are, to put
the matter simply, numerous. Figure 10.2 provides a generalized map
of the world, with the locations of some modern dog/pet cemeteries
indicated. Mary Thurston (1996), whose book on dogs covers several
topics, devotes the entire next-to-last chapter (Chapter 11) to "Saying
Good-Bye" to dogs. In fact, in describing several of these modern facilities she tells us that "Victorian pet mortuary rites could be as formal

---

[3] What people will do to accommodate dogs can be remarkable. John Grogan (2005)
has captured that reality wonderfully in an irresistibly engaging book about his own
dog, billed as "the world's worst dog." This dog, Marley, did everything from fail
obedience school at one point, with instructions not to return, to make a shambles
of parts of Grogan's house, including outright destruction. Additionally, Marley did
indeed "sniff genitals," and showed a special appetite for soiled diapers. In spite of
Marley's routine misdeeds he won the hearts of Grogan's entire family, and genuinely
poignant moments sometimes ensued.

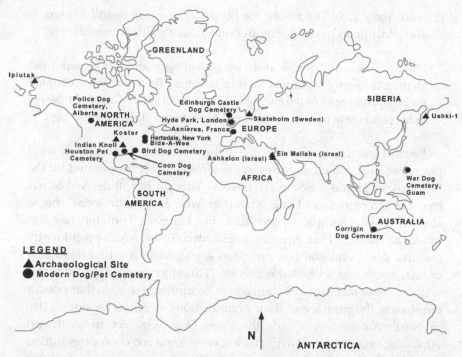

FIGURE 10.2. Generalized map of the world, showing the locations of some modern dog/pet cemeteries. Locations of some prehistoric dog burials, as indicated in Chapter 3 (Figure 3.8) are indicated as well, for reference. Such burials are dealt with more fully in Chapter 7, especially. Adapted and expanded from Morey (2006: 170, Figure 5).

as any concocted for humans" (Thurston 1996: 259). The oldest facility that is currently in operation in the United States is the Hartsdale Pet Cemetery in Hartsdale, New York, established in 1896 (Thurston 1996: 263). Significantly, this cemetery is also known as the Hartsdale Canine Cemetery, due to the prevalence of dogs that are buried there, including some that were associated with well-known people. This cemetery also has a war dog memorial, in the form of an inscribed stone monument, with a standing dog on its top. Unlike the war dog memorial in Figure 10.1, depicting a Doberman, this memorial was more generalized, the dog reportedly being modeled after a dog that passed by the designer, Walter A Buttendorf, almost every day, accompanied by its owner. In Guam, as a part of World War II activities, the Marines paid tribute to their war dog dead by means of a cemetery, developed in 1944

(Lemish 1996: 128). The reason for honoring war dog dead in Guam is that in addition to participating in hundreds of military patrols,

> The dogs also did double duty by providing nighttime security and alerting to enemy activity at least forty times. These actions accounted for an additional 66 of the enemy killed, as reported by Gen. A. H. Noble in his report of activities of the war dog platoons. (Lemish 1996: 127)

Beyond war dog memorials, one of the most prominent actively used pet cemeteries and crematories in the United States is operated by the Bide-A-Wee Home Association, in New York. In fact, Bide-A-Wee has two pet cemeteries on Long Island, as well as a "retirement" home for pets in Westhampton, New York. For her part, Thurston has also noted the Houston Pet Cemetery, in Houston, Texas, which prominently features dogs, and she has provided a picture of a fabric-lined dog casket, associated with that cemetery (Thurston 1996: 261). As of this writing, these facilities maintain active computer web sites that note or emphasize the primacy of their organizations' services for dogs. Also in North America, one finds the Coon Dog Cemetery in northwest Alabama, established in 1937, having now some 200 coon dogs buried there. As well, a bird dog cemetery in Waynesboro, Georgia, features some 75 headstones for field trial bird dogs. Meanwhile, in Canada, there is the Police Dog Cemetery in Innisfail, Alberta, located next to the facility where police dogs are trained. At that cemetery, the first official police dog of the Royal Canadian Mounted Police, called Dale of Cawsalta, is honored by means of a life-size dog statue in front of the kennels.

Outside of North America, the Hyde Park Dog Cemetery, in London, was established in 1888, and at one lavish pet cemetery, at Huntingdonshire, England, there were by 1926 more than 600 graves, including those for dogs (Thurston 1996: 256–257). One dog had been embalmed in Italy "and shipped back to England for interment in its own marble mausoleum complete with stained-glass windows" (Thurston 1996: 257). In addition, according to Thurston (1996: 256), the oldest public pet cemetery, established in 1896, is the dog cemetery in Asnières, France, and as of the mid-1990s, some 40,000 animals were buried there. There is also the Edinburgh Castle Dog Cemetery in Scotland, a small garden in use since the 1840s for the burial of officers' dogs, but it is not open to visitors. Meanwhile, the Corrigin Dog Cemetery was established in 1974, and is located in western Australia, not far from Perth.

But it is worth emphasizing again that the burial of dogs is not confined to those linked with the Victorian tradition. For example, Atsushi Nobayashi (2006) presented an ethnoarchaeological study of hunting activities carried out with dogs among the Tsou, an aboriginal group in Taiwan. Nobayashi related the following concerning the disposition of hunting dogs when they died:

> If the dog died, the hunter who owned it buried it somewhere near the village, usually in an area separate from the village cemetery. Pasuya [one of the hunters] stated that he had buried his dogs on his own premises. Other villagers did not bury hunting dogs in a single area. These hunters decided where they would bury the dogs on an individual basis. (Nobayashi 2006: 80)

Considering the attention commonly lavished on dogs when they die, a statement in a recent book is surprising: "Societal belief holds that candid expressive grief over a mere dog is improper. There are consequently few public rituals or customs to commemorate the loss" (Kuzniar 2006: 139). That statement comes in the final chapter, entitled "Mourning," in Kuzniar's book, *Melancholia's Dog*. One can only wonder which society she has in mind, since dogs are treated much like people at death in several societies in the world, and not just western societies. Certainly established customs are involved in most such practices, and Chapter 5 highlighted ongoing dog funeral service facilities in Korea. One can reasonably infer that Kuzniar is unaware of all this activity. That is especially surprising in view of her recognition, noted at the end of Chapter 9, that many modern people mourn the loss of a dog in much the same terms that they mourn the loss of a person. In any case, this focus on modern dog cemeteries leads rather directly to an important question.

ARE MODERN DOGS A RELIABLE GUIDE TO PREHISTORIC DOGS?

This question is legitimate to pose in bringing this volume to a close. At one level, the answer must be an unequivocal no. Modern dogs live under vastly different conditions than their predecessors, as known from the archaeological record. For that reason, modern dogs engage in many specific behaviors that were not known to prehistoric dogs. A conspicuous example is provided in this very chapter, namely the use of dogs in modern medicine. There was no modern medicine in the prehistoric past, by definition. This fact doesn't mean, though, that some past dogs might not have had an uncanny capacity for grasping when something was amiss with a person, for example, a debilitating illness like cancer. At the same time, and to reverse the equation, one can just as legitimately pose the question of whether or not modern people are a reliable guide to prehistoric people. Again, and for much the same reason, at one level the answer must also be an unequivocal no. Nevertheless, at another level, dogs are dogs, and people are people, a reality that seems to underlie one particular sense in which both modern dogs and people are legitimate guides to dogs and people of the past. That sense involves their mutual perceptions of each other. It is those enduring perceptions that underlie the careful burial of deceased dogs for thousands of years, in different parts of the world. To be sure, the coverage of dog burials in Chapter 7 revealed that dogs were buried under a variety of circumstances, sometimes even as sacrificial victims.

That particular reality, though, doesn't distinguish dogs from all other animals. Animals of different kinds, including some that apparently were used rather exclusively for economic purposes, have also been used in sacrificial capacities (e.g., Aaris-Sørensen 1981; Fadiman 1997: Chapter 9; Green 2001: Chapter 2).[4] Be that as it may, Chapter 7 covered numerous cases, in different parts of the world, where dogs were buried individually with great care. In the Old World, an especially dramatic example is Ashkelon, in present-day Israel, a cemetery area with more than 1,000 individually buried dogs. That enigmatic context (as well as Hatch, in Virginia) may represent a situation imbued with what was referred to at the close of Chapter 7 as "spirituality writ large." But many situations elsewhere, especially as highlighted in the New World, seem to be a matter of personal reverence. In those situations, modern dog burials, in cemeteries, are a valid guide to the past. They represent an enduring social bond between people and dogs, as is also evident in many settings in the past.

To simply anticipate the final words in this volume, the epilogue that follows takes a decidedly different approach from the preceding

---

[4] The sacrifice of other animals sometimes took on rather dramatic proportions. A conspicuous example is the Maussolleion at Halikarnassos, in present-day Turkey. Dating to more than 2,000 years ago, about 350 B.C. by the Christian calendar, this facility was an elaborate tomb for King Maussolos of Karia, in southwest Turkey. There, Aaris-Sørensen (1981: 94–9) inventoried skeletal remains, including complete skeletons, of 235 domestic animals, mostly sheep/goat, cattle, and chicken. From contexts there of unclear specific provenience, there were more such remains, "and dog, *Canis familiaris*, represented by 5 teeth" (Aaris-Sørensen 1981: 107). That was the sum total of dog remains listed.

chapters, which focused on shedding some light on the overall journey of the dog through time as a species. In contrast, the epilogue represents an individual dog's journey. The approach is personal, makes no pretense at scholarly standards of expression, and is, in my view, an ideal way to bring this volume to a close, given its overall emphasis. That said, I leave it for you, the reader, to encounter the final words.

EPILOGUE: ONE DOG'S JOURNEY

In these brief closing pages, I wish to highlight the amazingly strong bond that can exist between people and their dogs by relating a personal experience. The most effective way to do that, and keep the approach at this point truly informal, is by means of two personal letters that I wrote on the same day. In the fall of 2005 our dog Jingo, cherished by me and my wife Beth, died. It was a crushing blow, one from which we still haven't recovered fully. Two days after it happened, I composed the following letter to a former work associate and delivered it by way of email. I will let that letter speak for itself, and then comment briefly before sharing the one I wrote later that same day.

10/15/2005, Morning

Little Jingo's demise at the end is actually pretty touching, and that's what I'll tell you about today. On Tuesday, we got her home from the vet's in Lawrence, and Beth stayed with her all night downstairs. Jingo was just too weak to get up the stairs herself, and of course, she needed to be taken out periodically. She was regularly gasping for breath, and so weak it was painful to witness. On Wednesday, we got her through the morning, and in the afternoon, after she had managed to stagger around in the back yard a bit before she collapsed, it was a nice day, so both us of just sat out there with her, and had a sort of picnic, right there in the yard. There was a spot in the yard, where sometimes she'd stagger and plop down, "her spot," where I think she was trying to go to just die, as an animal will do. We'd just pick her up and bring her back in. That night, we both spread pads out on the downstairs floor, and stayed with her. She was in terrible discomfort, but was trying so hard to stay with us. Every now and then she'd yelp a little in pain, and we learned to take that as a cue that she needed to get outside for a few minutes. Basically, we lay there all night with her, all of us getting little sleep, and I just spent the night stroking her, holding her paw, and being there for her, showering her with love and affection. Bear in mind that I don't use the word "love" lightly. I loved that animal, and she loved being with us, and wanted so bad to stay with us. We honestly don't think she would have made it through the night without us being there, adoring her. Maybe a better way to say that is that she got through that night because we were there.

Anyway, she was supposed to get back to the Lawrence vet's in the morning, Thursday, and we dropped her off there, and Beth dropped me at the museum to do a make-up test thing. At the vet's they were going to do some more ultra-sound and X-ray things. When we finally got back there, in the early afternoon, they were all stumped. We had absolutely expected to hear that they'd found a huge growth, or something like that, that they couldn't do anything about, and we were ready to face having to put her down. But they still didn't know what it was. The vet, a guy we like a lot, said we should get her to this specialty animal hospital over in Overland Park. It's a 24-hour place, and he contacted them, with a specific guy in mind, and the idea was to get her there that day, and this guy would get her in the morning, when he came back in. So we drove over to K. C. right then, with little Jingo desperately hanging on in the back seat of our car, while I just kept my hand back there with her. She was pitifully weak, but every now and then she'd manage to raise her paw just a little, just to reach out to me, and tell me, in her dog way, that she wanted to stay with us. And as usual, we had to stop once on the way, when she yelped a little, and needed to get outside for a minute.

We got her there, and somebody outside picked her up and carried her in for us. They took her back to one of those kennels, put an oxygen tube in her, and started some IV nutrition. If she could get through the night, this specialist was supposed to see her in the morning. We insisted on seeing her before we left, and when we did, I just knew it was over. Despite the oxygen, her breathing was still horrible, and though we managed to get her to open an eye, there was nothing there. I managed to tap her paw and nose through the wire mesh, and there was zero response. She had given up, couldn't be with us, and life just wasn't worth trying to cling to anymore. On our way home, Beth tried to paint the best face on it, suggesting that she was just so relieved to finally be getting some oxygen, that she was just collapsed from exhaustion. I went along with that possibility, but I knew better, deep down.

About 11:30 at home that night, the dreaded phone call came. She was in cardiac arrest. They tried CPR for a few minutes, but called back and said it was not doing any good, and I just said to stop. Enough was enough. Let the poor animal die, and stop suffering. So she did, and we're going to have her cremated, and get some ashes back. One thing that was nice was that the vet clinic in Lawrence even sent us some flowers.

This was a once in a lifetime dog. There will never be another Jingo, and it'll be a long time before I have the heart to try bonding with another dog. She truly was our furry little baby, and for the first time in my life, I really know what true heartbreak is. For now, it's a different sign-off than usual.

Good-bye, Jingo
Darcy

This person responded, expressing an interest in knowing how it was that Beth and I came to have Jingo. So, I wrote and told her, as follows:

10/15/2005, Afternoon

Thanks for getting back to me so promptly, and I'll tell you some good things today, for a welcome change. Our furry little baby was literally delivered to us in our mailbox. Where we lived in Tennessee, outside of Knoxville, was in a really rural area, in a big trailer on quite a chunk of beautiful land. The rural road running close to it was a place that Beth used to jog. A standard roadside mailbox was at the end of our drive. One afternoon, she was coming back from her jog, heard a bit of noise from that direction, and looked inside. Somebody had driven by, and stuffed a helpless little puppy in our mailbox. That's how she became our mailbox dog. Beth pulled her out, and carried her up to our trailer. The little dog wriggled and whined, as you'd expect. That night, we at first tried a pad out in the main area, but the dog was inconsolable. So Beth went and got her, and carried her into her room, and put her on the bed. The dog was instantly comforted, and curled up next to her and slept. It was all over. The next day, Beth announced to me, in no uncertain terms, "I want to keep the puppy!" Who was I to argue, at least if I valued our marriage? We kept the puppy, who still needed a name. We fiddled with several possibilities, involving mailboxes and such, and ended up with Jingo, as the name of a horse that was in a story that Beth had read. And to add to the appropriateness of that name, in the story, the horse Jingo got its name as a result of the common exclamation "by jingo!" By jingo, our new dog became Jingo. So as you can see, our dear little Jingo was basically born in our mailbox, a little mutt that somebody didn't want and ditched. Well, we wanted her, and still do.

In some ways, today is harder than yesterday. Yesterday, we were basically in shock. Today, the reality is setting in. No longer will I be able to go in our living room, spot Jingo on the couch, and announce to her loudly, with her being deaf and unable to hear me, "Get your worthless butt of the couch, you little idiot, I want to lie down here!" Of course, what I'd do then is sit down next to her and cuddle her. This was the end of a huge era in our lives together. It's just so hard to picture life without her now, but that's the way it is.

I located the negative of the original picture that was used in the Jingo paper (you know which one), and we're going to have a large print made, and have it framed. Between that, and some of her ashes after she's cremated, a little of Jingo will always be with us. In fact, a lot of Jingo will always be with us.

That's all for now, and we're truly in mourning and trying to get over this huge obstacle. Beware if you're bonding like this with your dog, for she won't live forever, and it'll tear you up when she goes, however that happens. But, by all means, don't pass up the chance to have with her, if you can, what we had with our furry baby Jingo for 13 years, and still carry around with us. You're so right, it can't be replaced or recreated.

Missing Jingo,
Darcy

That, then, is one dog's journey, or rather its end point, and quite near its beginning. I left out the journey between those points, but you can surmise that it was a good journey for Jingo, and for us as well. Just after its end point, we did indeed have a large print made of her previously published picture, which is now displayed in our home next to a hand-carved box containing her ashes. Here we are a few years later, and we have once again opened our hearts to another little rescued mutt named Jezebel, who began her journey on a Tennessee back road.

Having shared that personal account, I now bring this volume to a close.

The End

APPENDIX A

Tables A.3 and A.4, comprising the bulk of this appendix, provide a compilation of raw data on canid specimens that serve as the basis for inferences covered especially in Chapter 3. Chapter 3 also covers the criteria used to generate metric observations, the bulk of this data base. Table A.3 is restricted to wild canid specimens, and Table A.4 provides the raw data on all archaeological dog specimens that are used. Many of the metric data are as appear in Morey (1990: Appendix A). There are, however, two exceptions, one of which is the domestic dogs (Table A.4). For reasons explained in the text of Chapter 3, a small part of these data was revised from those used in the 1990 study, and those revised data, presented here, are in Morey (1992: 188). The other data, on modern canids, were mostly collected in the late 1980s, and at several different institutions. As for the second exception alluded to earlier, some data on juvenile wolves recorded in the late 1980s were not part of Morey's (1990) earlier synthetic study and do not appear in that data base. They are, however, part of this study and appear in the present database. Several conventions for recording qualitative observations require advance explanation, starting with the provenience of the individual specimens themselves. For all the wild canids, a specimen is given a three letter acronym, followed by the six digit sequence that specifies its identity in a given collection. For example, case no. 1 in Table A.3, a wolf, is specimen JFB001360 in the ascending numerical sequence. In Table A.4, the archaeological domestic dogs also begin with a three letter acronym, followed by an identifier sequence, its format unique to each institution, and each series within that institution. The following is a key for the different three letter acronyms.

JFB   University of Minnesota, James Ford Bell Museum of Natural History, Minneapolis, Minnesota

SNM   Smithsonian Institution, National Museum of Natural History, Washington, D.C.

UIM   University of Illinois, Natural History Museum, Urbana, Illinois

KUM   University of Kansas, Natural History Museum, Lawrence, Kansas

ISM   Illinois State Museum, Springfield, Illinois

UKL   University of Kentucky, Museum of Anthropology, Lexington, Kentucky

MCL   University of Tennessee, Frank H. McClung Museum, Knoxville, Tennessee

IPA   Institute for Prehistory, University of Århus, Moesgård, Denmark

ZMC   University of Copenhagen, Zoological Museum, Denmark

The measurements are all in whole millimeters, except for the two dental measurements, which are to the nearest tenth of a millimeter. Under "Observations," *A* is an assigned age category, based especially on dental criteria. Table A.1 summarizes the five age categories, as adapted from Morey (1990: 81). The observational categories (descriptions) apply to all taxa, but the estimated chronological correlates (corresponding age) are just approximations pertaining directly only to the Gray Wolf, *Canis lupus*.

After age category, **S** is sex, for which M = male, F = female, and U = unknown. In fact, **U** in any column signifies unknown, unmeasured, or inapplicable, (in other words, no data). Finally, **SG** denotes subspecies or geographic region, the former applying to wild canids, the latter to archaeological domestic dogs. A number represents each one, and all are listed in that column for the sake of completeness, though in certain instances a single number applies to all examples of that species within an age category. They are simply written out in the Table A.1 heading in such cases, but for the other cases the changing numerical code prohibits that step. Accordingly Table A.2 presents the numerical code value for all others. Tables A.3 and A.4, the bulk of this appendix, follow immediately after Table A.2.

TABLE A.1. *Ontogenetic age categories for analyzed canid crania*

Age Category	Description	Corresponding Age
1: Puppy	Deciduous dentition unerupted, erupting, or in place	newborn–4 months
2: Juvenile	Deciduous dentition being replaced	4–6 months
3: Advanced Juvenile	Permanent dentition erupted Cranial sutures not fully sealed Bone very porous	6 months–1 year
4: Young Adult	Most sutures fully closed Most bone fully ossified No visible wear on teeth	1–2 years
5: Adult	All sutures fully closed Visible wear on teeth	Older than 2 years

TABLE A.2. *Numerical codes for certain taxa found in Tables A.3 and A.4*

SG Numerical Code	Taxon Common Name (wild canids), or Geographic Region/Country (Archaeological Dogs)	Taxon Latin Name
1	Unknown	
2	Gray Wolf	*Canis lupus lycaon*
3	Gray Wolf	*Canis lupus baylei*
4	Gray Wolf	*Canis lupus nubilus*
5	Gray Wolf	*Canis lupus irremotus*
6	Gray (Chinese) Wolf	*Canis lupus chanco*
7	Gray Wolf	*Canis lupus youngi*
8	Gray Wolf	*Canis lupus arctos*
21	Red Wolf	*Canis rufus rufus*
22	Coyote	*Canis latrans thamnos*
23	Golden Jackal	*Canis aureus indicus*
24	Golden Jackal	*Canis aureus lanka*
25	Golden Jackal	*Canis aureus lupaster*
26	Golden Jackal	*Canis aureus maroccanus*
27	Golden Jackal	*Canis aureus anthus*
28	Golden Jackal	*Canis aureus algirensis*
30	Kentucky	*Canis lupus familiaris* (Dog)
31	Alabama	*Canis lupus familiaris* (Dog)
32	Tennessee	*Canis lupus familiaris* (Dog)
33	Illinois	*Canis lupus familiaris* (Dog)
40	Northern Europe: Denmark	*Canis lupus familiaris* (Dog)
41	Northern Europe: Germany	*Canis lupus familiaris* (Dog)

TABLE A.3. *Raw data on all wild canid specimens used in analysis of metric data (Chapter 3)*

		Measurements (mm)								Observations		
Case	Specimen	CL	PL	PW	OI	MCW	IM2	P3	P4	A	S	SG
Canis lupus lycaon (**adults**)												
1	JFB001360	228	120	73	108	72	121	15.2	23.0	5	F	2
2	JFB001872	250	130	80	116	77	130	15.9	24.4	5	M	2
3	JFB012295	252	133	80	118	77	131	15.7	24.5	5	M	2
4	JFB013259	230	121	77	109	75	123	16.0	22.9	5	M	2
5	JFB005689	213	111	70	97	67	113	15.3	22.7	4	U	2
6	JFB012290	226	123	76	110	73	123	15.7	23.6	5	F	2
7	JFB001219	225	116	78	100	73	119	14.9	23.6	5	M	2
8	JFB012252	230	122	73	106	72	120	15.7	24.0	5	M	2
9	JFB001221	240	126	81	111	74	127	17.5	26.8	5	M	2
10	JFB013243	224	118	74	103	76	121	15.3	23.9	5	M	2
11	JFB012299	223	118	74	101	72	122	15.0	23.2	5	M	2
12	JFB010634	240	125	79	112	79	126	15.7	25.1	4	M	2
13	JFB012296	244	129	84	115	74	129	14.7	26.3	5	M	2
14	JFB010633	236	122	75	107	74	125	16.7	24.6	5	M	2
15	JFB013260	237	123	77	108	72	126	15.6	23.7	5	M	2
16	JFB012301	228	123	76	107	71	123	16.2	24.0	5	M	2
17	JFB001930	229	122	73	115	71	126	16.3	24.6	4	F	2
18	JFB003850	230	120	77	106	74	121	15.6	23.2	5	M	2
19	JFB001856	235	124	74	107	73	126	14.6	25.0	5	F	2
20	JFB012303	223	124	73	107	73	123	15.0	24.3	5	F	2
21	JFB001220	227	115	71	104	72	119	15.0	22.4	4	M	2
22	JFB013263	249	126	83	120	76	132	16.8	26.8	5	M	2
23	JFB013252	226	119	75	105	74	118	15.2	23.7	5	F	2
24	JFB013266	209	110	73	96	73	111	U	21.4	5	F	2
25	JFB013255	219	116	68	103	71	117	14.7	23.1	5	F	2
26	JFB013256	237	129	81	112	74	127	15.8	24.5	5	F	2
27	JFB012308	243	129	82	113	78	128	15.5	25.0	5	M	2
28	JFB012309	255	135	81	122	75	135	16.7	26.0	5	M	2
29	JFB013262	227	115	75	108	73	118	14.6	23.3	5	M	2
30	JFB013261	239	128	79	111	74	127	15.6	24.1	5	M	2
31	JFB013257	238	127	82	106	73	129	16.3	25.3	5	M	2
32	JFB010631	233	123	78	110	69	126	16.0	23.4	5	F	2
33	JFB013264	247	129	80	119	73	132	14.9	23.9	5	M	2
34	JFB013250	218	117	74	99	70	116	14.8	22.2	5	F	2
35	JFB010637	246	126	91	113	76	131	16.0	24.2	5	M	2
36	JFB012313	244	125	84	111	78	127	15.8	24.3	5	F	2
37	JFB012312	231	123	81	112	U	127	17.2	25.4	5	M	2
38	JFB001350	228	119	79	109	74	122	16.3	23.9	4	M	2
39	JFB012289	227	117	76	104	76	120	15.6	23.6	5	F	2
40	JFB012304	214	115	75	102	70	119	15.4	24.6	4	M	2
41	JFB012306	219	117	75	99	70	120	15.9	23.8	4	M	2

		Measurements (mm)								Observations		
Case	Specimen	CL	PL	PW	OI	MCW	IM2	P3	P4	A	S	SG
42	JFB013253	220	123	79	104	72	122	16.8	U	4	M	2
43	SNM289995	249	132	78	118	75	129	16.5	24.4	5	F	2
44	SNM265071	223	120	79	110	73	123	16.1	25.6	5	M	2
45	SNM258637	216	112	69	100	68	111	13.5	20.9	5	F	2
46	SNM243973	224	120	70	107	73	120	13.9	23.0	4	F	2
47	SNM243395	224	115	74	104	69	119	15.4	24.9	5	M	2
48	SNM242290	216	117	69	102	68	115	14.2	22.6	5	F	2
49	SNM170692	230	118	73	103	72	115	14.9	22.8	5	M	2
50	SNM530436	204	106	69	095	68	109	14.1	22.4	4	F	2
51	SNM530435	216	113	72	100	72	117	15.4	24.6	4	F	2
52	SNM513676	215	111	73	100	71	114	14.6	23.2	4	F	2
53	SNM529877	212	110	70	99	70	114	14.7	22.3	4	F	2
54	SNM512026	233	124	74	109	U	124	16.7	23.7	4	U	2
55	SNM512009	231	120	77	105	72	122	17.2	23.6	4	U	2
56	SNM512007	233	122	73	109	74	122	14.9	15.9	4	U	2
57	SNM347921	230	111	72	101	72	117	15.0	23.2	4	F	2

Canis lupus baylei (**adults**)

58	SNM224484	231	117	76	110	73	123	15.0	22.9	5	M	3
59	SNM224485	212	106	73	99	72	114	13.9	23.1	4	F	3
60	SNM225394	210	111	71	101	72	113	14.8	22.9	5	U	3
61	SNM228269	220	114	72	104	73	117	13.9	24.6	5	M	3
62	SNM231320	217	115	73	101	73	113	14.1	24.1	5	M	3
63	SNM231322	224	114	76	103	72	117	14.6	22.7	5	M	3
64	SNM231323	217	111	74	105	76	116	13.8	22.9	4	F	3
65	SNM231324	225	117	74	107	72	123	15.1	24.4	5	M	3
66	SNM231532	218	115	74	102	73	115	14.5	23.0	5	M	3
67	SNM231533	217	112	75	101	73	117	15.8	23.4	4	M	3
68	SNM231534	216	111	74	99	66	114	14.5	22.6	4	M	3
69	SNM231536	210	107	73	101	70	113	14.4	22.3	5	F	3
70	SNM232446	235	124	78	111	73	124	U	24.8	5	M	3
71	SNM002193	203	104	69	96	66	109	12.5	21.7	5	U	3
72	SNM285754	233	119	76	111	74	124	15.5	24.3	5	M	3
73	SNM094728	210	110	73	102	67	118	15.9	24.5	5	F	3
74	SNM003335	220	114	73	106	74	119	14.5	24.3	5	U	3
75	SNM167989	208	111	71	96	68	111	13.9	22.9	4	F	3
76	SNM095752	202	105	67	94	69	110	12.3	21.2	4	U	3
77	SNM098307	222	114	69	105	70	117	13.9	23.9	5	M	3
78	SNM098311	216	116	70	104	72	118	14.0	22.8	5	F	3
79	SNM098313	225	116	72	107	74	122	13.3	23.6	5	M	3
80	SNM099668	209	108	69	99	73	113	14.7	24.0	4	U	3

(*continued*)

TABLE A.3 *(continued)*

Case	Specimen	CL	PL	PW	OI	MCW	IM2	P3	P4	A	S	SG
											Measurements (mm) → Observations		
81	SNM117059	232	121	77	108	77	123	16.0	24.5	5	M	3
82	SNM117060	224	117	74	105	75	118	13.8	U	5	M	3
83	SNM117061	214	113	73	100	73	116	14.1	23.0	4	F	3
84	SNM117062	222	113	72	105	74	120	14.9	23.8	5	F	3
85	SNM117542	219	112	70	105	72	119	14.9	22.6	4	F	3
86	SNM170556	228	115	75	107	72	120	15.7	24.7	5	M	3
87	SNM235089	225	116	74	109	71	122	15.3	24.5	4	M	3
88	UIM001048	218	113	71	101	70	114	U	23.8	5	F	3
89	UIM001153	227	116	77	104	73	119	U	24.0	5	M	3
90	UIM001156	221	115	74	106	73	117	U	23.5	5	M	3
91	UIM001160	221	120	73	107	74	120	U	24.5	5	M	3
92	UIM001161	215	114	73	102	72	115	U	23.8	5	U	3
93	UIM001163	210	109	74	99	69	113	U	22.5	5	U	3
94	UIM001165	215	111	73	100	72	112	U	23.5	5	U	3
95	UIM001164	207	108	70	93	66	108	U	23.7	5	M	3
96	UIM??????	211	109	72	98	70	111	U	24.0	5	F	3
97	UIM001149	206	106	70	99	67	112	U	23.0	5	F	3
98	UIM004105	209	108	73	100	69	112	U	22.6	5	F	3
99	UIM004106	213	110	71	99	71	115	U	22.6	5	F	3
100	KUM076473	214	113	71	101	70	114	U	23.1	5	F	3
Canis rufus rufus (**adults**)												
101	SNM266506	193	102	57	91	61	105	12.8	19.6	4	F	21
102	SNM266173	189	97	59	86	58	101	12.0	19.9	5	U	21
103	SNM265645	206	108	63	96	64	109	12.6	20.2	5	M	21
104	SNM265599	192	99	55	89	59	103	12.3	19.9	4	M	21
105	SNM224531	209	107	60	98	64	112	12.6	22.3	4	M	21
106	SNM224972	204	106	61	96	63	108	13.7	21.2	5	M	21
107	SNM224973	195	103	56	91	60	104	12.5	20.0	4	M	21
108	SNM224974	198	103	59	94	62	106	13.0	21.1	4	M	21
109	SNM225366	196	103	60	95	59	108	11.8	20.4	5	M	21
110	SNM225367	204	107	58	96	61	109	12.3	20.3	5	M	21
111	SNM227899	202	108	58	96	61	111	12.2	19.3	5	M	21
112	SNM227900	199	103	61	U	60	106	13.0	21.4	5	M	21
113	SNM228069	202	108	58	95	61	110	13.2	21.6	5	M	21
114	SNM228089	195	100	61	93	62	103	12.3	20.5	4	F	21
115	SNM228239	207	108	58	101	61	113	12.4	21.8	5	M	21
116	SNM228517	198	103	62	93	60	108	12.7	21.3	5	M	21
117	SNM251084	218	110	63	100	63	114	13.9	21.4	5	M	21
118	SNM251085	225	113	63	U	63	118	13.0	22.0	5	M	21
119	SNM251086	206	105	62	93	64	109	13.3	21.6	5	F	21
120	SNM261609	204	106	56	96	60	111	12.8	22.4	4	M	21
121	SNM261753	201	104	62	96	61	109	12.5	20.7	5	M	21

Case	Specimen	Measurements (mm)								Observations		
		CL	PL	PW	OI	MCW	IM2	P3	P4	A	S	SG
122	SNM262105	207	106	61	93	61	107	13.1	19.7	5	F	21
123	SNM262106	200	103	59	90	60	107	12.4	19.1	4	F	21
124	SNM265458	209	108	59	98	62	110	12.7	19.4	5	F	21
125	KUM024879	198	103	59	92	61	106	U	21.0	5	M	21
128	KUM060148	201	105	62	95	62	109	U	21.4	5	F	21
127	KUM024878	202	105	65	98	62	110	U	21.9	5	M	21
128	KUM060149	194	107	61	94	58	106	U	20.9	5	M	21
129	KUM054820	207	106	63	96	61	112	U	21.8	5	M	21

Canis latrans thamnos (**adults**)

Case	Specimen	CL	PL	PW	OI	MCW	IM2	P3	P4	A	S	SG
130	ISM686535	180	96	57	87	58	100	12.8	20.7	5	M	22
131	ISM001631	182	94	58	83	58	98	11.7	19.0	5	U	22
132	ISM614755	174	88	53	78	57	95	11.8	19.6	5	F	22
133	ISM 683716	158	81	51	71	54	82	11.7	17.2	5	F	22
134	ISM614240	170	86	55	77	55	93	11.9	20.6	5	F	22
135	ISM687981	187	93	54	86	59	102	11.9	19.3	5	F	22
136	ISM614378	183	95	54	87	58	100	11.9	19.0	5	F	22
137	ISM614204	179	92	54	84	59	89	11.9	18.7	5	U	22
138	ISM614754	189	98	61	90	62	101	14.1	21.2	5	M	22
139	ISM688233	186	102	57	88	59	103	11.2	19.0	5	U	22
140	ISM614569	180	91	55	85	59	97	11.9	18.7	5	F	22
141	ISM614472	185	97	57	86	60	101	11.6	20.4	5	F	22
142	ISM614379	203	106	63	97	63	109	12.9	19.3	5	M	22
143	ISM683778	188	98	57	88	60	102	11.9	20.6	5	M	22
144	ISM687966	195	101	59	91	58	106	12.5	21.2	5	U	22
145	ISM614705	190	98	55	90	60	103	12.1	19.3	5	U	22
146	ISM687968	183	94	53	84	56	98	11.3	20.0	5	M	22
147	ISM614674	184	96	57	87	58	102	12.7	20.9	5	M	22
148	ISM687748	195	101	60	92	65	104	12.3	20.0	5	M	22
149	ISM690935	193	100	61	89	61	105	13.1	22.1	5	U	22
150	ISM614658	184	96	56	89	60	101	12.5	20.5	5	U	22
151	ISM614390	178	91	55	80	57	96	11.7	18.4	5	U	22
152	ISM614666	185	97	58	89	60	104	11.7	19.4	5	U	22
153	ISM688235	186	98	59	88	61	104	11.5	21.1	5	U	22
154	ISM614262	186	96	59	87	59	103	13.7	21.5	5	M	22
155	ISM614254	192	100	57	89	61	104	12.3	20.1	5	M	22
156	ISM614382	202	105	62	97	63	108	13.8	20.2	5	M	22
157	ISM614731	167	87	49	78	55	93	10.0	17.8	5	F	22
158	ISM614213	176	91	55	82	59	98	12.4	18.5	5	U	22
159	ISM614562	183	95	53	85	57	100	12.3	20.6	5	M	22
160	ISM614463	189	95	55	89	59	104	12.7	19.5	5	M	22

(*continued*)

TABLE A.3 *(continued)*

Case	Specimen	\[Measurements (mm)\] CL	PL	PW	OI	MCW	IM2	P3	P4	\[Observations\] A	S	SG
161	ISM687983	177	92	58	82	59	96	11.5	19.8	5	U	22
162	ISM614675	184	96	56	83	57	100	13.1	20.0	5	M	22
163	ISM614398	182	92	54	82	59	97	12.3	19.5	5	U	22
164	ISM614242	189	99	60	88	60	104	13.3	19.9	5	M	22
165	ISM614241	178	92	57	82	58	97	12.6	19.8	5	F	22
166	ISM614263	184	92	59	88	58	100	12.1	19.3	5	F	22
167	ISM687967	184	94	55	85	59	99	11.7	20.0	5	M	22
168	ISM614628	187	96	55	88	57	101	12.6	20.0	5	M	22
169	ISM001364	183	95	55	86	57	100	11.8	19.6	4	U	22
170	ISM001356	198	104	59	94	63	108	13.6	21.3	4	F	22
171	ISM001665	167	84	51	78	59	90	11.0	18.6	4	U	22
172	ISM001600	189	98	59	89	59	104	14.3	21.8	4	U	22
173	ISM001631	186	99	58	88	59	102	12.2	21.5	4	U	22
174	ISM000549	185	96	55	89	60	102	12.0	19.5	4	M	22
175	ISM001357	193	96	56	89	61	102	11.8	19.7	4	M	22
176	ISM614565	175	90	54	81	58	98	12.5	20.4	4	F	22
177	ISM614265	173	87	55	78	56	93	11.3	19.3	4	F	22
178	ISM614625	176	91	53	83	60	96	11.9	18.3	4	U	22
179	ISM614277	192	101	58	91	60	105	13.3	21.3	4	M	22
180	ISM614623	186	94	57	88	60	101	13.2	19.5	4	F	22
181	ISM685944	174	90	54	81	55	96	11.1	19.5	4	F	22
182	ISM684394	183	94	56	86	58	101	12.3	20.8	4	F	22
183	ISM614474	177	95	53	82	56	95	11.9	19.2	4	F	22
184	ISM614729	178	90	53	80	54	94	12.8	19.6	4	U	22
185	ISM614261	190	98	60	86	61	102	12.7	20.2	4	F	22
186	ISM614750	185	94	56	84	60	100	10.8	21.1	4	M	22
187	ISM614532	186	97	53	85	58	103	12.0	20.2	4	M	22
188	ISM686003	182	94	57	83	57	97	11.7	19.2	4	M	22
189	ISM614389	186	95	56	86	59	99	12.6	20.3	4	M	22
190	ISM614627	184	95	56	85	58	100	12.8	19.5	4	F	22
191	ISM689929	177	92	57	81	56	98	13.1	19.1	4	M	22

Canis aureus (**adults**)

Case	Specimen	CL	PL	PW	OI	MCW	IM2	P3	P4	A	S	SG
192	SNM290135	149	76	51	67	53	82	9.2	U	5	M	23
193	SNM173280	150	76	47	66	52	U	8.6	16.0	5	F	23
194	SNM173283	157	78	50	68	54	82	9.5	16.7	5	F	23
195	SNM173284	153	78	50	66	52	81	10.3	16.5	4	F	23
196	SNM256727	155	78	55	67	55	80	10.3	18.2	5	F	24
197	SNM321958	192	101	61	88	59	101	11.4	18.8	5	M	25
198	SNM321956	180	90	56	82	58	92	10.0	17.3	5	M	25
199	SNM321954	176	91	55	81	57	95	10.6	16.9	5	F	25
200	SNM322834	150	79	49	69	51	83	10.5	17.1	5	M	25
201	SNM399436	155	80	52	68	54	86	10.4	17.8	4	U	1

Case	Specimen	Measurements (mm)								Observations		
		CL	PL	PW	OI	MCW	IM2	P3	P4	A	S	SG
202	SNM322833	159	82	50	72	53	87	10.4	18.1	4	M	25
203	SNM399433	157	80	47	70	52	86	9.6	16.8	5	F	1
204	SNM399432	163	83	54	74	54	89	9.8	17.5	5	M	1
205	SNM410910	151	77	49	68	50	83	9.6	16.1	5	F	26
206	SNM410911	151	81	45	69	49	82	9.1	14.9	4	F	26
207	SNM476031	157	81	49	71	54	U	9.7	16.7	5	F	26
208	SNM486165	146	79	49	69	51	83	10.6	18.0	5	F	26
209	SNM486167	152	80	48	69	52	83	9.6	17.1	5	F	26
210	SNM476030	146	77	47	65	51	79	10.0	16.2	5	M	26
211	SNM378686	147	76	44	66	49	79	9.5	15.6	4	F	27
212	SNM476034	163	85	53	73	55	90	9.8	18.5	4	M	28
213	SNM378688	158	82	50	70	53	84	10.6	17.1	5	M	27
214	SNM378685	157	81	51	69	52	84	10.6	18.6	5	M	27
215	SNM378684	155	80	50	68	51	82	10.8	17.6	5	F	27
216	SNM378683	156	81	48	70	53	84	9.8	17.5	5	F	27
217	SNM476856	155	80	50	68	54	85	9.7	17.4	5	F	28
218	SNM399435	160	83	52	72	55	88	11.0	18.1	5	M	1
219	SNM321951	183	93	58	85	59	97	11.6	18.5	5	M	25
220	SNM399434	156	82	49	71	54	85	10.3	17.1	4	F	1
Canis lupus (**juveniles**)*												
221	JFB000091	217	115	74	104	72	118	15.3	23.1	3	F	2
222	JBF005019	216	114	72	101	71	115	15.8	24.0	3	F	2
223	JFB010632	182	98	69	83	66	101	U	22.6	3	F	2
224	JFB012726	203	113	71	95	66	116	15.8	22.9	3	F	2
225	JFB013254	187	100	73	87	65	109	U	U	2	M	2
226	JFB012302	207	110	68	99	67	115	16.0	24.6	3	U	2
227	JFB012305	194	104	68	89	65	111	13.9	22.4	3	F	2
228	JFB012307	188	103	71	88	64	107	U	U	2	F	2
229	JFB013249	187	103	70	86	62	109	U	U	2	F	2
230	JFB013258	184	98	70	83	63	107	U	U	2	M	2
231	SNM529878	194	103	64	93	66	108	U	U	2	F	2
232	SNM347920	196	104	65	91	69	108	14.4	21.5	3	F	2
233	SNM347916	208	111	72	96	72	112	13.8	22.5	3	M	2
234	SNM243394	191	102	67	90	67	106	U	U	2	F	2
235	SNM243393	194	107	63	92	67	108	15.0	22.1	3	F	2
236	SNM242291	217	114	68	103	70	119	15.8	23.8	3	F	2
237	SNM022371	206	110	72	95	69	112	13.5	24.3	3	F	2
238	SNM156838	230	115	74	106	70	118	14.8	22.7	3	U	2
239	SNM512021	218	113	72	100	71	115	14.5	21.8	3	M	2
240	SNM012314	215	115	72	102	72	100	15.3	21.9	3	F	2

(*continued*)

TABLE A.3 *(continued)*

Case	Specimen	CL	PL	PW	OI	MCW	IM2	P3	P4	A	S	SG
										Observations		
241	SNM347919	198	103	67	93	68	107	14.3	22.2	3	M	2
242	SNM170567	223	116	74	109	73	123	16.9	24.7	3	M	2
243	SNM224172	212	109	76	99	73	116	15.6	24.4	3	M	3
244	SNM224187	203	105	70	96	70	113	14.8	24.3	3	F	3
245	SNM117064	226	120	75	109	74	123	15.6	24.9	3	U	3
246	SNM098328	223	112	74	104	75	123	15.1	24.8	3	M	3
247	SNM168427	113	63	54	49	54	61	U	U	1	M	4
248	SNM036588	58	32	31	22	33	31	U	U	1	M	4
249	SNM036559	58	32	30	22	34	31	U	U	1	F	4
250	SNM036560	58	33	32	22	35	31	U	U	1	M	4
251	SNM036557	56	32	29	21	32	30	U	U	1	F	4
252	SNM036556	56	32	30	21	32	29	U	U	1	F	4
253	SNM147203	75	42	42	28	43	41	U	U	1	M	5
254	SNM147204	75	44	43	30	45	43	U	U	1	U	5
255	SNM147195	74	43	43	29	43	41	U	U	1	U	5
256	SNM147205	101	58	49	43	49	53	U	U	1	F	5
257	SNM198463	197	106	66	99	69	112	13.1	25.0	3	U	6
258	SNM232440	143	78	59	64	60	75	U	U	1	M	7
259	SNM232439	139	73	55	62	57	71	U	U	1	M	7
260	SNM232442	138	72	57	59	58	69	U	U	1	M	7
261	SNM232441	135	70	57	60	56	61	U	U	1	F	7
262	SNM231338	93	54	45	38	53	U	U	U	1	M	7
263	SNM231340	89	49	46	36	47	U	U	U	1	M	7
264	SNM231341	86	49	46	34	49	U	U	U	1	M	7
265	SNM301317	225	118	78	110	72	123	19.2	27.7	3	U	8
266	SNM291011	200	106	75	96	70	115	U	U	2	M	8
267	SNM291008	233	122	82	105	76	126	17.7	26.7	3	M	8

[*] Clearly, some of the measurement points designed around adults (see Chapter 3, especially Figure 3.1, page 32) are not directly applicable to some subadults. This is especially true for dimensions that utilize tooth locations as reference points. But subadults falling into age categories 1 and 2 have no permanent dentition, in some cases no dentition. Accordingly, defining criteria for measurements like PW or IM2 are not directly applicable. For PW, I took the measurement as the widest dimension of the palate, corresponding approximately to PW for adults. Similarly, IM2 for subadults was the maximum length of the lateral margin of the palate, where the teeth eventually would have erupted in a living animal.

TABLE A.4. *Data on all archaeological domestic dog specimens used in analysis of metric data (Chapter 3)*[a]

Case	Specimen	CL	PL	PW	OI	MCW	IM2	P3	P4	A	S	SG
						Measurements (mm)				Observations		

North America: Cases 268–314 (47 total)

Indian Knoll, Kentucky (15OH2)

Case	Specimen	CL	PL	PW	OI	MCW	IM2	P3	P4	A	S	SG
268	UKL 1–4	156	80	56	70	54	85	10.7	16.7	5	U	30
269	UKL 1–24	(142)	74	54	63	54	80	10.3	16.5	4	U	30
270	UKL 1–26	136	73	49	62	51	77	10.9	16.7	5	U	30
271	UKL 1–30	125	65	49	54	47	69	9.6	15.6	5	U	30
272	UKL 1–35	144	76	53	66	51	82	11.0	18.1	4	U	30
273	UKL 1–55	133	69	54	56	51	73	10.4	16.8	5	U	30
274	UKL 1–56	154	80	56	66	55	84	11.1	17.3	5	U	30
275	UKL 1–60	(136)	U	53	61	53	77	9.7	16.5	4	U	30
276	UKL 1–117	144	75	54	64	54	80	11.0	16.4	5	U	30
277	UKL 1–129	153	U	57	69	56	83	10.8	16.6	5	U	30
278	UKL 1–130	153	81	56	70	53	85	11.0	16.3	5	U	30
279	UKL 1–132	142	75	48	64	51	80	9.7	16.2	5	U	30
280	UKL 1–133	147	U	53	69	U	81	10.3	16.5	5	U	30
281	UKL 1–134	165	85	57	73	59	87	10.2	17.2	4	U	30

Carlston Annis, Kentucky (15BT5)

Case	Specimen	CL	PL	PW	OI	MCW	IM2	P3	P4	A	S	SG
282	UKL 1–146	143	75	51	67	53	80	10.2	15.6	5	U	30
283	UKL 1–148	157	83	54	72	56	87	10.6	16.6	5	U	30
284	UKL 1–150	141	76	53	64	51	(80)	10.9	17.4	5	U	30
285	UKL 1–151	154	79	55	69	55	84	10.4	16.0	5	U	30
286	UKL, no number	(153)	81	56	(70)	55	84	10.9	16.9	5	U	30

Ward, Kentucky (15MCL11)

Case	Specimen	CL	PL	PW	OI	MCW	IM2	P3	P4	A	S	SG
287	UKL 1–70	157	80	55	(72)	56	84	10.7	17.6	4	U	30
288	UKL 1–72	141	73	54	64	52	U	10.2	16.4	4	U	30
289	UKL 1–98	148	80	55	U	51	82	9.5	16.6	5	U	30
290	UKL 1–99	(163)	U	57	72	57	87	10.7	16.5	5	U	30

Chiggerville, Kentucky (15OH1)

Case	Specimen	CL	PL	PW	OI	MCW	IM2	P3	P4	A	S	SG
291	UKL 1–61	151	78	52	71	U	85	10.8	17.2	5	U	30

Read, Kentucky (15BT10)

Case	Specimen	CL	PL	PW	OI	MCW	IM2	P3	P4	A	S	SG
292	UKL 1–144	160	U	(59)	U	58	87	U	18.2	5	U	30

Perry, Alabama (1LU25)

Case	Specimen	CL	PL	PW	OI	MCW	IM2	P3	P4	A	S	SG
293	UKL 2–43	144	74	51	64	53	79	10.0	15.6	5	U	31
294	UKL 2–45	145	75	53	65	53	81	11.0	16.4	5	U	31
295	UKL 2–52	146	76	52	65	56	78	9.3	15.3	5	U	31
296	UKL 2–53	141	73	54	61	52	(80)	10.2	15.8	5	U	31
297	UKL 2–55	129	70	49	59	49	73	9.7	15.0	5	U	31

(continued)

TABLE A.4 *(continued)*

		Measurements (mm)								Observations		
Case	Specimen	CL	PL	PW	OI	MCW	IM2	P3	P4	A	S	SG
298	UKL 2–73	142	73	53	64	50	80	10.8	17.5	5	U	31
299	UKL 2–82	154	79	59	69	U	82	10.4	17.4	5	U	31
Whitesburg Bridge, Alabama (1MA10)												
300	UKL 40–25	156	U	57	72	55	86	11.3	17.6	5	U	31
Flint River, Alabama (1MA48)												
301	UKL 40–7	156	82	57	72	56	85	10.1	15.6	5	U	31
302	UKL 40–9	167	86	57	78	58	91	10.6	17.0	5	U	31
Little Bear Creek, Alabama (1CT8)												
303	UKL 2–93	148	U	(56)	70	52	84	9.8	16.6	5	U	31
304	UKL 2–97	156	83	58	71	U	86	10.9	17.1	5	U	31
Mulberry Creek, Alabama (1CT27)												
305	UKL 2–3	152	79	50	68	53	82	9.5	16.1	5	U	31
306	UKL 2–5	151	78	54	68	55	86	10.9	17.7	5	U	31
307	UKL 2–9	157	82	(56)	70	55	86	11.4	18.1	5	U	31
Bailey, Tennessee (40GL26)												
308	UTK 86–157	141	72	(52)	61	55	78	9.0	15.3	5	U	32
Cherry, Tennessee (40McL84)												
309	MCL 84–22	151	79	52	68	51	82	9.8	15.6	5	U	32
310	MCL 84–49	133	69	49	59	50	73	9.3	14.1	5	U	32
Eva, Tennessee (40McL6)												
311	MCL 6–16	149	80	57	67	53	82	10.9	U	5	U	32
312	MCL 6–49	153	82	57	72	54	83	10.3	16.9	5	U	32
Koster, Illinois												
313	ISM F2256	165	85	61	77	59	91	U	18.5	5	U	33
Modoc, Illinois (11R5)												
314	ISM B-2	162	86	59	75	56	90	11.8	17.9	5	U	33
Europe: Cases 315–332 (18 total)												
Ringkloster, Denmark												
315	IPA 1592AVEN	156	80	56	71	57	84	10.6	17.3	5	U	40
316	IPA 1592AYFG	158	80	57	71	58	84	10.1	15.5	5	U	40
Ertebølle, Denmark												
317	ZMC?	168	U	57	U	56	90	11.0	18.1	5	U	40
Vedbaek, Denmark												
318	ZMC 1944–45	161	84	61	74	59	88	11.3	18.7	5	U	40
Saltpetermosen, Denmark												
319	ZMC H.7–1	178	93	63	82	62	96	11.4	18.1	5	U	40
320	ZMC H.7–2	163	83	60	75	59	87	11.5	17.3	5	U	40

		Measurements (mm)								Observations		
Case	Specimen	CL	PL	PW	OI	MCW	IM2	P3	P4	A	S	SG
Bundsø, Denmark												
321	ZMC BII	136	72	52	61	54	76	10.3	15.1	5	U	40
322	ZMC F. A. 62	137	71	50	62	51	74	9.2	14.2	5	U	40
323	ZMC DS 3	154	76	53	69	56	79	8.5	14.3	5	U	40
324	ZMC KV A.1	160	84	57	73	60	87	10.3	16.9	5	U	40
325	ZMC KV B	158	83	57	71	56	84	10.9	17.5	5	U	40
326	ZMC BS 2	159	80	59	68	58	86	10.8	18.1	5	U	40
327	ZMC DS 5	149	77	51	66	53	80	10.4	16.9	5	U	40
Spodsbjerg, Denmark												
328	ZMC 9688:941	149	78	54	70	57	82	9.8	17.1	5	U	40
Lidsø, Denmark												
329	ZMC?	149	77	53	68	56	80	9.1	15.6	5	U	40
330	ZMC?	157	80	53	72	57	86	10.5	16.3	5	U	40
331	ZMC?	148	78	54	67	58	78	10.4	16.4	5	U	40
Senckenberg, Germany[b]												
332	ZMC?	178	93	63	81	59	99	12.8	19.0	5	U	41

[a] Parentheses around a measurement indicate that that the measurement was retained even though it was estimated on an incomplete specimen. As covered in the text of Chapter 3, some estimates were eliminated due to concerns in replicating them accurately, but those retained were judged to be sufficiently accurate.

[b] The Senckenberg specimen is a plaster cast; the original was lost or destroyed during World War II.

APPENDIX B

Table B.1 assembles an extensive series of documented nonmodern dog burials. In the Old World, I have included only summary entries for a series from Northern Europe (Scandinavia) as well as a series from continental Europe and England inventoried by Prummel (1992). That step is due to their quantity, along with the fact that most primary sources are not in English and I have not seen them. Beyond those, there are certainly even more from the Old World, but I have only passing familiarity with much of the literature pertaining to that side of the world. For example, Trantalidou (2006: 100–102, Tables 3, 4, 5) has inventoried a series of sites in ancient Greece where dog remains are found buried with people. Similarly, Mazzorin & Minniti (2006: 63–64) have called attention to settings in both ancient Greece and Italy, more than 2,000 years old, where dogs were buried with people, or by themselves. Likewise, Wilkens (2006: 134–135) has also focused on several such sites in Italy. In addition, Nikolova (2005: 105) has indicated some from the Balkan region of Europe. As well, besides the dramatic number of dog burials at Ashkelon, in Israel, Wapnish & Hesse (1993: 67–70) have indicated others from that region.

The list is more extensive for the New World, due simply to my greater familiarity with that setting, though certainly, there are more. For example, lacking from this series are some burials indicated by Schwartz (1997: Table 4.1). Also, there are more from northeastern North America indicated by Handley (2000), many of those being in older or more obscure sources that I have not attempted to obtain. Similarly, additional examples from the late prehistory in North American desert southwest can be found in E. Hill (2000: 379–387) and also in Lang & Harris (1984: 90). Also from the desert southwest, much more recently Dody Fugate is quoted by Anne Casselman (2008: 1) as saying: "I have a database now of almost 700 dog burials, and a large number of them are either buried in groups in places of ritual or they're buried with individual human beings." They are apparently most common between 400 B.C. and 1100 A.D., but no specific examples are provided. As well, additional dog burials from late prehistoric or historic times in the present state of California can be found by consulting Hale & Salls (2000).

For many sites only a time range is offered. In some of those cases, especially in the New World, it is highly likely, and perhaps known, that the dog burials date to a restricted span within that general range. For example, 8000–3000 B.P. is a time range that often appears with Archaic Period sites in the midwestern and interior eastern United States. That span subsumes, by convention, what are known as the Middle and Late Archaic Periods (see Chapter 1, Figure 1.1). The dog burials may be concentrated in or confined to a portion of that span, usually the latter portion (e.g., the Green River Valley in Kentucky), but it seems best to indicate the general span, rather than risk assigning any burials to a more restricted span that could be inaccurate.

The reality of truly high quantities of burials is underscored by the existence of nearly 200 Archaic Period dog burials from a specific locality where I have considerable first-hand background, the Green River Valley of Kentucky. The total documented from there exceeds the known quantities from other restricted North American localities, a pattern that can be verified by collapsing some of the sites according to their specific regions. In fact, the Green River region gets a mere summary entry for this table, as that region gets separate treatment in Chapter 7. To be sure, the Green River Valley is noteworthy for its frequency of dog burials, though part of that pattern may well stem from my enhanced familiarity with that region.

The single most dramatic case of dog burials, from any setting, is clearly Ashkelon. As covered in the text of Chapter 7, though, Ashkelon represents a distinctive order of business. The same is also likely true of Hatch, in Virginia, though on a lesser scale of magnitude, and perhaps Yin in China, as well. Overall, given the uncertain quantities associated with any number of cases, there seemed to be little point in suggesting a grand total.

TABLE B.1: *Numbers of dogs buried with people and total number of buried dogs for different sites/complexes around the world, prehistoric in most cases[a]*

Site/Complex, Country (Date)	Reference(s)	# with Humans	Total (dogs)
(A) Old World			
Bonn-Oberkassel, Germany (ca. 14,000 B.P.)	Nobis 1979, 1986, 1996; Benecke 1987; Street 2002	1	1
Ein Mallaha, Israel (12,000–11,000 B.P.)	Davis & Valla 1978	1	1
Kamikuroiwa, Japan (12,000 – 10,000 B.P.?; problematical dates, Incipient Jomon Period)	Kipfer 2000: 269	0	2
Hayonim Terrace, Israel (11,000–10,500 B.P.)	Tchernov & Valla 1997	2	2
Ushki-1, Siberia (ca. 10,650 B.P.)	Dikov 1996: 245; Vasil'evskiy 1998: 291; Goebel & Slobodkin 1999: 137	0	1
Almeö, Sweden (ca. 9000 B.P.)	Arnesson-Westerdahl 1985; Larsson 1990: 158	0	4?
Ust'-Belaia, Siberia (ca. 9000 B.P.?)	Chard 1974: 58; J.W. Olsen 1985: 66	0	1
Iron Gates Area, former Yugoslavia (8500–8000 B.P.)	Radovanović 1999 (see also Bökönyi 1970: 1703)	3	4
Polderweg, Netherlands (7500–7000 B.P.)	Van de Noort 2007: 85	0	3
Vedbaek, Denmark (7300–6500 B.P.)	Nielsen & Petersen 1993	0	1
Skateholm, Sweden (6500–5500 B.P.)	Larsson 1990, 1995; Fahlander 2008	4?	14?
Botai, Kazakhstan (5700–5100 B.P.)	S. L. Olsen 2000	0	43?
4 sites, Poland (5500–4200 B.P.)	Poznań 2006: 4 (Table 2)	0	4
Tell Brak, Iraq (5000–4000 B.P.)	Clutton-Brock 2001	0	1
Cemetery C, El Kadada, Sudan (4840–4630 B.P.)	Reinold 2005: 108, 113	16	16
Esbjerg, Denmark (4800–4400 B.P.)	Lauenborg 1982	0	1

Site/Context	Reference		
Classic Greek Contexts (4300–1300 B.P.)	L. P. Day 1984	19	?
Unar 2, U. Arab Emirates (ca. 4200 B.P.)	Blau & Beech 1999	1	1
Sintashta, Russia (4000–3600 B.P.)	Jones-Bley 2000: 129	0	"a number"
Tagara, Japan (4000–2300 B.P., Jomon Period)	Shigehara & Hongo 2000: 62–63	0	22
Chin Tafidet, Niger (3900–3300 B.P.)	Paris 2000: 114, 117	0	3
Apatheia, Greece (3700–3400 B.P.)	Konsolaki-Yannopoulou 2001: 218	1	3
Kerameikos Cemetery, Athens, Greece (3500–2750 B.P.)	Closterman 2007: 639	2?	2
Borger, Netherlands (3430–3210 B.P.)	Prummel 2006: 68–70	1	2
Yin, China (ca. 3380–3100 B.P.)	J. W. Olsen 1985: 60–61	439?	439
Dimini, Greece (3340–3200 B.P.)	Prummel 2006: 70–75	?	8
Drama–"Kajrjaka," Bulgaria (3100–2800 B.P.)	Benecke & Lichardus 2007	0	1
Van-Yoncatepe, Anatolia (ca. 3000 B.P. or slightly later)	Onar 2005; Onar & Belli 2005	"many"	at least 50
Jarlshof, Scotland (2800–2200 B.P.)	Curle & Scot 1934: 255	0	1
Duzerra Cave, Austria (2700–2450 B.P.)	Galik 2000 (possibly burials)	?	at least 45
Sindos, Greece (2700–2300 B.P.)	Antikas 2008: 24–25	2?	2
Metaponto, Italy (2700–2100 B.P.)	Carter 2003: 15–16	?	"several"
Siracusa, Sicily (2550–2400 B.P.)	Chilardi 2006: 33–35	0	2
Ashkelon, Israel (2550–2300 B.P.)	Stager 1991; Wapnish & Hesse 1993; Halpern 2000	0	1200+
Lismullin 1, Ireland (2520–2370 B.P.)	O'Connell 2007: 54	0	1
Site at Ely, England (2500–2200 B.P.)	Atkins & Mudd 2003	0	1
Eretria, Greece (2336–2146 B.P.)	Chenal-Velarde 2006: 25–30	at least 26	at least 26
Ein Tirghi, Egypt (2200–1200 B.P.)	Churcher 1993	0	16
Côte-d'Or, France (Vertault) (ca. 2000 B.P.)	Horard-Herbin 2000: 115	0	150
Herefordshire, England (2000–1900 B.P.; Roman Period)	Sherlock & Pikes 2002: 32	?	2
Dryburn Bridge, Scotland (2000–1600 B.P.; Roman Period)	Thoms 2005: 90	0	1

(continued)

Site/Complex, Country (Date)	Reference(s)	# with Humans	Total (dogs)
Nomentana, Italy (1850–1800 B.P.)	Gräslund 2004: 170	0	4
Barcombe Villa, England (1800–1700 B.P.; Roman Period)	Rudling & Butler 2004: 3	0	2
Yasmina, Tunisia (1800–1500 B.P.)	MacKinnon & Belanger 2006	1	1
York Road, Leicester, England (1700–1600 B.P.; Roman Period)	Baxter 2006	1?	1
Duzerra Cave, Austria (1600–1500 B.P.)	Galik 2000	?	unspec.
55 cemeteries, Continental Europe and England (1600–1200 B.P.)	Prummel 1992: 135, 139	Some	114
5 Anglo-Saxon settlements, United Kingdom (1600–1200 B.P.)	Hamerow 2006: 4–7	0	8
48 cemeteries, Scandinavia (1600–900 B.P.)	Prummel 1992: 135, 139	Some	246
Rickeby, Sweden (1400–1300 B.P.)	Gräslund 2004: 168	1	4
Machrins, Colonsay, Scotland (ca. 1200 B.P.)	Ritchie 1981: 8	1	1
Cape St. Francis, South Africa (ca. 1200 B.P.)	Voigt 1983: 67	0	1
Old Uppsala, Sweden (1200–1100 B.P.); boat grave	Gräslund 2004: 167	1	1
Oseberg, Norway (1200–1100 B.P.); boat grave	Gräslund 2004: 169	2	2
Visegrád, Hungary (1000–900 B.P.)	Daróczi-Szabó 2006: 86	6	6
Houtskär (Forunabb), Finland (ca. 150 B.P.?: "of recent age")	Tuovinen 2002: 49	0	1
(B) Polynesia			
Misc. Contexts, Australia (prehistoric/early historic, dingoes)	Corbett 1995: 21	?	unspec.
Hane, Ua Huka Island (ca. 1300 B.P.)	Sinoto 1966; G.Clark 1996:33	?	3
Palliser Bay, New Zealand (ca. 770 B.P.)	Leach 1979: 85	?	1

Site	Reference		
Shag Mouth, New Zealand (ca. 700 B.P.)	G. Clark 1996: 33	?	1
Bellows Beach, Hawaii (ca. 500 B.P.)	G. Clark 1996: 33	?	1
False Island, New Zealand (470+ B.P.)	Lockerbie 1959: 90	?	1
Lotofaga, Samoa (ca. 450 B.P.)	Green & Davidson 1969: 239	?	1
Lanai, Mamaki, Hawaii (400–200 B.P.)	Titcomb 1969: 19–20	?	1
Nuolo Flat, Hawaii (ca. 300 B.P.)	Wood-Jones 1931: 40	?	2
Afareaitu, Moorea Island (ca. 300 B.P.)	Titcomb 1969: 26	?	1
Halekulani Hotel Site, Hawaii (300–200 B.P.)	Jourdane & Dye 2006: 5	0	2
Pukapuka, Northern Cooks (ca. 240 B.P.)	Shigehara et al. 1993	?	1
(C) New World			
Koster, Illinois (ca. 8500 B.P.)	Morey & Wiant 1992	0	3
Dust Cave, Alabama (8400–5600 B.P.)	Morey 1994b; Walker et al. 2005	0	4
Duncan Tract, Tennessee (8200–1600 B.P.)	Breitburg 1983: 392	0	3
Green River Valley, Kentucky (8000–3000 B.P.)	See Chapter 7, Table 7.1	31+	192
Cherry, Tennessee (8000–3000 B.P.)[3]	Magennis 1977: 80	4	4?
Bailey, Tennessee (8000–3000 B.P.)	Bentz 1988	0	1
Eva, Tennessee (8000–3000 B.P.)	Lewis & Lewis 1961: 144	4	at least 18
40MY105, Tennessee (8000–3000 B.P.)	S. D. Moore 1991: 43	0	1
O'Neal, Alabama (8000–3000 B.P.)	Webb & DeJarnette 1942: 135	0	1
Long Branch, Alabama (8000–3000 B.P.)	Webb & DeJarnette 1942: 183	?	"occasional"
Mulberry Creek, Alabama (8000–3000 B.P.)	Webb & DeJarnette 1942: 246	1	3
Perry, Alabama (8000–3000 B.P.)[2]	Webb & DeJarnette 1942: 68–69, 84; 1948a: 22	13+	55
Flint River, Alabama (8000–3000 B.P.)[4]	Webb & DeJarnette 1948b: 37	0	19
Whitesburg Bridge, Alabama (8000–3000 B.P.)[4]	Webb & DeJarnette 1948c: 16	0	9
Little Bear Creek, Alabama (8000–3000 B.P.)[4]	Webb & DeJarnette 1948d: 21	0?	unspec.
Rodgers Shelter, Missouri (ca. 7500 B.P.)	R. B. McMillan 1970	0	1
Anderson, Tennessee (ca. 7000 B.P.)	Dowd 1989: 60, 122	0	1
Modoc Shelter, Illinois (ca. 7000 B.P.)	Parmalee 1959: 63	0	2

(continued)

TABLE B.1 *(continued)*

Site/Complex, Country (Date)	Reference(s)	# with Humans	Total (dogs)
Koster, Illinois (ca. 7000 B.P.)	F. Hill 1972	0	1
Braden, Idaho (ca. 6600 B.P.)	Yohe & Pavesic 2000	2	2
Several sites, Ecuador (5600–3800 B.P.)	Stahl 2003: 187	?	"infrequent"
Ricker, Alabama (5600–3000 B.P.)	Hale 1983: 332–334	0	1
Gaston, North Carolina (5500–4300 B.P.?)	VanDerwarker 2001: 4; South 2005: 59–60	1+?	"several" "a number"
Bluegrass, Indiana (5300–5000 B.P.)	Stafford et al. 2000: 320	?	12
Bible Site, Tennessee (5000–3000 B.P.)	Parmalee 1966	0	2
Westmoreland-Barber, Tennessee (5000–1200 B.P.)	Guilday & Tanner 1966: 145	0	2
Turner Farm, Maine (4500–4000 B.P.)	Bourque 1995: 86	0	6
Real Alto, Ecuador (4500–3750 B.P.?)	Wing 1986: 262	?	"a number"
SJo-68, California (ca. 4050 B.P.)	Haag & Heizer 1953	0	1
Russell Cave, Alabama (4000+ B.P.)	Miller 1956: 554, 556	0	1
Port au Choix, Newfoundland (4000–3700 B.P.)	Tuck 1976: 77–78, 202	4	4
East Steubenville, West Virginia (ca. 3850–3400 B.P.)	D. MacDonald 2003: 62	0	2
Boardwalk, NW coast, Canada (3850–1500 B.P.)	Stewart & Stewart 1996: 41; Ames 2001: 7 (the dating)	?	"occasionally"
several sites, NW Coast, North America (3800–1500 B.P.)	Ames & Maschner 1999: 187–188	?	unspec.
Robinson, Tennessee (3300–2500 B.P.)	Morse 1967: 20; Guilday 1967	0	2
40HW45, Tennessee (3000–2500 B.P.)	Curren 1981: 387–394	0	18
La Playa, Mexico (ca. 3000–1800 B.P.)	Carpenter et al. 2005: 27	0	1
Costello-King, Arizona (ca. 2700 B.P.)	Ezzo & Stiner 2000	0	1
Watmough Bight, Washington (2650–120 B.P.?)	Barsh et al. 2006: 1–2	1	1
Chupicuaro, Mexico (2600–2100 B.P.)	Adams 2005: 126	46	46
White's Mound, Georgia (ca. 2500 B.P.)	S. J. Olsen 1970	0	2
Rosamachay, Peru (2400–1700 B.P.)	Wing 1986: 262	0	1

Site	Reference		
Broad Reach, North Carolina (ca. 2400–400 B.P.)	Millis 2010	0	13
several sites, West Indies (2250–450 B.P.)	Wing 2001: 493; Wing 2007: 416	"many"	unspec.
14 sites, Illinois (2200–1600 B.P.)	Cantwell 1980: Table 1	most	at least 34
Monte Albán, Mexico (2150–1800 B.P.)	Licón 2003: 162	1	30
Scioto Cavern, Ohio (2100–1300 B.P.)	Potter & Baby 1964	?	1
Crooks, Louisiana (2000–1600 B.P.)	Ford 1940: 41	3	25?
Sorcé, Puerto Rico (2000–1500 B.P.)	Wing 1991: 380	0	3
Trowbridge, Kansas (1800–1400 B.P.)	S. Collins 1999	0	7
Palmer Burial Mound, Florida (1800–1100 B.P.)	Bullen & Bullen 1976: 44–46	0	1
Wildcat Canyon, Oregon (1800–800 B.P.)	Dumond & Minor 1983: 116	0	4
Sipán, Peru (1700–1200 B.P.)	Alva & Donnan 1993: 123, 159	2	9–10
Bayshore Homes, Florida (1700–850 B.P.)	Sears 1960	1	2
James Village, Alabama (1600–1000 B.P.)	Walthall 1980: 168	0	1
McCulloch, Missouri (1550–800 B.F.)	Darwent & Gilliland 2001: 151	0	5
Kersey, Missouri (1550–800 B.P.)	Darwent & Gilliland 2001: 159	0	1
Buck Burial Mound, Florida (1500–1200 B.P.)	Lazarus 1979: 7	0	1
Ipiutak, Alaska (1500–1150 B.P.)	Larsen & Rainy 1948: 225–250	5	6
Smith Mound 4, Minnesota (1500–1100 B.P.)	Wilford 1950: 169	0	1
Ancon, Peru (1400–550 B.P.)	Nehring 1887	unspec.	unspec. (many)
Tula, Mexico (1350–1250 B.P.)	Valadez 1996: 50–51; Valadez et al. 1999	?	27
El Riego Cave, Mexico (1300 B.P.–1521 A.D.)	Flannery 1967: 168	0	2
CA-Ora-849, California (1250–750 B.P.?)	Langenwalter 2005	0	1
Mancos Canyon, Colorado (1200–850 B.P.)	Emslie 1978	0	20
Morell-Sheets, Indiana (1200–800 B.P.)	McCord 2005: 173	0	1
Lambert Farm, Rhode Island (1150–500 B.P.)	Kerber 1997: 66–78	0	3
Ausmus Farm Mounds, Tennessee, Site 10 (ca. 1100 B.P.?)	Webb 1938: 109–110	2	2
Teotihuacan, Mexico (ca. 1100 B.P.)	Valadez et al. 2006: 125	0	1

(continued)

270

Site/Complex, Country (Date)	Reference(s)	# with Humans	Total (dogs)
Dadasta óðir, Iceland (1100–800 B.P.)	McGovern 2004: 3	1	"common"
Iceland, multiple contexts (1100–800 B.P.)	Perdikaris et al. 2002: 7	?	3
Hiwassee Island, Tennessee (1100–700 B.P.)	Lewis & Kneberg 1946: 23, 24	2	40+
Chiribaya Baja, Peru (1100–650 B.P.)	Dittmar et al 2003; de Pastino 2006; Lange 2007	?	1
Norris Farms Cemetery, Illinois (ca. 1100–400 B.P.)	Santure & Esarey 1990: 93	1	2
Miller Cave, Missouri (ca. 1000 B.P.?)	Darwent & Gilliland 2001: 160	0	2
Bessemer, Alabama (1000–800 B.P.)	DeJarnette & Wimberly 1941: 22–23	0	1
Vir 150, North Carolina (1000–600 B.P.)	VanDerwarker 2001: 5	0	3
Archery Range, New York City (1000–400 B.P.)	Cantwell & Wall 2001: 101	?	unspec.
College Point, New York City (1000–400 B.P.)	Cantwell & Wall 2001: 104	?	13
Unnamed sites, New York City (1000–400 B.P.)	Cantwell & Wall 2001: 105–106	0	1
Strawtown Enclosure, Indiana (900–600 B.P.)	McCullough et al. 2004: 14	0	2
Tule Creek Village, California (800–600 B.P.)	Vellanoweth et al. 2008	0	38
Hatch, Virginia (800–400 B.P.)	Gregory 1979	?	105
	Boyd & Boyd 1992: 263	?	1
John Green, Virginia (800–400 B.P.)	MacCord 1970	1	2 or 3
Fisher, Illinois (800–400 B.P.)	Parmalee 1962: 406	0	1
Unnamed site, Greenland (ca. 800–200 B.P.)	Nyegaard 1995: 100, 103	1	1
Lone Tree Cove, Michigan (750–600 B.P.)	C.P. Clark 1990	0	1
Winslow, Maryland (ca. 700 B.P.)	J. Dent 2003: 3	0	1
Grasshopper Ruin. Arizona (700–600 B.P.)	S. J. Olsen 1968: 2	1	2
Josey Farm, Mississippi (700–150 B.P.)	Hogue 2003: 186	0	2
Quaker Creek, North Carolina (600–400 B.P.)	Davis et al. 1998: 5	0	1
Madisonville, Ohio (600–350 B.P.)	Purtill 2000	0	3
Buena Vista Lake, California (600–250 B.P.)	Wedel 1941: 35	0	

Site	Source		
			unspec.
Lo Dimás, Peru (520–460 B.P.)	Sandweiss & Wing 1997: 54–55	0	1
SNI-25, San Nicolas Island, California (ca. 510 B.P.)	Kerr et al. 2002: 33	0	1
Cleveland Site, Ontario (500–420 B.P.)	Bathurst & Barta 2004: 918	0	3
Armorel, Arkansas (500–300 B.P.?)	Pavao-Zuckerman 2001	?	1
Blood Run, Iowa (ca. 500–300 B.P.)	Harvey 1979: 137	?	1
Eschelman, Pennsylvania (400–375 B.P.)	Guilday et al. 1962: 64–65	0	2
Ibaugh, Pennsylvania (400–375 B.P.?)	Witthoft et al. 1959:115	0	1
Frank Bay, Ontario (400–300 B.P.)	Ridley 1954: 49; Quimby 1966: 112	0	3
Munsee Cemetery, New Jersey (400–200 B.P.)	Heye & Pepper 1915: 59	0	2
Eel Point C, San Clemente Island, California (400–200 B.P.?)	Hardy 2000: 87	?	unspec.
22OK904, Mississippi (400–180 B.P.)	Hogue 2003: 187	0	1
Fig Springs Mission, Florida (392–344 B.P.)	Weisman 2000	0	1
Roseborough Lake, Texas (300–200 B.P.)	Yates & Koler-Matznick 2006: 143–144	0	2
Meadowcroft, Pennsylvania (ca. 225 B.P.)	Stuckenrath et al. 1982: 79	0	1
Ryan, Nebraska (ca. 210–160 B.P.?)	O'Shea 1984: 57	0	1
Blount and Monroe Counties, Tennessee (Late Prehistoric/Proto-Historic/Historic Cherokee)	Parmalee & Bogan 1978; Schroedl & Breitburg 1986: 178; Polhemus 1969: 100	1	8
Big Dog Cave, San Clemente Island, California (Historic Period)	Hardy 2000: 91	0	1
Chattooga, South Carolina (Historic Cherokee)	Schroedl & Parmalee 1997	0	3
7S-F-68, a Euro-American cemetery, Delaware (1750–1800 A.D.)	LeeDecker et al. 1995: 37, 40	0	3
Los Angeles, California (1800s A.D.)	Costello 2004: 17	0	1
Poplar Forest, Virginia (ca. 1870–1900 A.D.; originally Thomas Jefferson's plantation)	Heath 2003: 12	0	1
Scranage Enclosure, Indiana	White et al. 2002: 70 ("recent")	0	1

[a] Most dates given are based on radiocarbon determinations. Given the volume of information, no attempt is made to distinguish between uncalibrated versus calibrated dates, since there is no consistency among the sources in reporting these data.

Note: For footnotes, see Chapter 7.

REFERENCES

Aaris-Sørensen, K.
(1977). Vedbaek – Jaegeren og hans Hunde. In *Vedbaekprojektet – I Marken og i Museerne*, ed. E. B. Petersen, J. H. Jønsson, P. V. Petersen & K. Aaris-Sørensen, pp. 170–176. Søllerød Kommune, Denmark: *Søllerødbogen*, 1977 (in Danish).

Aaris-Sørensen, K.
(1981). A zoological analysis of the osteological material from the sacrificial layer at the Maussolleion at Hallikarnassos. In *The Mausolleion at Hallikarnassos*, Volume 1: *The Sacrificial Deposit*, ed. K. Jeppesen, pp. 91–110 (plus two plates). Aarhus, Denmark: Publications of the Jutland Archaeological Society.

Aaris-Sørensen, K.
(1985). Den terrestriske pattedyrfauna i det sydfynske øhav gennem Atlantikum og Tidlig Subboreal. In *Yngre stenalder på øerne syd for Fyn*, ed. J. Skaarup, pp. 458–466. Rudkøbing, Denmark: *Meddelelser fra Langelands Museum* (in Danish).

Aaris-Sørensen, K.
(1988). *Danmarks Forhistoriske Dyreverden – Fra Istid til Vikingetid*. Copenhagen, Denmark: Gyldendal (in Danish).

Aaris-Sørensen, K.
(2001). *The Danish Fauna throughout 20,000 Years from Mammoth Steppe to Cultural Steppe – A Guide to an Exhibition about the Changeability of Nature*. Copenhagen, Denmark: Zoological Museum, University of Copenhagen.

Aaris-Sørensen, K. & Petersen, E. B.
(1986). The Prejlerup Aurochs – An archaeolozoological discovery from boreal Denmark. *Striae*, 24, 111–117.

References

Adams, R. E. W.
  (2005). *Prehistoric Mesoamerica* (3rd edition). Norman, Oklahoma: University of Oklahoma Press.

Ahler, S. A., Thiessen, T. D. & Trimble, M. K.
  (1991). *People of the Willows – The Prehistory and Early History of the Hidatsa Indians*. Grand Forks, North Dakota: University of North Dakota Press.

Albert, A. & Bulcroft, K.
  (1987). Pets and urban life. *Anthrozoös*, 1, 9–23.

Albert, A. & Bulcroft, K.
  (1988). Pets, families, and the life course. *Journal of Marriage and the Family*, 50, 543–552.

Albrethsen, S. E. & Petersen, E. B.
  (1976). Excavation of a Mesolithic cemetery at Vedbæk, Denmark. *Acta Archaeologica*, 47, 1–28.

Allen, G. M.
  (1920). Dogs of the American aborigines. *Bulletin of the Museum of Comparative Zoology, Harvard University*, 63 (9), 431–517.

Allis, S. A.
  (1887). Forty years among the Indians, on the eastern borders of Nebraska. *Transactions and Reports of the Nebraska State Historical Society* (Lincoln, Nebraska), 2, 133–166.

Alva, W. & Donnan, C. W.
  (1993). *Royal Tombs of Sipán*. Los Angeles, California: Fowler Museum of Cultural History.

American Museum of Natural History.
  (1971). *The Language and Music of the Wolves*. Auburn, California: The Audio Partners, Inc. (a recorded soundtrack, narrated by Robert Redford, issued again in 1986).

American Pet Products Manufacturers Association.
  (1988). *A Nationwide Survey of Pet Owners*. American Pet Products Manufacturers Association, Inc., 60 East 42nd Street, New York, NY 10165.

Ames, K. M.
  (2001). Slaves, chiefs and labour on the northern northwest coast. *World Archaeology*, 33 (1), 1–17.

Ames, K. M. & Maschner. H. D. G.
  (1999). *Peoples of the Northwest Coast – Their Archaeology and Prehistory*. New York City: Thames and Hudson.

Andersen, A. C., & Wooten, E.
  (1959). The estrous cycle of the dog. In *Reproduction in Domestic Animals*, ed. H. H. Cole & P. T. Cupps, pp. 359–397. New York City: Academic Press.

Andersen, S. H.
(1975). *Ringkloster, en jysk inlandsboplads med Ertebøllekultur. KUML*, 1973–1974, 11–108 (in Danish).

Andersen, S. H., & Johansen, E.
(1986). Ertebølle revisited. *Journal of Danish Archaeology*, 5, 31–61.

Andresen, J. M., Byrd, B. F., Elson, M. D., McGuire, R. H., Mendoza, R. G., Staski, E. & White, J. P.
(1981). The deer hunters: Star Carr revisited. *World Archaeology*, 13(1), 31–46.

Antikas, T. G.
(2008). They didn't shoot horses: Fracture management in a horse of the 5th century BCE from Sindos, Central Macedonia, Greece. *Veterinarija ir Zootechnika*, 42(64), 24–27.

Arnesson-Westerdahl, A.
(1985). *Djuren vid Hornborgasjön för 9000 år sedan. Sveriges Natur*, 76(3), 40–43 (in Swedish).

Arnold, C. D.
(1979). Possible evidence of domestic dog in a Paleoeskimo context. *Arctic*, 32(3), 263–265.

Ash, R. A.
(2005). Hound dog blues: Depression in dogs. www.doggiefun.com: one page (may be defunct).

Ashby, J. R.
(1974). Food Habits and Distribution of the Red Fox (*Vulpes fulva* Desmarest) and the Gray Fox (*Urocyon cinereoargenteus* Schreber) on the WMA, Tennessee. Cookeville, Tennessee: Unpublished M.S. Thesis, Department of Biology, Tennessee Technological University.

Atkins, D. L. & Dillon, L. S.
(1971). Evolution of the cerebellum in the genus *Canis. Journal of Mammalogy*, 52(1), 96–107.

Atkins, R. & Mudd, A.
(2003). An Iron Age and Romano-British settlement at Prickwillow Road, Ely, Cambridgeshire: Excavations 1999–2000. *Proceedings of the Cambridge Antiquarian Society*, 92, 5–55.

Audubon, M. R.
(1960). *Audubon and His Journals*. New York City: Dover (originally published in 1897 by Charles Scribner's Sons).

Aydinlioğlu, A., Arslan, K., Erdoğan, A. R., Rağbetli, M. Ç., Keleş, P. & Diyarbakirli, S.
(2000). The relationship of callosal anatomy to paw preference in dogs. *European Journal of Morphology*, 38(2), 128–133.

Ballantine, R.
(1996). *The Sawtooth Wolves*. Bearsville, New York: Rufus Publications, Inc. (Photographs by Jim Dutcher).

Balseiro, S. C. & Correia, H. R.
(2006). Is olfactory detection of human cancer by dogs based on major histocompatibility complex-dependent odour components? – A possible cure and a precocious diagnosis of cancer. *Medical Hypotheses*, 66(2), 270–272.

Barja, I. & Rosinelli, S.
(2008). Does habitat type modify group size in roe deer and red deer under predation risk by Iberian wolves? *Canadian Journal of Zoology*, 86(3), 170–176.

Barrett, N. A., Large, M. M., Smith, G. L., Karayanidis, F., Michie, P. T., Kavanagh, D. J., Fawdry, R., Henderson, D. & O'Sullivan, B. T.
(2003). Human brain regions required for the dividing and switching of attention between two features of a single object. *Cognitive Brain Research*, 17(1), 1–13.

Barsh, R. L., Jones, J. M. & Suttles, W.
(2006). History, ethnography, and archaeology of the Coastal Salish woolydog. In *Dogs and People in Social, Working, Economic or Symbolic Interaction*, ed. L. M. Snyder & E. A. Moore, pp. 1–11. Oxford, England: Oxbow Books (Proceedings of the 9th ICAZ Conference, Durham, England, 2002).

Bathurst, R. R. & Barta, J. L.
(2004). Molecular evidence of tuberculosis induced hypertrophic osteopathy in a 16th-century Iroquoian dog. *Journal of Archaeological Science*, 31, 917–925.

Bauer, N. K.
(2006). *Dog Heroes of September 11th: A Tribute to America's Search and Rescue Dogs*. Allenhurst, New Jersey: Kennel Club Books.

Baus de Czitrom, C.
(1988). *Los perros de la antigua provincia de Colima*. Roma, México: Instituto Nacional de Antropología e Historía (in Spanish).

Baxter, I. L.
(2006). A dwarf hound skeleton from a Romano-British grave at York Road, Leicester, England, U.K., with a discussion of other Roman small dog types and speculation regarding their respective aetiologies. In *Dogs and People in Social, Working, Economic or Symbolic Interaction*, ed. L. M. Snyder & E. A. Moore, pp. 12–23. Oxford, England: Oxbow Books (Proceedings of the 9th ICAZ Conference, Durham, England. 2002).

Bazaliiskiy, V. I. & Savelyev, N. A.
(2003). The wolf of Baikal: The "Lokomotiv" early Neolithic cemetery in Siberia (Russia). *Antiquity*, 77, 20–30.

Beck, A. M.
(1996). Animal contact and the older person: Companionship, health, and the quality of life. Presented at the AARP Biennial Convention, Denver, Colorado, May 23, 1996.

Beitz, A. J. & Fletcher, T. F.
(1993). The brain. In *Miller's Anatomy of the Dog* (3rd edition), ed. H. E. Evans, pp. 894–952. Philadelphia, Pennsylvania: W. B. Saunders.

Bellars, A. R. M.
(1969). Veterinary studies of the British Antarctic Survey's sledge dogs: I. Survey of diseases and accidents. *British Antarctic Survey Bulletin*, 21, 1–18.

Bellars, A. R. M. & Godsal, M. F.
(1969). Veterinary studies on the British Antarctic survey's sledge dogs: II. Occupational osteoarthritis. *British Antarctic Survey Bulletin*, 22, 15–38.

Belyaev, D. K.
(1979). Destabilizing selection as a factor in domestication. *Journal of Heredity*, 70, 301–308.

Belyaev, D. K. & Trut, L. N.
(1975). Some genetic and endocrine effects of selection for domestication in silver foxes. In *The Wild Canids*, ed. M. W. Fox, pp. 416–426. New York City: Van Nostrand Reinhold.

Belyaev, D. K. & Trut, L. N.
(1982). Accelerating Evolution. *Science in the U.S.S.R.*, 5, 24–29, 60–64.

Benecke, N.
(1987). Studies on early dog remains from Northern Europe. *Journal of Archaeological Science*, 14, 31–49.

Benecke, N. & Lichardus, J.
(2007). The "poor" dog from Drama: Observations on an early Iron Age dog skeleton from south-east Bulgaria. In *Internationale Archäologie – Studia Honoraria*, ed. C. Dobiat & K. Leidorf, pp. 67–77. Rahden, Germany: Marie Leidorf.

Benecke, N. & Hanik, S.
(2002). Dogs for the living and dogs for the dead – On the exploitation of dogs in Mesolithic Europe. Abstract of paper presented at the 9th ICAZ conference, 23–28 August, University of Durham, England, ICAZ Durham Scientific Programme, 2002, pp. 20–21.

Bentz, C.
(1988). The Bailey Site: Late Archaic, Late Woodland, and Historic Settlement and Subsistence in the Lower Elk River Drainage of Tennessee. Nashville, Tennessee: Report submitted to the Tennessee Department of Transportation.

References

Berdan, F. F.
   (1982). *The Aztecs of Central Mexico – An Imperial Society*. New York City: Holt, Rinehart and Winston.

Bierce, A.
   (2000). *The Unabridged Devil's Dictionary*. Athens, Georgia: University of Georgia Press (originally published in 1911 by the Neale Publishing Company).

Binford, L. R.
   (1981). *Bones – Ancient Men and Modern Myths*. New York City: Academic Press.

Bingham, P. M.
   (1999). Human uniqueness: A general theory. *The Quarterly Review of Biology*, 74(2), 133–169.

Björnerfeldt, S.
   (2007). *Consequences of the Domestication of Man's Best Friend, the Dog*. Uppsala, Sweden: Digital Comprehensive Summaries of Uppsala Dissertations from the Faculty of Science and Technology, 289.

Björnerfeldt, S., Webster, M. T. & Vilà, C.
   (2006). Relaxation of selective constraint on dog mitochondrial DNA following domestication. *Genome Research*, 16(7), 1–5 (DOI 10.1101/gr.5117706: June 29, 2006).

Blanco, J. C. & Cortés, Y.
   (2007). Dispersal patterns, social structure and mortality of wolves living in agricultural habitats in Spain. *Journal of Zoology*, 273(1), 114–124.

Blanco., J. C., Cortés, Y. & Virgós, E.
   (2005). Wolf response to two kinds of barriers in an agricultural habitat in Spain. *Canadian Journal of Zoology*, 82(2), 312–323.

Blau, S. & Beech, M.
   (1999). One woman and her dog: An Umm an-Nar example from the United Arab Emirates. *Arabian Archaeology and Epigraphy*, 10, 34–42.

Bloom, P.
   (2004). Can a dog learn a word? *Science*, 304, 1605–1606.

Boas, F.
   (1964). *The Central Eskimo*. Lincoln, Nebraska: University of Nebraska Press (originally published in 1888, as part of the sixth Annual Report of the Bureau of Ethnology, Smithsonian Institution, Washington, D.C.).

References

Bodson, L.

(2000). Motivations for pet-keeping in Ancient Greece and Rome: A preliminary survey. In *Companion Animals and Us – Exploring the Relationships between People and Pets*, ed. A. L. Podberscek, E. S. Paul & J. Serpell, pp. 27–41. Cambridge, England: Cambridge University Press.

Boesch, C. & Tomasello, M.

(1998). Chimpanzee and human cultures. *Current Anthropology*, 19(5), 591–614.

Bohlman, S.

(1999). Ontologies of music. In *Rethinking Music*, ed. N. Cook & M. Everist, pp. 17–34. Oxford, England: Oxford University Press.

Boitani, L.

(2003). Wolf conservation and recovery. In *Wolves – Behavior, Ecology, and Conservation*, ed. L. D. Mech & L. Boitani, pp. 317–340. Chicago, Illinois: University of Chicago Press.

Boitani, L., Ciucci, P. & Ortolani, A.

(2007). Behaviour and social ecology of free-ranging dogs. In *The Behavioural Biology of Dogs*, ed. P. Jensen, pp. 147–165. Wallingford, England, and Cambridge, Massachusetts: CAB International.

Boitani, L., Francisci, F., Ciucci, P. & Andreoli, G.

(1995). Population biology and ecology of feral dogs in central Italy. In *The Domestic Dog: Its Evolution, Behaviour and Interactions with People*, ed. J. Serpell, pp. 217–244. Cambridge, England: Cambridge University Press.

Bonner, J. T.

(1980). *The Evolution of Culture in Animals*. Princeton, New Jersey: Princeton University Press.

Bourque, B. J.

(1995). *Diversity and Complexity in Prehistoric Maritime Societies – A Gulf of Maine Perspective*. New York City: Plenum.

Bower, B.

(1997). Ancient human saunters into limelight. *Science News*, 152, 117.

Bowron, E. P., Rebbert, C. R., Rosenblum, R. & Secord, W.

(2006). *Best in Show – The Dog in Art from the Renaissance to Today*. New Haven, Connecticut: Yale University Press.

Boyd, D. C. & Boyd, C. C.

(1992). Late Woodland mortuary variability in Virginia. In *Middle and Late Woodland Research in Virginia: A Synthesis*, ed. T. R. Reinhart & M. E. N. Hodges, pp. 249–275. Archaeological Society of Virginia, Special Publication, 29.

Boyko, A. R., Boyko, R. H., Boyko, C. M., Parker, H. G., et al.
(2009). Complex population structure in African village dogs and its implications for inferring dog domestication history. *Proceedings of the National Academy of Sciences*, 106(33), 13903–13908.

Bozell, J. R.
(1988). Changes in the role of the dog in protohistoric-historic Pawnee culture. *Plains Anthropologist*, 33, 95–111.

Bökönyi, S.
(1969). Archaeological problems and methods of recognizing animal domestication. In *The Domestication and Exploitation of Plants and Animals*, ed. P. J. Ucko & G. W. Dimbleby, pp. 219–229. Chicago, Illinois: Aldine.

Bökönyi, S.
(1970). Animal remains from Lepinski Vir. *Science*, 167, 1702–1704.

Bökönyi, S.
(1974). *History of Domestic Animals in Central and Eastern Europe*. Budapest, Hungary: Akadémiai Kiadó.

Bökönyi, S.
(1983). Domestication, dispersal, and use of animals in Europe. In *Domestication, Conservation and Use of Animal Resources*, ed. L. Peel & D. E. Tribe, pp. 1–20. Amsterdam, the Netherlands: Elsevier.

Brackenridge, H. M.
(1904). Journal of a voyage up the River Missouri performed in eighteen hundred and eleven. In *Early Western Travels 6*, ed. R. G. Thwaites, pp. 10–166. Cleveland, Ohio: Arthur H. Clark.

Brackman, B. & Brackman, J.
(2002). *The Dog in the Picture*. Lawrence, Kansas: Sirius Press.

Bradshaw, J. W. S. & Nott, H. M. R.
(1995). Social and communication behaviour of companion dogs. In *The Domestic Dog: Its Evolution, Behaviour and Interactions with People*, ed. J. Serpell, pp. 115–130. Cambridge, England: Cambridge University Press.

Branson, N. J. & Rogers, L. J.
(2006). Relationship between paw preference strength and noise phobia in *Canis familiaris*. *Journal of Comparative Psychology*, 120(3), 176–183.

Bräuer, J., Kaminski, J., Riedel, J., Call, J. & Tomasello, M.
(2006). Making inferences about the location of hidden food: Social dog, causal ape. *Journal of Comparative Psychology*, 120(1), 38–47.

Breitburg, E.
(n.d.). The Anderson Site (40WM9): An analysis of faunal remains recovered from a Tennessee Central Basin Middle Archaic site. Unpublished manuscript (as listed in Dowd 1989: 193).

Breitburg, E.

(1983). An analysis of faunal remains recovered from the Duncan Tract site (40TR27), Trousdale County, Tennessee. In *The Duncan Tract Site (40TR27) Trousdale County, Tennessee*, ed. C. H. McNutt & G. Weaver, pp. 357–400. Chattanooga, Tennessee: Tennessee Valley Authority, Publications in Anthropology, 33.

Brewer, D. J.

(2001a). The evolution of the modern dog. In *Dogs in Antiquity: Anubis to Cerberus – The Origins of the Domestic Dog*, ed. D. J. Brewer, T. Clark & A. Phillips, pp. 1–20. Warminster, England: Aris & Phillips.

Brewer, D. J.

(2001b). The path to domestication. In *Dogs in Antiquity: Anubis to Cerberus – The Origins of the Domestic Dog*, ed. D. J. Brewer, T. Clark & A. Phillips, pp. 21–27. Warminster, England: Aris & Phillips.

Brothwell, D., Malega, A. & Burleigh, R.

(1979). Studies on Amerindian dogs, 2: Variation in early Peruvian dogs. *Journal of Archaeological Science*, 6, 139–161.

Bruford, M. W., Bradley, D. G. & Luikart, G.

(2003). DNA markers reveal the complexity of livestock domestication. *Nature Reviews, Genetics*, 4, 900–910.

Bubna-Littitz, H.

(2007). Sensory physiology and dog behaviour. In *The Behavioural Biology of Dogs*, ed. P. Jensen, pp. 91–104. Wallingford, England, and Cambridge, Massachusetts: CAB International.

Budiansky, S.

(2002). *The Character of Cats*. New York City: Penguin Putnam.

Bullen, P. & Bullen, A. K.

(1976). The Palmer Site. *Florida Anthropologist*, 29(2), 1–55.

Burleigh, R. & Brothwell, D.

(1978). Studies on Amerindian dogs, 1: Carbon isotopes in relation to maize in the diet of domestic dogs from early Peru and Ecuador. *Journal of Archaeological Science*, 5, 355–362.

Burrows, R.

(1968). *Wild Fox*. New York City: Taplinger.

Burt, W. H. & Grossenheider, R. P.

(1976). *A Field Guide to the Mammals* (3rd edition). Boston: Houghton Mifflin.

Call, J., Bräuer, J., Kaminski, J. & Tomasello, M.

(2003). Domestic dogs (*Canis familiaris*) are sensitive to the attentional state of humans. *Journal of Comparative Psychology*, 117(3), 257–263.

Cantwell, A.-M.
> (1980). Middle Woodland dog ceremonialism in Illinois. *Wisconsin Archeologist*, 61(4), 480–496.

Cantwell, A.-M. & Wall, D.
> (2001). *Unearthing Gotham – The Archaeology of New York City*. New Haven, Connecticut: Yale University Press.

Carpenter, J. P., Sánchez, G. & Villalpando, C. M. E.
> (2005). The Late Archaic/early agricultural period in Sonora, Mexico. In *The Late Archaic across the Borderlands – From Foraging to Farming*, ed. B. J. Vierra, pp. 13–40. Austin, Texas: University of Texas Press.

Carter, J. C.
> (2003). The Chora of Metaponto: The animal inhabitants and their environment. In *Living off the Chora – Diet and Nutrition at Metaponto*, pp. 13–20 (editor not given). Austin, Texas: Institute of Classical Archaeology, University of Texas.

Case, L. P.
> (1999). *The Dog: Its Behavior, Nutrition and Health*. Ames, Iowa: Iowa State University Press.

Casselman, A.
> (2008). Buried dogs were divine "escorts" for ancient Americans. *National Geographic News*, 23 April (2 pages, electronic) http://news.nationalgeographic.com/news/2008/04/080423-dog-burial.html.

Chapman, J. & Watson, P. J.
> (1993). The Archaic Period and the flotation revolution. In *Foraging and Farming in the Eastern Woodlands*, ed. C. M. Scarry, pp. 27–38. Gainesville, Florida: University Press of Florida.

Chard, C. S.
> (1974). *Northeast Asia in Prehistory*. Madison, Wisconsin: University of Wisconsin Press.

Chauvet, J.-M., Deschamps, E. B. & Hillaire, C.
> (1996). *The Dawn of Art: The Chauvet Cave*. New York City: Harry N. Abrams.

Chavez, A. S. & Gese, E. M.
> (2006). Landscape use and movements of wolves in relation to livestock in a wildland-agriculture matrix. *Journal of Wildlife Management*, 70(4), 1079–1086.

Chen, S. H. & Desmond, J. E.
> (2005). Cerebrocerebellar networks during articulatory rehearsal and verbal working memory tasks. *Neuroimage*, 24, 332–338.

References

Chenal-Velarde, I.
(2006). Food, rituals? The exploitation of dogs from Eretria (Greece) during the Helladic and Hellenistic Periods. In *Dogs and People in Social, Working, Economic or Symbolic Interaction*, ed. L. M. Snyder & E. A. Moore, pp. 24–31. Oxford, England: Oxbow Books (Proceedings of the 9th ICAZ Conference, Durham, England, 2002).

Chilardi, S.
(2006). Artemis pit? Dog remains from a well in the ancient town of Siracusa (Sicily). In *Dogs and People in Social, Working, Economic or Symbolic Interaction*, ed. L. M. Snyder & E. A. Moore, pp. 32–37. Oxford, England: Oxbow Books (Proceedings of the 9th ICAZ Conference, Durham, England, 2002).

Churcher, C. S.
(1993). Dogs from Ein Tirghi cemetery, Balat, Dakhleh Oasis, western desert of Egypt. In *Skeletons in Her Cupboard: Festschrift for Juliet Clutton-Brock*, ed. A Clason, S. Payne & H.-P. Uerpmann, pp. 39–59. Oxford, England: Oxbow Monograph, 34.

Churchland, P. S.
(1986). *Neurophilosophy – Toward a Unified Science of the Mind/Brain*. Cambridge, Massachusetts: MIT Press.

Churchland, P. S.
(2002). *Brain-Wise – Studies in Neurophilosophy*. Cambridge, Massachusetts: MIT Press.

Cipponeri, T. & Verrell, P.
(2003). An uneasy alliance: Unequal distribution of affiliative interactions among members of a captive wolf pack. *Canadian Journal of Zoology*, 81(10), 1763–1766.

Ciucci, P., Lucchini, V., Boitani, L. & Randi, E.
(2003). Dewclaws in wolves as evidence of admixed ancestry with dogs. *Canadian Journal of Zoology*, 81(12), 2077–2081.

Clark, C. P.
(1990). A dog burial from Isle Royale, Lake Superior: An example of household ritual sacrifice in the Terminal Woodland Period. *Midcontinental Journal of Archaeology*, 15(2), 265–278.

Clark, G.
(1996). Animal burials from Polynesia. *Archaeology in New Zealand*, 39, 30–38.

Clark, J. G. D.
(1954). *Excavations at Star Carr, an Early Mesolithic Site at Seamer, Near Scarborough, Yorkshire*. Cambridge, England: Cambridge University Press (reprinted in 1971).

Cleland, C. E.
(1973). Appendix I: Notes on the dog skull from Nanook component 2. In *Archaeology of the Lake Harbour District, Baffin Island*, ed. M. S. Maxwell, pp. 353–356. Ottawa, Canada: National Museum of Man, Mercury Series, Archaeological Survey of Canada, Paper, 6.

Closterman, W. E.
(2007). Family ideology and family history: The function of funerary markers in Classical Attic Peribolos tombs. *American Journal of Archaeology*, 111, 633–652.

Clottes, J.
(2003). *Chauvet Cave – The Art of Earliest Times*. Salt Lake City, Utah: University of Utah Press.

Clutton-Brock, J.
(1970). The origins of the dog. In *Science in Archaeology*, ed. D. Brothwell & E. S. Higgs, pp. 303–309. New York City: Praeger.

Clutton-Brock, J.
(1977). Man-made dogs. *Science*, 197, 1340–1342.

Clutton-Brock, J.
(1981). *Domesticated Animals from Early Times*. Austin, Texas: The University of Texas Press.

Clutton-Brock, J.
(1984). Dog. In *Evolution of Domesticated Animals*, ed. I. L. Mason, pp. 198–211. London, England, and New York City: Longman.

Clutton-Brock, J.
(1995). Origins of the dog: Domestication and early history. In *The Domestic Dog: Its Evolution, Behaviour and Interactions with People*, ed. J. Serpell, pp. 7–20. Cambridge, England: Cambridge University Press.

Clutton-Brock, J.
(1999). *A Natural History of Domesticated Mammals* (2nd edition). Cambridge, England: Cambridge University Press.

Clutton-Brock, J.
(2001). Ritual burials of a dog and six domestic donkeys. In *Excavations at Tell Brak*, Volume 2: *Nagar in the Third Millennium B.C.*, ed. D. Oates, J. Oates. & H. McDonald, pp. 327–338. London and Cambridge, England: British School of Archaeology in Iraq, and McDonald Institute for Archaeological Research.

Clutton-Brock, J. & Hammond, N.
(1994). Hot dogs: Comestible canids in preclassic Maya culture at Cuello, Belize. *Journal of Archaeological Science*, 21, 819–826.

Clutton-Brock, J. & Jewell, P.
(1993). Origin and domestication of the dog. In *Miller's Anatomy of the Dog* (3rd edition), by H. E. Evans, pp. 21–31. Philadelphia, Pennsylvania: W. B. Saunders.

Clutton-Brock, J. & Kitchener, A. C.
(2000). An anomalous wolf, *Canis lupus arctos*, from Ellesmere Island and the problem of hybridisation between wild and domestic canids. In *Dogs through Time: An Archaeological Perspective*, ed. S. J. Crockford, pp. 257–268. Oxford, England. BAR International Series, 889.

Clutton-Brock, J. & Noe-Nygaard, N.
(1990). New osteological and C-isotope evidence on Mesolithic dogs: Companions to hunters and fishers at Star Carr, Seamer Carr, and Kongemose. *Journal of Archaeological Science*, 17, 643–653.

Cock, A. G.
(1966). Genetical aspects of metrical growth and form in animals. *The Quarterly Review of Biology*, 41(2), 131–190.

Cohen, S. P.
(2002). Can pets function as family members? *Western Journal of Nursing Research*, 24(6), 621–638.

Colcord, J. C.
(1938). *Songs of American Sailormen*. New York City: W. W. Norton.

Cole, L. C.
(1954). The population consequences of life history phenomena. *The Quarterly of Review of Biology*, 29, 103–137.

Coleman, J. T.
(2004). *Vicious: Wolves and Men in America*. New Haven, Connecticut: Yale University Press.

Collins, B. J.
(1990). The puppy in Hittite ritual. *Journal of Cuneiform Studies*, 42, 211–226.

Collins, S.
(1999). "On Land Where the Indians Lived" – Harry Martin Trowbridge a Wyandotte County avocationalist. *Bulletin* (Society for American Archaeology), 17(5).

Coppinger, R. & Coppinger, L.
(2001). *Dogs – A Startling New Understanding of Canine Origin, Behavior, & Evolution*. New York City: Scribner.

Coppinger, R. & Feinstein, M.
(1991). 'Hark! Hark! The dogs do bark...' and bark and bark. *Smithsonian*, 21(10), 119–129.

Coppinger, R. & Schneider, R.
(1995). Evolution of working dogs. In *The Domestic Dog: Its Evolution, Behaviour and Interactions with People*, ed. J. Serpell, pp. 21–47. Cambridge, England: Cambridge University Press.

Corbett, L.
(1995). *The Dingo in Australia and Asia*. London, England, and Ithaca, New York: Comstock/Cornell.

Coren, S.
(1994). *The Intelligence of Dogs: Canine Consciousness and Capabilities*. New York City: Free Press.

Costello, J. C.
(2004). The Chinese in Gum San ("Golden Mountain"). *The SAA Archaeological Record*, 4(5), 14–17.

Crabtree, P. J.
(1993). Early animal domestication in the Middle East and Europe. In *Archaeological Method and Theory* (Vol. 5), ed. M. B. Schiffer, pp. 201–245. New York City: Academic Press.

Cridlebaugh, P. A., Breitberg, E. & Jones, D.
(1986). Inhumations and human skeletal remains. In *Penitentiary Branch: A Late Archaic Cumberland River Shell Midden in Middle Tennessee*, by P. A. Cridlebaugh, pp. 43–65. Nashville, Tennessee: Tennessee Department of Conservation, Division of Archaeology, Report of Investigations, 4.

Crockford, S. J.
(1997). *Osteometry of Makah and Coast Salish Dogs*. Burnaby, British Columbia: Simon Fraser University Archaeology Press.

Crockford, S. J.
(2000a, editor). *Dogs through Time: An Archaeological Perspective*. Oxford, England: BAR International Series, 889.

Crockford, S. J.
(2000b). Dog evolution: A role for thyroid hormone physiology in domestication changes. In *Dogs through Time: An Archaeological Perspective*, ed. S. J. Crockford, pp. 11–20. Oxford, England: BAR International Series, 889.

Crockford. S. J.
(2000c). A commentary on dog evolution: Regional variation, breed development and hybridisation with wolves. In *Dogs through Time: An Archaeological Perspective*, ed. S. J. Crockford, pp. 295–312. Oxford, England: BAR International Series, 889.

Crockford, S. J.
(2002). Animal domestication and heterochronic speciation: The role of thyroid hormone. In *Human Evolution through Developmental Change*, ed.

N. Minugh-Purvis & K. J. McNamara, pp. 122–153. Baltimore, Maryland: Johns Hopkins University Press.

Crockford, S. J.
(2006). *Rhythms of Life: Thyroid Hormones & the Origin of Species.* Victoria, British Columbia: Trafford.

Crockford, S. J. & Pye, C. J.
(1997). Forensic reconstruction of prehistoric dogs from the northwest coast. *Canadian Journal of Archaeology*, 21(2), 149–153.

Cross, E. C.
(1940). Arthritis among wolves. *Canadian Field-Naturalist*, 54, 2–4.

Cross, I.
(2001). Music, mind and evolution. *Psychology of Music*, 29(1), 95–102.

Curle, A. O. & Scot, F. S. A.
(1934). An account of further excavations at Jarlshof, Sumburgh, Shetland, in 1932 and 1933, on behalf of H. M. Office of Works. *Proceedings of the Society of Antiquaries of Scotland*, March 12, 224–319.

Curren, C. B., Jr.
(1981). The dog burials at Phipps Bend. In *The Phipps Bend Archaeological Project*, ed. R. H. Lafferty III, pp. 387–396. Tuscaloosa, Alabama and Chattanooga, Tennessee: University of Alabama, Office of Archaeological Research, Research Series, 4, and Tennessee Valley Authority, Publications in Anthropology, 26.

Dahr, E.
(1942). *Über die Variation der Hirnschale bei wilden und zahmen Caniden. Arkiv für Zoologi*, 33A(16), 1–56 (in German, with an English summary).

Damas, D.
(1984). Copper Eskimo. In *Handbook of North American Indians*, Volume 5: *Arctic*, ed. D. Damas, pp. 397–414. Washington, D. C.: Smithsonian Institution.

Darimont, C. T., Reimchen, T. E. & Paquet, P. C.
(2003). Foraging behavior by gray wolves on salmon streams in coastal British Columbia. *Canadian Journal of Zoology*, 81(2), 349–353.

Dark, P.
(2003). Dogs, a crane (not duck) and diet at Star Carr: A response to Schulting and Richards. *Journal of Archaeological Science*, 30(10), 1353–1356.

Daróczi-Szabó, M.
(2006). Variability in medieval dogs from Hungary. In *Dogs and People in Social, Working, Economic or Symbolic Interaction*, ed. L. M. Snyder & E. A. Moore, pp. 85–94. Oxford, England: Oxbow Books (Proceedings of the 9th ICAZ Conference, Durham, England, 2002).

Darwent, C. M.
(2004). The highs and lows of Arctic mammals: Temporal change and regional variability in Paleoeskimo subsistence. In *Colonisation, Migration, and Marginal Areas: A Zooarchaeological Approach*, ed. M. Mondini, S. Muñoz & S. Wickler, pp. 62–73. Oxford, England: Oxbow Books (Proceedings of the 9th ICAZ Conference, Durham, England, 2002).

Darwent, C. & Gilliland, J. E.
(2001). Osteological analysis of domestic dogs from burials in southern Missouri. *Missouri Archaeologist*, 62, 149–169.

Darwin, C.
(1868). *The Variation of Animals and Plants under Domestication*, Volume 1. New York City: Appleton.

Darwin, C.
(1962). *The Origin of Species by Means of Natural Selection or the Preservation of Favoured Races in the Struggle For Life*. New York City: MacMillan (originally published in 1859).

Davis, R. P. S., Jr., Lambert, P. M., Steponaitis, V. P., Larsen, C. S. & Ward, H. T.
(1998). *An Abbreviated NAGPRA Inventory of the North Carolina Archaeological Collection*. Chapel Hill, North Carolina: The University of North Carolina, Research Laboratories of Archaeology.

Davis, S. J. M.
(1987). *The Archaeology of Animals*. New Haven, Connecticut: Yale University Press.

Davis, S. J. M. & Valla, F.
(1978). Evidence for domestication of the dog 12,000 years ago in the Natufian of Israel. *Nature*, 276, 608–610.

Day, L. P.
(1984). Dog burials in the Greek world. *American Journal of Archaeology*, 88, 21–32.

Day, S. P.
(1996). Dogs, deer and diet at Star Carr: a reconsideration of C-isotope evidence from early Mesolithic dog remains from the Vale of Pickering, Yorkshire, England. *Journal of Archaeological Science*, 23, 783–787.

Dayan, T.
(1994). Early domesticated dogs of the Near East. *Journal of Archaeological Science*, 21, 633–640.

Degerbøl, M.
(1927). Uber prähistorische dänische Hunde. *Videnskabelige Meddelelser fra Dansk Naturhistorisk i København*. Copenhagen, Denmark: Selskabets Bestyrelse (in German).

Degerbøl, M.
   (1934). Zoological Appendix – Animal bones from the Eskimo settlement in
   Dødemandsbugten, Clavering Island. In Dødemandbugten – An Eskimo
   Settlement on Clavering Island, by H. Larsen, pp. 173–181. Copenhagen,
   Denmark: *Meddelelser om Grønland*, 102(1).

Degerbøl, M.
   (1939). *Dyreknogler*. In *Bundsø, en yngre stenalders Boplads paa Als*, by T. Mathiassen, pp. 85–198. Copenhagen, Denmark: *Aarbøger for Norkisk Oldkyn-
   dighed og Historie*, 1939 (in Danish).

Degerbøl, M.
   (1946). Untitled description of animal bones from Vedbaek. In *En Boplads fra
   Aeldre Stenalder ved Vedbaek Boldbaner*, by T. Mathiassen, pp. 32–33. Søllerød
   Kommune, Denmark: *Søllerødbogen*, 1946 (in Danish).

Degerbøl, M.
   (1961). On a find of a preboreal domestic dog (*Canis familiaris* L.) from Star
   Carr, Yorkshire, with remarks on other Mesolithic dogs. *Proceedings of the
   Prehistoric Society*, 27, 35–55.

DeJarnette, D. L.
   (1952). Alabama archaeology: A summary. In *Archaeology of the Eastern United
   States*, ed. J. B. Griffin, pp. 272–284. Chicago, Illinois: University of Chicago
   Press.

DeJarnette, D. L. & Wimberly, S. B.
   (1941). *The Bessemer Site: Excavation of Three Mounds and Surrounding Village
   Areas Near Bessemer, Alabama*. Tuscaloosa, Alabama: Geological Survey of
   Alabama, *Museum Paper*, 17.

Denig, E. T.
   (2000). *The Assiniboine*. Norman, Oklahoma: University of Oklahoma Press.

Dent, J.
   (2003). Fully intact dog skeleton is the first for Winslow site. *ASM Ink*, 29(9),
   3 (Newsletter of the Archaeological Society of Maryland, Inc.).

Dent, R. J.
   (1995). *Chesapeake Prehistory – Old Traditions, New Directions*. New York City:
   Plenum.

de Pastino, B.
   (2006). Photo in the news: Dog mummies found in ancient Peru pet cemetery.
   National Geographic News, 25 September, 2 pages (electronic) http://news.
   national/geographic.com/news/2006/09/060925-dog-mummy.html.

Derr, M.
   (1996). The making of a marathon mutt. *Natural History*, 105(3), 34–41.

References

Derr, M.
(2004a). *Dog's Best Friend: Annals of the Dog–Human Relationship*. Chicago, Illinois: University of Chicago Press (originally published in 1997).

Derr, M.
(2004b). *A Dog's History of America: How our Best Friend Conquered, and Settled a Continent*. New York City: North Point Press.

De Waal, F.
(2001). *The Ape and the Sushi Master – Cultural Reflections of a Primatologist*. New York City: Basic Books.

Dewey, T. & Smith, J.
(2002). Canis lupus (On-line). Animal Diversity Web, accessed 27 May 2006 (http://animaldiversity.ummz.umich.edu/site/accounts/information/Canis_lupus.html.).

Diamond, J.
(1987). The worst mistake in the history of human race. *Discover Magazine*, May, 80–84.

Diamond, J.
(1998). Ants, crops, and history. *Science*, 281, 1974–1975.

Diamond, J.
(2002). Evolution, consequences and future of plant and animal domestication. *Nature*, 418, 700–707.

Dikov, N. N.
(1996). The Ushki sites, Kamchatka Peninsula. In *American Beginnings – The Prehistory and Paleoecology of Beringia*, ed. F. H. West, pp. 244–250. Chicago: University of Chicago Press.

Dittmar, K., Ludewig, E., Conlogue, J. & Guillen, S.
(2003). Bone-pathologies in Peruvian dog mummies from the Chiribaya Culture, southern Peru, determined by X-ray, pp. 113–15. Copenhagen, Denmark: University of Copenhagen Press (Proceedings of the 4th World Congress on Mummy Studies, Nuuk, Greenland 2001).

Dixon, W. J.
(1981). BMDP Statistical Software. Los Angeles, California: University of California Press.

Dobney, K. & Larson, G.
(2006). Genetics and animal domestication: New windows on an elusive process. *Journal of Zoology*, 269(2), 261–271.

Dolan R. J.
(1998). A cognitive affective role for the cerebellum. *Brain*, 121, 545–546.

References

Donn, J.
(2007). War dogs win hearts – Handlers point to raw courage, devotion of pawed comrades in arms. *Memphis Commercial Appeal* (newspaper): 12 August, A10 (www.commercialappeal.com, an Associated Press feature).

Douglas, K.
(2000). Mind of a dog. *New Scientist*, March, 22–27.

Dowd, J. T.
(1989). *The Anderson Site: Middle Archaic Adaptation in Tennessee's Central Basin*. Knoxville, Tennessee: Tennessee Anthropological Association, *Miscellaneous Paper*, 13.

Drake, A. G. & Klingenberg, C. P.
(2008). The pace of morphological change: Historical transformation of skull shape in St. Bernard dogs. *Proceedings of the Royal Society* B, 275, 71–76.

Ducos, P.
(1978). "Domestication" defined and methodological approaches to its recognition in faunal assemblages. In *Approaches to Faunal Analysis in the Middle East*, ed. R. H. Meadow & M. A. Zeder, pp. 53–56. Cambridge, Massachusetts: Peabody Museum of Archaeology and Ethnology, Bulletin, 2.

Dukas, R.
(2004). Causes and consequences of limited attention. *Brain, Behavior and Evolution*, 63(4), 197–210.

Dumond, D.
(1984). Prehistory of the Bering Sea region. In *Handbook of North American Indians*, Volume 5: *Arctic*, ed. D. Damas, pp. 94–105. Washington, D.C.: Smithsonian Institution.

Dumond, D.
(1987). *The Eskimos and Aleuts* (2nd edition). New York: Thames and Hudson.

Dumond, D. E. & Minor, R.
(1983). *Archaeology in the John Day Reservoir: The Wildcat Canyon Site 35-GM-9*. Eugene, Oregon: University of Oregon, Anthropological Papers, 30.

Dunbar, J. B.
(1880). The Pawnee Indians – Their habits and customs. *Magazine of American History*, 5(5), 321–345.

Dunnell, R. C.
(1971). *Systematics in Prehistory*. New York City: The Free Press.

Eichhorn, G. E. & Jones, S. B.
(2000). *The Dog Album – Studio Portraits of Dogs and Their People*. New York: Stewart, Tabori and Chang.

Emslie, S. D.
    (1978). Dog burials from Mancos Canyon, Colorado. *The Kiva*, 43 (3–4), 167–182.

Epstein, H.
    (1955). Domestication features in animals as functions of human society. *Agricultural History*, 29 (4), 137–146.

Epstein, H.
    (1971). *Origins of the Domestic Animals of Africa*, Volume 1. New York City: Africana.

Evans, H. E.
    (1993). *Miller's Anatomy of the Dog* (3rd edition). Philadelphia, Pennsylvania: W. B. Saunders.

Ezzo, J. A. & Stiner, M. C.
    (2000). A Late Archaic dog burial from the Tucson Basin, Arizona. *The Kiva*, 66(2), 291–305.

Fadiman, A.
    (1997). *The Spirit Catches You and You Fall Down*. New York City: Farrar, Strauss and Giroux.

Fagan, B. M.
    (1991). *Ancient North America – The Archaeology of a Continent*. New York City: Thames and Hudson.

Fagan, B. M.
    (2004). *The Great Journey – The Peopling of Ancient America*. Gainesville, Florida: University Press of Florida.

Fahlander, F.
    (2008). A piece of the Mesolithic – Horizontal stratigraphy and bodily manipulations at Skateholm. In *The Materiality of Death: Bodies, Burials, Beliefs*, ed. F. Fahlander & T. Oestigaard, pp. 29–45. Oxford, England: Archaeopress.

Feddersen-Petersen, D. U.
    (2004). Communication in wolves and dogs. In *Encyclopedia of Animal Behavior*, ed. M. Bekoff, pp. 385–394. Westport, Connecticut: Greenwood Press.

Feddersen-Petersen, D. U.
    (2007). Social behaviour of dogs and related canids. In *The Behavioural Biology of Dogs*, ed. P. Jensen, pp. 105–119. Wallingford, England, and Cambridge, Massachusetts: CAB International.

Feder, K. L.
    (2007). *The Past in Perspective – An Introduction to Human Prehistory* (4th edition). New York City: McGraw Hill.

Fentress, J. C.
> (1967). Observations on the behavioral development of a hand-reared male timber wolf. *American Zoologist*, 7, 339–351.

Fiedel, S. J.
> (2005). Man's best friend – Mammoth's worst enemy? A speculative essay on the role of dogs in Paleoindian colonization and megafaunal extinction. *World Archaeology*, 37(1), 11–25.

Fiedel, S. J. & Haynes, G.
> (2004). A premature burial: Comments on Grayson and Meltzer's "requiem for overkill." *Journal of Archaeological Science*, 31(1), 121–131.

Fiennes, R.
> (1976). *The Order of Wolves*. Indianapolis, Indiana: Bobbs-Merrill.

Fischer, A., Olsen, J, Richards, M., Heinemeier, J., Sveinbjörnsdóttir, Á. E. & Bennike, P.
> (2007). Coast-inland mobility and diet in the Danish Mesolithic and Neolithic: Evidence from stable isotope values of humans and dogs. *Journal of Archaeological Science*, 34, 2125–2150.

Fitzhugh, W. H.
> (1984). Paleoeskimo cultures of Greenland. In *Handbook of North American Indians*, Volume 5: *Arctic*, ed. D. Damas, pp. 528–539. Washington, D.C.: Smithsonian Institution.

Flannery, K. V.
> (1967). Vertebrate fauna and hunting patterns. In *The Prehistory of the Tehuacan Valley*, Volume 1, ed. D. S. Byers, pp. 132–177. Austin, Texas: University of Texas Press.

Ford, J. A.
> (1940). *Crooks Site, a Marksville Period Burial Mound in La Salle Parish, Louisiana*. New Orleans, Louisiana: Louisiana Department of Conservation, Anthropological Study, 3.

Fox, M. W.
> (1971). *Integrative Development of Brain and Behavior in the Dog*. Chicago, Illinois: University of Chicago Press.

Fox, M. W.
> (1978). *The Dog: Its Domestication and Behavior*. New York City: Garland STPM Press.

Fox-Dobbs, K., Bump, J. K., Peterson, R. O., Fox, D. L. & Koch. P. L.
> (2007). Carnivore-specific stable isotope variables and variation in the foraging ecology of modern and ancient wolf populations: Case studies from Isle Royale, Minnesota, and La Brea. *Canadian Journal of Zoology*, 85(4), 458–471.

Frame, P. F., Cluff, H. D. & Hik, D. S.
(2008). Wolf reproduction in response to caribou migration and industrial development on the Central Barrens of mainland Canada. *Arctic*, 61(2), 134–142.

Freuchen, P.
(1935). Part II. Field notes and biological observations. In *Report of the Fifth Thule Expedition 1921–1924*, Volume 2: 4–5, pp. 68–278. Copenhagen, Denmark: Gyldendal.

Friedmann, E. & Thomas, S. A.
(1998). Pet ownership and one-year survival after acute myocardial infarction in the Cardiac Arrhythmia Suppression Trial (CAST). In *Companion Animals in Human Health*, ed. C. C. Wilson & D. C. Turner, pp. 187–201. Thousand Oaks, California: Sage Publications.

Friesen, T. M. & Arnold, C. D.
(2008). The timing of the Thule migration: New dates from the western Canadian Arctic. *American Antiquity*, 73(3), 527–538.

Fritz, S. H., Stephenson, R. O., Hayes, R. D. & Boitani, L.
(2003). Wolves and humans. In *Wolves – Behavior, Ecology, and Conservation*, ed. L. D. Mech & L. Boitani, pp. 289–316. Chicago, Illinois: University of Chicago Press.

Fukuzawa, M., Mills, D. S. & Cooper, J. J.
(2005). The effect of human command phonetic characteristics on auditory cognition in dogs (*Canis familiaris*). *Journal of Comparative Psychology*, 119(1), 117–120.

Fuller, T. K., Mech, L. D. & Cochrane, J. F.
(2003). Wolf population dynamics. In *Wolves – Behavior, Ecology, and Conservation*, ed. L. D. Mech & L. Boitani, pp. 161–191. Chicago, Illinois: University of Chicago Press.

Funk, H.
(2008). Introduction. In *On the Skulls and Lower Jaws of the Japanese Stone Age Dog Races*, by K. Hasebe, pp. 5–99. Paderborn, Germany: Lykos Press.

Furst, P. F.
(1998). Shamanic symbolism, transformation, and deities in west Mexican funerary art. In *Ancient West Mexico – Art and Archaeology of the Unknown Past*, ed. R. F. Townsend, pp. 169–189. Chicago, Illinois: The Art Institute of Chicago, Thames and Hudson.

Gácsi, M., Miklósi, Á., Varga, O., Topál, J. & Csányi, V.
(2004). Are readers of our faces readers of our minds? Dogs (*Canis familiaris*) show situation-dependent recognition of human's attention. *Animal Cognition*, 7, 144–153.

Galik, A.
  (2000). Dog remains from the Late Hallstatt period (500–400 BC) of the Chimney Cave Durezza, near Villach, Austria. In *Dogs through Time: An Archaeological Perspective*, ed. S. J. Crockford, pp. 129–37. Oxford, England: BAR International Series, 889.

Galton, F.
  (1865). The first steps towards the domestication of animals. *Transactions of the Ethnological Society of London*, 3, 122–138.

Galton, F.
  (1973). *Inquiries into Human Faculty and Its Development* (2nd edition). New York City: AMC Press (reprint of the 1907 edition, published by J. M. Dent & Sons, London, England, and by E. P. Dutton & Company, New York City; first edition published in 1883, by MacMillan Publishers, London, England).

Garcia, M. A.
  (2005). *Ichnologie générale de la grotte Chauvet. Bulletin de la Société Préhistorique Française*, 102, 103–108 (in French).

Gautier, A.
  (1990). *La Domestication. Et L'Homme Créa Ses Animaux*. Paris, France: Editions Errance (in French).

Gautier, A.
  (1992a). Domestication animale et théorie de l-évolution. *Revue des Questions Scientifiques*, 163(2), 147–160 (in French).

Gautier, A.
  (1992b). Domestication animale et animaux domestiques prétendument oubliés. *Colloques D'Histoire des Connaissances Zoologiques*, 3, 31–36 (in French).

Gazzola, A., Bertelli, I., Avanzinelli, E., Tolosano, A., Bertotto, P. & Appolonio, M.
  (2005). Predation by wolves (*Canis lupus*) on wild and domestic ungulates of the western alps, Italy. *Journal of Zoology*, 266(2), 205–213.

Gazzola, A., Capitani, C., Mattioli, L. & Apollonio, M.
  (2008). Livestock damage and wolf presence. *Journal of Zoology*, 274(3), 261–269.

Gentry, A., Clutton-Brock, J. & Groves, C. P.
  (2004). The naming of wild animal species and their domestic derivatives. *Journal of Archaeological Science*, 31, 645–651.

Germonpré, M., Sablin, M. V., Stevens, R. E., Hedges, R. E. M., Hofreiter, M., Stiller, M. & Després, V. R.
  (2009). Fossil dogs and wolves from Paleolithic sites in Belgium, the Ukraine and Russia: Osteometry, ancient DNA and stable isotopes. *Journal of Archaeological Science*, 36(2), 473–490.

Giesel, J. T.
(1976). Reproductive strategies as adaptations to life in temporally het-
erogeneous environments. *Annual Review of Ecology and Systematics*, 7,
57–79.

Gilbert, M. T. P., Kivisild, T., Grønnow, B. & twenty others
(2008). Paleo-Eskimo MtDNA genome reveals matrilineal discontinuity in
Greenland. *Science*, 320(5884), 1787–1789.

Gittleman, J. L.
(1986). Carnivore brain size, behavioral ecology, and phylogeny. *Journal of
Mammalogy*, 67(1), 23–36.

Giuffra, E., Kijas, J. M. H., Amarger, V., Carlborg, Ö, Jeon, J.-T. & Andersson, L.
(2000). The origin of the domestic pig: Independent domestication and subsequent introgression. *Genetics*, 154, 1785–1791.

Glob, P. V.
(1935). Eskimo Settlements in Kempe Fjord and King Oscar Fjord. Copenhagen, Denmark: *Meddelelser om Grønland*, 102(2).

Glover, J. M.
(2003). Olaus Murie's spiritual connection with wilderness. *International Journal of Wilderness*, 9(1), 4–8.

Goebel, T. & Slobodkin, S. B.
(1999). The colonization of western Berengia – Technology, ecology, and
adaptations. In *Ice Age People of North America – Environments, Origins,
and Adaptations*, ed. R. Bonnischen & K. Turnmire, pp. 104–155. Corvallis,
Oregon: Oregon State University Press.

Gotfredsen, A. B.
(1996). The fauna from the Saqqaq site of Nipisat 1, Sisimiut District, west
Greenland: Preliminary results. In *The Paleo-Eskimo Cultures of Greenland:
New Perspectives in Greenlandic Archaeology*, ed. B. Grønnow & J. Pind,
pp. 97–110. Copenhagen, Denmark: The Danish Polar Center.

Gotfredsen, A. B.
(2004). Mammals. In Nipisat – A Saqqaq Culture Site in Sisimiut, Central West
Greenland, by A. B. Gotfredsen & T. Møbjerg, pp. 142–191. Copenhagen,
Denmark: *Meddelelser om Grønland*, Man & Society, 31.

Gotfredsen, A. B., Gravlund, P. & Rosenlund, K.
(1994). *I Skjoldungen spiste de isaer fisk og saeler. Forskning i Grønland/
Tusaat*, 1994(1–2), 46–55 (in Danish and Greenlandic).

Gotfredsen, A. B., Gulløv, H. C. & Rosing, M.
(1992). *Østgrønlandske ressourcer – rapport fra projekt Skjoldungen. Forskning i
Grønland/Tusaat*, 1992–1, 19–27 (in Danish and Greenlandic).

References

Gottwald, B., Mihajlovic, Z., Wilde, B. & Mehdorn, H. M.
  (2003). Does the cerebellum contribute to specific aspects of attention? *Neuropsychologia*, 41(11), 1452–1460.

Gould, S. J.
  (1966). Allometry and size in ontogeny and phylogeny. *Cambridge Philosophical Society, Biological Reviews*, 41(4), 587–640.

Gould, S. J.
  (1977). *Ontogeny and Phylogeny*. Cambridge, Massachusetts: Harvard University Press.

Graham, H. & Graham, J.
  (1990). After the flood: The Shadyside, Ohio, disaster. *Dog Sports*, October (two electronic pages).

Grandin, T. & Johnson, C.
  (2005). *Animals in Translation – Using the Mysteries of Autism to Decode Animal Behavior*. Orlando, Florida: Harcourt.

Grandin, T. & Johnson, C.
  (2009). *Animals Make Us Human – Creating the Best Life for Animals*. Boston, Massachusetts: Houghton Mifflin Harcourt.

Gräslund, A.-S.
  (2004). Dogs in graves – A question of symbolism? In *Pecus. Man and Animal in Antiquity*, ed. B. S. Frizell, pp. 167–176. Rome, Italy: Proceedings of the Conference at the Swedish Institute in Rome, September 9–12, 2002.

Grayson, D. K.
  (1988). Danger Cave, Last Supper Cave, and Hanging Rock Shelter: The Faunas. New York City: American Museum of Natural History, *Anthropological Papers*, 66(1).

Grayson, D. K.
  (2006). Review of: *Twilight of the Mammoths: Ice Age Extinctions and the Rewilding of America*, by P. S. Martin, University of California Press, Berkeley, California (2005). *The Quarterly Review of Biology*, 81(3), 259–264.

Grayson, D. K. & Meltzer, D. J.
  (2003). A requiem for North American overkill. *Journal of Archaeological Science*, 30, 585–593.

Green, M.
  (2001). *Dying for the Gods – Human Sacrifice in Iron Age & Roman Europe*. Charleston, South Carolina: Tempus.

Green, R. C. & Davidson, J. M.
  (1969). *Archaeology in Western Samoa*, Volume 1. Aukland, New Zealand: Aukland Institute and Museum, Bulletin, 6.

Gregory, L. B.
(1980). The Hatch Site: A preliminary report. *Quarterly Bulletin of the Archaeological Society of Virginia*, 34(4), 239–248.

Griffin, J. B.
(1967). Eastern North American archaeology: A summary. *Science*, 156(3772), 175–191.

Grogan, J.
(2005). *Marley & Me – Life and Love with the World's Worst Dog*. New York City: Harper.

Grønnow, B.
(1988). Prehistory in permafrost – Investigations at the Saqqaq site, Qeqertasussuk, Disco Bay, west Greenland. *Journal of Danish Archaeology*, 7, 24–39.

Grønnow, B.
(1994). Qeqertasussuk: The archaeology of a frozen Saqqaq site in Disko Bugt, west Greenland. In *Threads of Arctic Prehistory: Papers in Honour of William E. Taylor, Jr.*, ed. D. Morrison & J.-L. Pilon, pp. 197–238. Hull, Quebec: Canadian Museum of Civilization, Mercury Series, Archaeological Survey of Canada, Paper, 149.

Grønnow, B. & Meldgaard, M.
(1988). *Boplads i Dybfrost. Naturens Verden*, 11–12, 409–440 (in Danish).

Grün, R., Shackleton, N. J. & Deacon, H. J.
(1990). Electron-spin resonance dating of tooth enamel from Klasies River Mouth Cave. *Current Anthropology*, 31(4), 427–432.

Guilday, J. E., Parmalee, P. W. & Tanner, D. P.
(1962). Aboriginal butchering techniques at the Eschelman site (36La12), Lancaster County, Pennsylvania. *Pennsylvania Archaeologist*, 32, 59–83.

Guilday, J. E. & Tanner, D. P.
(1966). Appendix A – Animal Remains from the Westmoreland-Barber site (40Mi-11), Marion County, Tennessee. In *Westmoreland-Barber Site (40Mi-11) Nickajack Reservoir Season II*, ed. C. H. Faulkner & J. B. Graham, pp. 138–148. Knoxville, Tennessee: University of Tennessee, Department of Anthropology, report submitted to the U. S. National Park Service.

Haag, W. G.
(1948). *An Osteometric Analysis of Some Aboriginal Dogs*. Lexington, Kentucky: The University of Kentucky, Department of Anthropology, Reports in Anthropology, 7(3).

Haag, W. G. & Heizer, R. F.
(1953). A dog burial from the Sacramento Valley. *American Antiquity*, 18(3), 263–265.

References

Hale, A. & Salls, R. A.
   (2000). The canine ceremony: Dog and fox burials on San Clemente Island. *Pacific Coast Archaeological Society Quarterly*, 36(4), 80–94.

Hale, H. S.
   (1983). Chapter IX – Analysis of faunal material from site 1FR310. In *Archaeological Investigations in the Cedar Creek and Upper Bear Creek Reservoirs*, ed. E. M. Futato, pp. 313–334. Tuscaloosa, Alabama, and Chattanooga, Tennessee: University of Alabama, Office of Archaeological Research, Report of Investigations, 13, and Tennessee Valley Authority, Publications in Anthropology, 32.

Hall, L.
   (2000). *Prince and Other Dogs, 1850–1940*. London: Bloomsbury.

Halpern, B.
   (2000). The canine conundrum at Ashkelon: A classical connection? In *The Archaeology of Jordan and Beyond: Essays in Honor of James A . Sauer*, ed. L. E. Stager, J. A. Greene & M. D. Coogan, pp. 133–144. Winona Lake, Indiana: Eisenbrauns.

Hamerow, H.
   (2006). 'Special Deposits' in Anglo-Saxon settlements. *Medieval Archaeology*, 50, 1–30.

Hancock, J. L. & Rowlands, I. W.
   (1949). The physiology of reproduction in the dog. *The Veterinary Record*, 61, 771–779.

Handley, B. M.
   (2000). Preliminary results in determining dog types from prehistoric sites in the northeastern United States. In *Dogs through Time: An Archaeological Perspective*, ed. S. J. Crockford, pp. 205–215. Oxford, England: BAR International Series, 889.

Haraway, D.
   (2003). For the love of a good dog: Webs of action in the world of dog genetics. In *Genetic Nature/Culture: Anthropology and Science between the Two-Culture Divide*, ed. A. Goodman, M. S. Lindee & D. Heath, pp. 111–131. Berkeley, California: University of California Press.

Harcourt, R. A.
   (1974). The dog in prehistoric and early historic Britain. *Journal of Archaeological Science*, 1, 151–175.

Hardy, A. T.
   (2000). Religious aspects of the material remains from San Clemente Island. *Pacific Coast Archaeological Society Quarterly*, 36(1), 78–96.

References

Hare, B.
(2007). From nonhuman to human mind: What changed and why? *Current Directions in Psychological Science*, 16(2), 60–64.

Hare, B., Brown, M., Williamson, C., & Tomasello, M.
(2002). The domestication of social cognition in dogs. *Science*, 298(5598), 1634–1636.

Hare, B., Plyusnina, I., Ignacio, N., Schepina, O., Stepika, A., Wrangham, R. & Trut, L.
(2005). Social cognitive evolution in captive foxes is a correlated by-product of experimental domestication. *Current Biology*, 15, 226–230.

Hare, B. & Tomasello, M.
(2005). Human-like social skills in dogs? *Trends in Cognitive Sciences*, 9(9), 439–444.

Harp, E.
(1984). History of archaeology after 1945. In *Handbook of North American Indians*, Volume 5: *Arctic*, ed. D. Damas, pp. 17–22. Washington, D.C.: Smithsonian Institution.

Harrington, F. H.
(1986). Timber wolf howling playback studies: Discrimination of pup from adult howls. *Animal Behavior*, 34, 1575–1577.

Harrington, F. H.
(1989). Chorus howling by wolves: Acoustic structure, pack size and the Beau Geste effect. *Bioacoustics*, 2, 117–136.

Harrington, F. H. & Asa, C. S.
(2003). Wolf communication. In *Wolves – Behavior, Ecology, and Conservation*, ed. L. D. Mech & L. Boitani, pp. 66–103. Chicago, Illinois: University of Chicago Press.

Harrington, F. H. & Mech, L. D.
(1978). Wolf vocalization. In *Wolf and Man: Evolution in Parallel*, ed. R. L. Hall & H. S. Sharp, pp. 109–132. New York City: Academic Press.

Harris, S.
(1977). Spinal arthritis (*Spondylosis deformans*) in the red fox *Vulpes vulpes* with some methodology of relevance to zooarchaeology. *Journal of Archaeological Science*, 4, 183–195.

Hart, L.
(1995). Dogs as human companions: A review of the relationship. In *The Domestic Dog: Its Evolution, Behaviour and Interactions with People*, ed. J. Serpell, pp. 161–178. Cambridge, England: Cambridge University Press.

References

Harvey, A. E.
  (1979). *Oneota Culture in Northwestern Iowa*. Iowa City, Iowa: The University of Iowa, Office of the State Archaeologist, Report, 12.

Hasebe, K.
  (2008). *On the Skulls and Jaws of the Japanese Stone Age Dog Races*. Paderborn, Germany: Lykos Press (edited, translated, and introduced by Holger Funk).

Hatting, T.
  (1978). Lidsø: Zoological remains from a Neolithic settlement. In *The Final TRB Culture in Denmark*, ed. K. Davidsen, pp. 193–207. Copenhagen, Denmark: Arkaeologiske Studier, 5.

Heath, B.
  (2003). Archaeology at Thomas Jefferson's Poplar Forest. *Newsletter* (Council for Northeast Historical Archaeology, Boston, Massachusetts), 56, 11–13.

Hedhammar, Å. & Hultin-Jäderlund, K.
  (2007). Behaviour and disease in dogs. In *The Behavioural Biology of Dogs*, ed. P. Jensen, pp. 243–261. Wallingford, England, and Cambridge, Massachusetts: CAB International.

Helmer, J. W.
  (1994). Resurrecting the spirit(s) of Taylor's "Carlsberg Culture": Cultural traditions and cultural horizons in Eastern Arctic prehistory. In *Threads of Arctic Prehistory: Papers in Honour of William E. Taylor, Jr.*, ed. D. Morrison & J.-L. Pilon, pp. 15–34. Hull, Quebec: Canadian Museum of Civilization, Mercury Series, Archaeological Survey of Canada, Paper, 149.

Hemmer, H.
  (1976). Man's strategy in domestication – A synthesis of new research trends. *Experientia, Basel*, 32, 663–666.

Henderson, N.
  (1994). Replicating dog travois travel on the Northern plains. *Plains Anthropologist*, 39(148), 145–159.

Heye, G. F. & Pepper G. H.
  (1915). *Exploration of a Munsee Cemetery near Montague, New Jersey*. New York City: Contributions from the Museum of the American Indian, Heye Foundation, 2(1).

Higham, C. F. W., Kijngam, A. & Manly, B. J. F.
  (1980). An analysis of prehistoric canid remains from Thailand. *Journal of Archaeological Science*, 7, 149–165.

Hill, E.
  (2000). The contextual analysis of animal interments and ritual practice in southwestern North America. *Kiva*, 65(4), 361–398.

References

Hill, F. C.
(1972). *A Middle Archaic Dog Burial in Illinois*. Evanston, Illinois: Foundation for Illinois Archaeology (an eight-page pamphlet).

Ho, S. Y. W., Phillips, M. J., Cooper, A. & Drummond, A. J.
(2005). Time dependency of molecular rate estimates and systematic overestimation of recent divergence times. *Molecular Biology & Evolution*, 22(7), 1561–1568.

Hoeflich, M.
(2005). Dogs are special friends. Lawrence, Kansas: *Lawrence Journal-World* (newspaper), 17 August, 9B (www.ljworld.com).

Hoffecker, J. F.
(2005). *A Prehistory of the North – Human Settlement of the Higher Latitudes*. New Brunswick, New Jersey: Rutgers University Press.

Hogue, S. H.
(2003). Corn dogs and hush puppies: Diet and domestication at two protohistoric farmsteads in Oktibbeha County, Mississippi. *Southeastern Archaeology*, 22(2), 185–195.

Hohmann, G. & Fruth, B.
(2003). Culture in Bonobos? Between-species and within-species variation in behavior. *Current Anthropology*, 44(4), 563–571.

Horard-Herbin, M.-P.
(2000). Dog management and use in the late Iron Age: The evidence from the Gallic site of Levroux (France). In *Dogs through Time: An Archaeological Perspective*, ed. S. J. Crockford, pp. 115–121. Oxford, England: BAR International Series, 889.

Horsburgh, K. A.
(2008). Wild or domesticated? An ancient DNA approach to canid species identification in South Africa's Western Cape Province. *Journal of Archaeological Science*, 35(6), 1474–1480.

Houpt, K. A.
(2007). Genetics of canine behavior. *Acta Veterinaria Brunensis*, 76, 431–444.

Houpt, K. A. & Wolski, T. R.
(1982). *Domestic Animal Behavior for Veterinarians and Animal Scientists*. Ames, Iowa: Iowa State University Press.

Hsu, W. H.
(1983). Effect of yohimbine on xylazine induced central nervous system depression in dogs. *Journal of the American Veterinary Medical Association*, 182(7), 698–699.

Hubrecht, R.
  (1995). The welfare of dogs in human care. In *The Domestic Dog: Its Evolution, Behaviour and Interactions with People*, ed. J. Serpell, pp. 179–198. Cambridge, England: Cambridge University Press.

Hullinger, R. L.
  (1993). The endocrine system. In *Miller's Anatomy of the Dog* (3rd edition), by H. E. Evans, pp. 559–585. Philadelphia, Pennsylvania: W. B. Saunders.

Jeffries, R. W.
  (1988a). Archaic Period research in Kentucky: Past accomplishments and future directions. In *Paleoindian and Archaic Research in Kentucky*, ed. C. D. Hockensmith, D. Pollack & T. N. Sanders, pp. 85–126. Frankfort, Kentucky: Kentucky Heritage Council.

Jeffries, R. W.
  (1988b). The Archaic in Kentucky: New Deal archaeological investigations. In *New Deal Era Archaeology and Current Research in Kentucky*, ed. D. Pollack & M. L. Powell, pp. 14–45. Frankfort, Kentucky: Kentucky Heritage Council.

Jenkins, T. W.
  (1978). *Functional Mammalian Neuroanatomy – With Emphasis on the Dog and Cat, Including an Atlas of the Central Nervous System of the Dog* (2nd edition). Philadelphia, Pennsylvania: Lea and Febiger.

Jenks, S. M. & Ginsburg, B. E.
  (1987). Socio-sexual dynamics in a captive wolf pack. In *Man and Wolf: Advances, Issues and Problems in Captive Wolf Research*, ed. H. Frank, pp. 375–399. Dordrecht, the Netherlands: Dr. W. Junk.

Jensen, P.
  (2007a, editor). *The Behavioural Biology of Dogs*. Wallingford, England, and Cambridge, Massachusetts: CAB International.

Jensen, P.
  (2007b). Mechanisms and function in dog behaviour. In *The Behavioural Biology of Dogs*, ed. P. Jensen, pp. 61–75. Wallingford, England, and Cambridge, Massachusetts: CAB International.

Jeppson, I.
  (2008). *Grønlands bedste slaedekusk. Sermitsiak*, 29 March (Greenland newspaper, one partial electronic page, periodically updated, in Danish: www.sermitsiak.gl).

Johnson, C. M.
  (2007). *A Chronology of Middle Missouri Plains Village Sites*. Washington, D.C.: Smithsonian Institution, *Contributions to Anthropology*, 47 (Scholarly Press).

Johnston, S. D., Larsen, R. E. & Olson, P. S.
(1982). Canine theriogenology. *Journal of the Society for Theriogenology*, 11, 1–50.

Jones-Bley, K.
(2000). Sintashta burials and their western European counterparts. In *Kurgans, Ritual Sites, and Settlements Eurasian Bronze and Iron Age*, ed. J. Davis-Kimball, E. M. Murphy, L. Koryakova & L. T. Yablonsky, pp. 126–133. Oxford, England: BAR International Series, 890.

Jourdane, E. H. R. & Dye, T. S.
(2006). Archaeological Monitoring for the Best Bridal Wedding Chapel at Hilton Hawaiian Village. Land of Waikiki, Honolulu (Kona) District, Island of O'ahu (TMK: 2–6-008:034). Honolulu, Hawaii: Report prepared by T. S. Dye & Colleages, Archaeologists, Inc., for Paul. H. Rosenthal, Ph.D., Inc, Hilo, Hawaii.

Kaminski, J.
(2008). The domestic dog: A forgotten star rising again (book review). *Trends in Cognitive Sciences*, 12(6), 211–212.

Kaminski, J., Call, J. & Fischer, J.
(2004). Word learning in a domestic dog: Evidence for "fast mapping." *Science*, 304, 1682–1683.

Kaminski, J., Call, J. & Tomasello, M.
(2004). Body orientation and face orientation: Two factors controlling apes' begging behavior from humans. *Animal Cognition*, 7(4), 216–223.

Kerber, J. E.
(1997). *Lambert Farm: Public Archaeology and Canine Burials along Narragansett Bay*. Fort Worth, Texas: Harcourt Brace College Publishers.

Kerr, S. L., Walker, P. L., Hawley, G. M. & Yoshida, B. Y.
(2002). Physical anthropology. In *The Ancient Mariners of San Nicolas Island – A Bioarchaeological Analysis of the Burial Collections*, ed. J. A. Ezzo, pp. 25–55. Tucson, Arizona, and Redlands, California: Statistical Research, Inc., Technical Report, 01–64.

Keyser, J. D.
(2007). The warrior as wolf: War symbolism in prehistoric Montana rock art. *American Indian Art Magazine*, 32(3), 62–69.

Kim, J. J., Andreasen, N. C., O'Leary, D.S., Wiser, A. K., Ponto, L. L., Watkins, G. L. & Hichwa, R. D.
(1999). Direct comparison of the neural substrates of recognition memory for words and faces. *Brain*, 122, 1069–1083.

Kipfer, B. A.
   (2000). *Encyclopedic Dictionary of Archaeology*. New York City: Springer.

Kirkwood. J. K.
   (1985). The influence of size on the biology of the dog. *Journal of Small Animal Practice*, 26, 97–110.

Kirschen, M. P., Chen, S. H. A., Schraedly-Desmond, P. & Desmond, J. E.
   (2005). Load- and practice-dependent increases in cerebro-cerebellar activation in verbal working memory: An fMRI study. *Neuroimage*, 24, 462–472.

Kirton, A., Wirrell, E., Zhang, J. & Hamiwka, L.
   (2004). Seizure-alerting and –response behaviors in dogs living with epileptic children. *Neurology*, 62, 2303–2305.

Kneberg, M. D.
   (1952). The Tennessee Area. In *Archaeology of Eastern United States*, ed. J. B. Griffin, pp. 190–198. Chicago, Illinois: University of Chicago Press.

Knuth, E.
   (1966/67). The ruins of the Musk-ox way. *Folk*, 8–9, 191–219.

Knuth, E.
   (1967). Archaeology of the Musk-Ox Way. *Contributions du Centre d'Études Arctiques et Finno-Scandinaves*, 5. Paris, France: École Pratique Des Hautes Études-Sorbonne.

Koch, C.
   (2004). *The Quest for Consciousness – A Neurobiological Approach*. Englewood, Colorado: Roberts and Company.

Koler-Matznick, J.
   (2002). The origin of the dog revisited. *Anthrozoös*, 15(2), 98–118.

Koler-Matznick, J., Brisbin, I. L. Jr. & McIntyre, J. K.
   (2000). The New Guinea singing dog: A living primitive dog. In *Dogs through Time: An Archaeological Perspective*, ed. S. J. Crockford, pp. 239–247. Oxford, England: BAR International Series, 889.

Konsolaki-Yannopoulou. E.
   (2001). New evidence for the practice of libations in the Aegean Bronze Age. *Aegaeum*, 22, 213–220 (plus 24 photographic plates).

Korea Animal Protection Society.
   (2003). From dog farm to soup bowl. Seoul, South Korea: International Aid for Korean Animals (on-line: www.koreananimals.org/animals/dogs.htm, Animal Issues, Dogs).

Korea Herald.
   (2000). Dog craze spawns diverse services in Korea. (an article in the August 5 digital edition of the Society and Arts Column in the Korea Now section of the newspaper).

Koster, J. M.
(2008). Hunting with dogs in Nicaragua: An optimal foraging approach. *Current Anthropology*, 49(5), 935–944.

Kramer, F. E.
(1996). Akia and Nipisat 1: Two Saqqaq sites in Sisimiut District, west Greenland. In *The Paleo-Eskimo Cultures of Greenland: New Perspectives in Greenlandic Archaeology*, ed. B. Grønnow & J. Pind, pp. 65–96. Copenhagen, Denmark: The Danish Polar Center.

Kreeger, T. J.
(2003). The internal wolf: Physiology, pathology, and pharmacology. In *Wolves – Behavior, Ecology, and Conservation*, ed. L. D. Mech & L. Boitani, pp. 192–217. Chicago, Illinois: University of Chicago Press.

Kruger, K. A. & Serpell, J. A.
(2006). Animal-assisted interventions in mental health: Definitions and theoretical foundations. In *Handbook on Animal-Assisted Therapy: Theoretical Foundations and Guidelines for Practice* (2nd edition), ed. A. H. Fine, pp. 21–38. New York City: Academic Press.

Kubinyi, E., Virányi. Z. & Miklósi, Á.
(2007). Comparative Social Cognition: From wolf to dog to humans. *Comparative Cognition & Behavior Reviews*, 2, 26–46.

Kukekova, A. V., Trut, L. N., Oskina, I. N., Kharlamova, A. V., Shikhevich, S. G., Kirkness, E. F., Aguirre, G. D. & Acland, G. M.
(2004). A marker set for construction of a genetic map of the silver fox (*Vulpes vulpes*). *Journal of Heredity*, 95(3), 185–194.

Kukekova, A. V., Acland, G. M., Oskina, I. N., Kharlamova, V, Trut, L. N., Chase, K., Lark, K. G., Erb, H. N. & Aguirre, G. D.
(2006). The genetics of domesticated behavior in canids: What can dogs and silver foxes tell us about each other? In *The Dog and its Genome*, ed. E. A. Ostrander, U. Giger & K. Lindblad-Toh, pp. 515–537. Woodbury, New York: Cold Spring Harbor Laboratory Press.

Kukekova, A. V., Trut, L. N., Chase, K., Shepeleva, D. V., & seven others.
(2008). Measurement of segregating behaviors in experimental silver fox pedigrees. *Behavioral Genetics*, 38(2), 185–194.

Kunkel, K. E., Pletscher, D. H., Boyd, D. K., Ream, R. R. & Fairchild, M. W.
(2004). Factors correlated with foraging behavior of wolves in and near Glacier National Park, Montana. *Journal of Wildlife Management*, 68(1), 167–178.

Kurz, R.
(1937). *Journal of Rudolph Friedrich Kurz*. Washington, D.C.: Smithsonian Institution, Bureau of American Ethnology, Report, 115.

References

Kuzniar, A. A.
  (2006). *Melancholia's Dog: Reflections on our Animal Kinship*. Chicago, Illinois: University of Chicago Press.

LaBarbera, M.
  (1989). Analyzing body size as a factor in ecology and evolution. *Annual Review of Ecology and Systematics*, 20, 97–117.

LaFrance, C., Garcia, L. J. & Labreche, J.
  (2007). The effect of a therapy dog on the communication skills of an adult with aphasia: Case report. *Journal of Communication Disorders*, 40(3), 215–224.

Landsberg, G., Hunthausen, W. & Ackerman, L.
  (2007). *Handbook of Behavior Problems of the Dog and Cat* (2nd edition). Edinburgh, Scotland: Elsevier Saunders.

Lang, R. W. & Harris, A. H.
  (1984). *The Faunal Remains from Arroyo Hondo Pueblo, New Mexico – A Study in Short-term Subsistence Change*. Santa Fe, New Mexico: School of American Research Press. Arroyo Hondo Archaeological Series, 5.

Lange, K. E.
  (2007). Hair of the dog. *National Geographic*, 211, May, 31.

Langenwalter, P. E.
  (2005). A late prehistoric dog burial associated with human graves in Orange County, California. *Journal of Ethnobiology*, 25(1), 25–37.

Langley, J. N.
  (1883). The structure of the dog's brain. *Journal of Physiology*, 4(4–5), 248–285, and two illustrative plates.

Larsen, H.
  (1934). Dødemandsbugten – An Eskimo Settlement on Clavering Island. Copenhagen, Denmark: *Meddelelser om Grønland* (Copenhagen, Denmark), 102(1).

Larsen, H. & Rainey, F.
  (1948). *Ipiutak and the Arctic Whale Hunting Culture*. New York City: American Museum of Natural History, *Anthropological Papers*, 42.

Larsson, L.
  (1990). Dogs in fraction – Symbols in action. In *Contributions to the Mesolithic in Europe*, ed. P. M. Vermeersch & P. Van Peer, pp. 153–160. Leuven, Belgium: Leuven University Press.

References

Larsson, L.
(1995). Pratiques mortuaires et sépultures de chiens dans les sociétés mésolithiques de Scandinavie méridionale. *L'Anthropologie*, 98(4), 562–575 (in French).

Lauenborg, M.
(1982). Hundens Grav. *Skalk*, 1982(1), 3–6 (in Danish).

Lawrence, B.
(1944). Appendix C – Bones from the Governador Area. In *Early Stockaded Settlements in the Governador, New Mexico*, ed. E. T. Hall, pp. 73–78. New York City: Columbia University, *Studies in Archaeology and Ethnology*, 2(1).

Lawrence, B.
(1967). Early domestic dogs. *Zeitschrift für Säugetierkunde*, 32, 44–59.

Lawrence, B.
(1968). Antiquity of large dogs in North America. *Tebiwa*, 11(2), 43–49.

Lazarus, Y. W.
(1979). *The Buck Burial Mound: A Mound of the Weeden Island Culture*. Fort Walton Beach, Florida: Temple Mound Museum.

Leach, B. F.
(1979). Excavations in the Washpool Valley, Palliser Bay. In *Prehistoric Man in Palliser Bay*, ed. B. F. Leach & H. Leach, pp. 67–136. Wellington, New Zealand: National Museum of New Zealand, Bulletin, 21.

LeeDecker, C. H., Bloom, J., Wuebber, I., Pipes, M.-L. & Rosenberg, K. R.
(1995). *Final Archaeological Excavations at a Late 18th-Century Family Cemetery for the U.S. Route 113 Dualization Milford to Georgetown Sussex County, Delaware*. Dover, Delaware: Delaware Department of Transportation, Archaeology Series, 134.

Lefebvre, D., Diederich, C., Delcourt, M. & Giffroy, J.-M.
(2007). The quality of the relation between handler and military dogs influences efficiency and welfare of dogs. *Applied Animal Behaviour Science*, 104(1–2), 49–60.

Lemish, M. G.
(1996). *War Dogs: Canines in Combat*. Washington, D.C.: Brassey's.

Leonard, J. A., Wayne, R. K., Wheeler, J., Valadez, R., Guillen, S. & Vilà, C.
(2002). Ancient DNA evidence for Old World origin of New World dogs. *Science*, 298, 1613–1616.

Leon-Portilla, M.
(1992). *The Broken Spears: The Aztec Account of the Conquest of Mexico*. Boston, Massachusetts: Beacon Press (originally published in 1962).

Lewis, M. & Clark, W.
(1904). *Original Journals of the Lewis and Clark Expedition, 1804–1806*, ed. R. G. Thwaites, 8 volumes (1904, Volumes 1 and 2; 1905, Volumes 3–8). New York City: Dodd, Mead & Co.

Lewis, T. N. M. & Kneberg M.
(1946). *Hiwassee Island: An Archaeological Account of Four Tennessee Indian Peoples*. Knoxville, Tennessee: University of Tennessee Press.

Lewis, T. N. M. & Lewis, M. K.
(1961). *Eva, an Archaic Site*. Knoxville, Tennessee: University of Tennessee Press.

Lewontin, R. C.
(1965). Selection for colonizing ability. In *The Genetics of Colonizing Species*, ed. H. G. Baker & G. L. Stebbins, pp. 77–91. New York City: Academic Press.

Licón, E. G.
(2003). Social Inequality at Monte Alban, Oaxaca: Household Analysis from Terminal Formative to Early Classic. Pittsburgh, Pennsylvania: Ph.D. Dissertation, Department of Anthropology, University of Pittsburgh (available electronically).

Lindberg, J., Björnerfeldt, S., Saetre, P., Svartberg, K., Seehuus, B., Bakken, M., Vilà, C. & Jazin, E.
(2005). Selection for tameness has changed brain gene expression in silver foxes. *Current Biology*, 15(22), R915–R916.

Lindblad-Toh, K., Wade, C. M., Mikkelson, T. S., et al.
(2005). Genome sequence, comparative analysis and haplotype structure of the domestic dog. *Nature*, 438(8 December), 803–819.

Litt, B. & Krieger, A.
(2007). Of seizure prediction, statistics, and dogs: A cautionary tale. *Neurology*, 68, 250–251.

Lockerbie. L.
(1959). From Moa-hunter to Classic Maori in southern New Zealand. In *Anthropology in the South Seas*, ed. J. D. Freeman & W. R. Geddes, pp. 75–110. New Plymouth, New Zealand: Avery.

Lockwood, R.
(1995). The ethology and epidemiology of canine aggression. In *The Domestic Dog: Its Evolution, Behaviour and Interactions with People*, ed. J. Serpell, pp. 131–138. Cambridge, England: Cambridge University Press.

Lopez, R. M.
(1978). *Of Wolves and Men*. New York City: Charles Scribner's Sons.

Lorenz, K.
(1954). *Man Meets Dog*. London, England: Methuen.

Lorenz, K.
(1975). Foreword. In *The Wild Canids*, ed. M. W. Fox, pp. vii–xii. New York City: Van Nostrand Reinhold.

Lovari, S., Sforzi, A., Scala, C. & Fico, R.
(2007). Mortality patterns of the wolf in Italy: Does the wolf keep himself from the door? *Journal of Zoology*, 272(2), 117–124.

Lyman, R. L.
(1994). *Vertebrate Taphonomy*. Cambridge, England: Cambridge University Press.

Lyons, L. A.
(2007). Review of: *The Dog and Its Genome*, ed. E. A. Ostrander, U. Giger & K. Lindblad-Toh, Cold Spring Harbor Laboratory Press, Cold Spring Harbor, New York, 2006. *The Quarterly Review of Biology*, 82(1), 63–64.

Lyras, G. A. & Van Der Geer, A. A. E.
(2003). External brain anatomy in relation to the phylogeny of the Caninae (Carnivora: Canidae). *Zoological Journal of the Linnean Society*, 138(4), 505–522.

MacArthur, R. W. & Wilson, E. O.
(1967). *The Theory of Island Biogeography*. Princeton, New Jersey: Princeton University Press.

MacCord, H. A.
(1970). The John Green Site, Greensville County, Virginia. *Quarterly Bulletin of the Archaeological Society of Virginia*, 25, 97–138.

MacDonald, D. H.
(2003). Pennsylvania Archaeological Data Synthesis: The Raccoon Creek Watershed. Monroeville, Pennsylvania, GAI Consultants, Inc.: Report submitted to the Pennsylvania Department of Transportation, Engineering District 11–0.

MacDonald, D. W. & Carr, G. M.
(1995). Variation in dog society: Between resource dispersion and social flux. In *The Domestic Dog: Its Evolution, Behaviour and Interactions with People*, ed. J. Serpell, pp. 199–216. Cambridge, England: Cambridge University Press.

MacDonald, D. W. & Reynolds, J. C.
(2004). 5.3 Red fox *Vulpes vulpes* Linnaeus, 1758 Least Concern (2004). In *Canids: Foxes, Wolves, Jackals and Dogs – Status Survey and Conservation Action Plan*, ed. C. Sillero-Zubiri, M. Hoffman & D. W. Macdonald, pp. 129–136. Gland, Switzerland, and Cambridge, England: The World Conservation Union.

References

MacDonald, K.
(1987). *Development and stability of personality characteristics in pre-pubertal wolves: Implications for pack organization and behavior.* In *Man and Wolf:* Advances, Issues, and Problems in Captive Wolf Research, ed. M. Frank., pp. 375–399. Dordrecht, the Netherlands: Dr. W. Junk.

MacIntosh, N. W. G.
(1975). The origin of the dingo: An enigma. In *The Wild Canids*, ed. M. W. Fox, pp. 87–106. New York City: Van Nostrand Reinhold.

MacKinnon, M. & Belanger, K.
(2006). In sickness and in health: Care for an arthritic Maltese dog from the Roman cemetery of Yasmina, Carthage, Tunisia. In *Dogs and People in Social, Working, Economic or Symbolic Interaction*, ed. L. M. Snyder & E. A. Moore, pp. 38–43. Oxford, England: Oxbow Books (Proceedings of the 9th ICAZ Conference, Durham, England, 2002).

Macpherson, K. & Roberts, W. A.
(2006). Do dogs (*Canis familiaris*) seek help in an emergency? *Journal of Comparative Psychology*, 120(2), 113–119.

Madsen, A. P., Müller, S., Neergaard, C., Petersen, C. G., Rostrop, E., Steenstrup, K. J. V. & Winge, H.
(1900). *Affaldsdynger fra Stenalderen i Danmark.* Copenhagen, Denmark: National Museum of Denmark. (in Danish).

Magennis, A., L.
(1977). Middle and Late Archaic Mortuary Patterning: An Example from the Western Tennessee Valley. Knoxville, Tennessee: Unpublished M.A. Thesis, Department of Anthropology, University of Tennessee.

Maines, S.
(2006). Step by step, professor perseveres after accident. *Lawrence Journal-World* (Lawrence, Kansas, newspaper), 17 April, 1A, 7A (www.ljworld .com).

Market Research Corporation of America.
(1987). *Anthrozoös*, 1, 123.

Marquardt, W. H. & Watson, P. J.
(2005a). Regional survey and testing. In *Archaeology of the Middle Green River Region, Kentucky*, ed. W. H. Marquardt & P. J. Watson, pp. 41–70. Gainesville, Florida: University of Florida, Institute of Archaeology and Paleoenvironmental Studies, Monograph, 5.

Marquardt, W. H. & Watson, P. J.
(2005b). SMAP investigations at the Carlston Annis site, 15BT5. In *Archaeology of the Middle Green River Region, Kentucky*, ed. W. H. Marquardt & P. J. Watson, pp. 87–120. Gainesville, Florida: University of Florida, Institute of Archaeology and Paleoenvironmental Studies, Monograph, 5.

Martin, J.
(2004). Seizure-alert dogs – Just the facts, hold the media hype. Electronic research article, two pages (www.epilepsy.com).

Martin, P. S.
(1967). Prehistoric overkill. In *Pleistocene Extinctions: The Search for a Cause*, ed. P. S. Martin & H. E. Wright, pp. 75–120. New Haven, Connecticut: Yale University Press.

Martin, P. S.
(1973). The discovery of America. *Science*, 179, 969–974.

Martin, P. S.
(1984). Prehistoric overkill: The global model. In *Quaternary Extinctions: A Prehistoric Revolution*, ed. P. S. Martin & R. G. Klein, pp. 354–403. Tucson, Arizona: University of Arizona Press.

Martin, P. S.
(2005). *Twilight of the Mammoths: Ice Age Extinctions and the Rewilding of America*. Berkeley, California: University of California Press.

Mathiassen, T.
(1927). *Archaeology of the Central Eskimos, the Thule Culture and Its Position within the Eskimo Culture. Report of the Fifth Thule Expedition 1921–1924*, Volume 4. Copenhagen, Denmark: Gyldendal.

Mathiassen, T.
(1930). Inugsuk, a Mediaeval Eskimo Settlement in Upernivik District, west Greenland. *Meddelelser om Grønland* (Copenhagen, Denmark), 77(4).

Mathiassen, T.
(1931). Ancient Eskimo Settlements in the Kangamiut Area. *Meddelelser om Grønland* (Copenhagen, Denmark), 91(1).

Mathiassen, T.
(1933). Prehistory of the Angmagssalik Eskimos. *Meddelelser om Grønland* (Copenhagen, Denmark), 92(4).

Mathiassen, T.
(1934). Contributions to the Archaeology of Disko Bay. *Meddelelser om Grønland* (Copenhagen, Denmark), 93(2).

Mathiassen, T.
(1936). The Former Eskimo Settlements on Frederik VI's Coast. *Meddelelser om Grønland* (Copenhagen, Denmark), 109(2).

Mathiassen, T.
(1958). The Sermermiut Excavations 1955. *Meddelelser om Grønland* (Copenhagen, Denmark), 161(3).

Mathiassen, T. & Holtved, E.
 (1936). The Eskimo Archaeology of Julianehaab District, with a Brief Summary of the Prehistory of the Greenlanders. *Meddelelser om Grønland* (Copenhagen, Denmark), 118(1).

Mattioli, L., Capitani, C., Avanizelli, E., Bertelli, I, Gazzola, A. & Appollonio, M.
 (2004). Predation by wolves (*Canis lupus*) on roe deer (*Capreolus capreolus*) in north-eastern Apennine, Italy. *Journal of Zoology*, 264(3), 249–258.

Maximilian, A. P.
 (1906). Travels in the interior of North America. In *Early Western Travels 22*, ed. R. G. Thwaites, pp. 5–393. Cleveland, Ohio: Arthur H. Clark.

Maxwell, M. S.
 (1984). Pre-Dorset and Dorset prehistory of Canada. In *Handbook of North American Indians*, Volume 5: *Arctic*, ed. D. Damas, pp. 359–368. Washington, D.C.: Smithsonian Institution.

Maxwell, M. S.
 (1985). *Prehistory of the Eastern Arctic*. Orlando, Florida: Academic Press.

Mayer, K. O.
 (2007). Grace and Garbo. *Memphis Commercial Appeal* (Memphis, Tennessee, newspaper): Monday, 26 November, M1, M6 (www.commercialappeal.com).

Mayr, E.
 (1988). *Toward a New Philosophy of Biology: Observations of an Evolutionist.* Cambridge, Massachusetts: Harvard University Press.

Mazzorin, J. D. G. & Minniti, C.
 (2006). Dog sacrifice in the ancient world: A ritual passage? In *Dogs and People in Social, Working, Economic or Symbolic Interaction*, ed. L. M. Snyder & E. A. Moore, pp. 62–66. Oxford, England: Oxbow Books (Proceedings of the 9th ICAZ Conference, Durham, England, 2002).

McAllister, I.
 (2007). *The Last Wild Wolves – Ghosts of the Rain Forest*. Berkeley, California: University of California Press.

McCabe, B. W., Baun, M. M., Speich, D. & Agrawal, S.
 (2002). Resident dog in the Alzheimer's special care unit. *Western Journal of Nursing Research*, 24(6), 684–696.

McCord, B. K.
 (2005). *Investigations in the Upper White River Drainage: The Albee Phase and Late Woodland /Prehistoric Settlement*. Muncie, Indiana: Ball State University, Archaeological Resources Management Service.

McCullough, K. M.
  (1989). *The Ruin Islanders: Early Thule Culture Pioneers in the Eastern High Arctic*. Hull, Quebec: Canadian Museum of Civilization, Mercury Series, Archaeological Survey of Canada, Paper, 141.

McCullough, R. G., White, A. A., Strezewski, M. R. & McCullough D.
  (2004). *Frontier Interaction during the Late Prehistoric Period: A Case Study from Central Indiana*. Fort Wayne, Indiana: Indiana University/Purdue University at Fort Wayne, Report of Investigations, 401.

McDougall, I., Brown, F. H. & Fleagle, J. G.
  (2005). Stratigraphic placement and age of modern humans from Kibish, Ethiopia. *Nature*, 433, 733–736.

McGhee R.
  (1974). *Beluga Hunters – An Archaeological Reconstruction of the History and Culture of the Mackenzie Delta Kittegaryumiut*. St. John's, Newfoundland: Memorial University of Newfoundland.

McGhee, R.
  (1984). *The Thule Village at Brooman Point, High Arctic Canada*. Ottawa, Ontario: National Museum of Man, Mercury Series, Archaeological Survey of Canada, Paper, 125.

McGhee, R.
  (1996). *Ancient People of the Arctic*. Vancouver, British Columbia: University of British Columbia Press.

McGovern, T. H.
  (2004). *Report of Bones from Dadastaðir, Mývatn District, Northern Iceland*. New York City: City University of New York, Northern Science and Education Center (NORSEC), Zooarchaeology Laboratory Report, 19.

McGreevy, P., Grassi, T. D. & Harman, A. M.
  (2004). A strong correlation exists between the distribution of retinal ganglion cells and nose length in the dog. *Brain, Behavior and Evolution*, 63(1), 13–22.

McKeown, M.
  (1975). Craniofacial variability and its relationship to disharmony of the jaws and teeth. *Journal of Anatomy*, 119, 579–588.

McLachlan, C.
  (2002). Gihli, the dog in Cherokee thought. *The Journal of Cherokee Studies*, 23, 4–18.

McMillan, F. D.
  (2004). *Unlocking the Animal Mind*. Rodale, New York: Holtzbrinck.

McMillan, R. B.
  (1970). Early canid burial from the western Ozark highland. *Science*, 167, 1246–1247.

Meats, A.

(1971). The relative importance to population increase of fluctuations in mortality, fecundity and the time variables of the reproductive schedule. *Oecologia*, 6, 223–237.

Mech, L. D.

(1970). *The Wolf*. Garden City, New York: Natural History Press.

Mech, L. D.

(1977). Productivity, mortality, and population trends of wolves in northeastern Minnesota. *Journal of Mammalogy*, 58, 559–574.

Mech, L. D.

(2007). Possible use of foresight, understanding, and planning by wolves hunting muskoxen. *Arctic*, 60(2), 145–149.

Mech, L. D. & Boitani, L.

(2003). Wolf social ecology. In *Wolves – Behavior, Ecology, and Conservation*, ed. L. D. Mech & L. Boitani, pp. 1–34. Chicago, Illinois: University of Chicago Press.

Mech, L. D. & Boitani. L.

(2004). 5.2 Grey wolf *Canis lupus* Linnaeus, 1758 Least Concern (2004). In *Canids: Foxes, Wolves, Jackals and Dogs – Status Survey and Conservation Action Plan*, ed. C. Sillero-Zubiri, M. Hoffman & D. W. Macdonald, pp. 124–29. Gland, Switzerland and Cambridge, England: The World Conservation Union.

Meldgaard, J.

(1960). Prehistoric culture sequences in the Eastern Arctic as elucidated by stratified sites at Igloolik. In *Men and Cultures: Selected Papers of the Fifth International Congress of Anthropological and Ethnological Sciences*, ed. A. F. C. Wallace, pp. 588–595. Philadelphia, Pennsylvania: University of Pennsylvania Press.

Meldgaard, J.

(1983). Qajâ, en køkkenmødding i dybfrost: Feltrapport fra arbejdsmarken i Grønland. Copenhagen, Denmark: *Nationalmuseets Arbejdsmark*, pp. 83–96 (in Danish).

Merriam Co., G. & C. (1974). *Webster's New Collegiate Dictionary*. Springfield, Massachusetts.

Mertens, R.

(1936). Der Hund aus dem Senckenberg-Moor, ein Begleiter des Ur's. *Natur und Volk*, 66, 506–510 (in German).

Miklósi, Á.

(2007a). *Dog Behaviour, Evolution, and Cognition*. Oxford, England: Oxford University Press.

References

Miklósi, Á.
(2007b). Human-animal interactions and social cognition in dogs. In *The Behavioural Biology of Dogs*, ed. P. Jensen, pp. 207–222. Wallingford, England, and Cambridge, Massachusetts: CAB International.

Miklósi, Á. & Topál, J.
(2005). Is there a simple recipe for how to make friends? *Trends in Cognitive Sciences*, 9 (10), 463–464.

Miklósi, Á., Kubinyi, E., Topál, J., Gásci, M., Virányi, Z. & Csányi, V.
(2003). A simple reason for a big difference: Wolves do not look back at humans, but dogs do. *Current Biology*, 13, 763–766.

Miklósi, Á., Pongrácz, P., Lakatos, G., Topál, J. & Csányi, V.
(2005). A comparative study of the use of visual communication signals in interactions between dogs (*Canis familiaris*) and humans and cats (*Felis catus*) and humans. *Journal of Comparative Psychology*, 119(2), 179–186.

Miller, C. F.
(1956). Life 8,000 years ago uncovered in an Alabama cave. *National Geographic*, 150(10), 542–558.

Millis, H.
(2010). *Archaeological Evaluation and Data Recovery Investigations of Sites 31CR218, 31CR258, 31CR259, and 31CR260 to be Adversely Affected by Construction for the Proposed Cannonsgate Subdivision, Carteret County, North Carolina. TRC, Chapel Hill*, North Carolina: Report to be submitted to R. A. North Development Inc., Matthews, North Carolina (in preparation).

Moehlman, P. D.
(1989). Interspecific variation in canid social systems. In *Carnivore Behavior, Ecology, and Evolution*, ed. J. L. Gittleman, pp. 143–163. Ithaca, New York: Cornell University Press.

Moore, C. B.
(1916). Some aboriginal sites on Green River, Kentucky. *Journal of the Academy of Natural Sciences of Philadelphia* (2nd series), 16, 431–487.

Moore, S. D.
(1991). Limited testing at site 40MY105: A multi-component accretionary mound, McNary County, Tennessee. In *The Archaic Period in the Midsouth*, ed. C. H. McNutt, pp. 40–45. Jackson Mississippi and Memphis, Tennessee: Mississippi Department of Archives and History, Archaeological Report, 24, and Memphis State University, Archaeological Research Center, Occasional Papers, 16.

Morey, D. F.
(1986). Studies on Amerindian dogs: Taxonomic analysis of canid crania from the northern plains. *Journal of Archaeological Science*, 13, 119–145.

Morey, D. F.
  (1990). Cranial Allometry and the Evolution of the Domestic Dog. Knoxville,
  Tennessee: Unpublished Ph.D. Dissertation, Department of Anthropology, University of Tennessee (Ann Arbor, Michigan: University Microfilms
  International, Order Number 9112869).

Morey, D. F.
  (1992). Size, shape and development in the evolution of the domestic dog.
  Journal of Archaeological Science, 19, 181–204.

Morey, D. F.
  (1994a). The early evolution of the domestic dog. *American Scientist*, 82, 336–
  347.

Morey, D. F.
  (1994b). *Canis* remains from Dust Cave. *Journal of Alabama Archaeology*, 40(1–
  2), 163–172.

Morey, D. F.
  (1995). The early evolution of the domestic dog. In *Exploring Evolutionary
  Biology*, ed. M. Slatkin, pp. 140–151. Sunderland, Massachusetts: Sinauer
  Associates (republication of Morey 1994a).

Morey, D. F.
  (1996). L'origine du plus vieil ami de l'homme. *La Recherche*, 288(Juin), 72–77
  (in French).

Morey, D. F.
  (1997). Review of: *The Domestic Dog: Its Evolution, Behaviour and Interactions
  with People* (ed. J. Serpell), Cambridge, England: Cambridge University
  Press, 1995. *The Quarterly Review of Biology*, 72(1), 87–88.

Morey, D. F.
  (2000). Review of: *The Dog: Its Behavior, Nutrition and Health* (L. P. Case),
  Ames, Iowa: Iowa State University Press, 1999. *The Quarterly Review of
  Biology*, 75(3), 325–326.

Morey, D. F.
  (2006). Burying key evidence: The social bond between dogs and people.
  Journal of Archaeological Science, 33(2), 158–175.

Morey, D. F. & Aaris-Sørensen, K.
  (2002). Paleoeskimo dogs of the Eastern Arctic. *Arctic*, 55(1), 44–56.

Morey, D. F. & Crothers, G. M.
  (1998). Clearing up clouded waters: Paleoenvironmental analysis of fresh-
  water mussel assemblages from the Green River shell middens, western
  Kentucky. *Journal of Archaeological Science*, 25, 907–926.

Morey, D. F., Crothers, G. M., Stein, J. K., Fenton, J. P. & Herrmann, N. P.
(2002). The fluvial and geomorphic context of Indian Knoll, an Archaic shell midden in west-central Kentucky. *Geoarchaeology*, 17(6), 521–553.

Morey, D. F. & Klippel, W. E.
(1991). Canid scavenging and deer bone survivorship at an Archaic Period site in Tennessee. *Archaeozoologia*, 4(1), 11–28.

Morey, D. F., McClellan, E. A., Lindor, N, Kellogg, G. & Lindor, R.
(2004). Why some people recover better from traumatic brain injury better than others: The ambidexterity theory. *Journal of Cognitive Rehabilitation*, 22(3), 12–19.

Morey, D. F. & Wiant, M. D.
(1992). Early Holocene domestic dog burials from the North American midwest. *Current Anthropology*, 33(2), 224–229.

Morrison, D. A.
(1983). *Thule Culture in the Western Coronation Gulf, N. W. T.* Ottawa, Ontario: National Museum of Man, Mercury Series, Archaeological Survey of Canada, Paper, 116.

Morrison, D. A.
(1994). An archaeological perspective on Neoeskimo economies. In *Threads of Arctic Prehistory: Papers in Honour of William E. Taylor, Jr.*, ed. D. Morrison & J.-L. Pilon, pp. 311–324. Hull, Quebec: Canadian Museum of Civilization, Mercury Series, Archaeological Survey of Canada, Paper, 149.

Morrison, D. G.
(1974). Discriminant analysis. In *Handbook of Marketing Research*, ed. R. Ferber, pp. 2-442–2-547. New York City: McGraw Hill.

Morse, D. F.
(1967). The Robinson Site and the Shell Mound Archaic Culture in the Middle South. Ann Arbor, Michigan: Unpublished Ph.D. dissertation, University of Michigan.

Mosimann, J. E. & Martin, P. S.
(1975). Simulating overkill by Paleoindians. *American Scientist*, 63, 304–313.

Mountjoy, J. B.
(1998). The evolution of complex societies in West Mexico: A comparative perspective. In *Ancient West Mexico – Art and Archaeology of the Unknown Past*, ed. R. F. Townsend, pp. 251–265. Chicago, Illinois: The Art Institute of Chicago, Thames and Hudson.

Mueller, U. G. & Gerardo, N.
(2002). Fungus-farming insects: Multiple origins and diverse evolutionary histories. *Proceedings of the National Academy of Sciences*, 99(24), 15247–15249.

Mueller, U. G., Rehner, S. A. & Schultz, T. R.
(1998). The evolution of agriculture in ants. *Science*, 281, 2034–2038.

Mugford, R. A.
(1995). Canine behavioural therapy. In *The Domestic Dog: Its Evolution, Behaviour and Interactions with People*, ed. J. Serpell, pp. 139–152. Cambridge, England: Cambridge University Press.

Mugford, R. A.
(2007). Behavioural disorders of dogs. In *The Behavioural Biology of Dogs*, ed. P. Jensen, pp. 225–242. Wallingford, England, and Cambridge, Massachusetts: CAB International.

Murie, A.
(1944). *The Wolves of Mount McKinley*. Washington, D.C.: U.S. National Park Service, Fauna Series, 5.

Murie, O. J.
(1954). *A Field Guide to Animal Tracks*. Boston, Massachusetts: Houghton Mifflin.

Musil R.
(1974). Tiergesellschaft der Kniegrotte. In *Die Kniegrotte, eine Magdalénian-Station in Thüringen*, ed. R. Fuestel, pp. 30–72. Wiemar, Germany: Veröffentlichungen des Museum für Ur- und Frühgeschicte Thüringens, 5, 30–95 (in German).

Musil, R.
(1984). The first known domestication of wolves in central Europe. In *Animals and Archaeology*, Volume 4: *Husbandry in Europe*, ed. C. Grigson & J. Clutton-Brock, pp. 23–25. Oxford, England: BAR International Series, 227.

Musil, R.
(2000). Evidence for the domestication of wolves in Central European Magdalenian sites. In *Dogs through Time: An Archaeological Perspective*, ed. S. J. Crockford, pp. 21–28. Oxford, England: BAR International Series, 889.

Møbjerg, T.
(1998). The Saqqaq culture in the Sisimiut municipality elucidated by the two sites Nipisat and Asummiut. In *Man, Culture and Environment in Ancient Greenland*, ed. J. Arneborg & H. C. Gulløv, pp. 98–118. Copenhagen, Denmark: The Danish Polar Center.

Møhl, J.
(1979). Description and analysis of the bone material from Nugarsuk: An Eskimo settlement representative of the Thule culture in west Greenland. In *Thule Eskimo Culture: An Anthropological Retrospective*, ed. A. P. McCartney, pp. 380–394. Ottawa, Ontario: National Museum of Man Mercury Series, Archaeological Survey of Canada, Paper, 88.

Møhl, J.
(1986). Dog Remains from a Paleoeskimo settlement in west Greenland. *Arctic Anthropology*, 23(1,2), 81–89.

Necker, C.
(1977). *The Natural History of Cats*. New York City: Delta.

Nehring, A.
(1887). Untitled two-page report, and accompanying photographic plate descriptions (no page numbers). In *The Necropolis of Ancon in Peru*, by W. Reiss & A. Stübel. Berlin, Germany: A. Asher & Co. (originally in German, reprinted, English translation by A. H. Keane).

Newsome, A. E., Corbett, L. K. & Carpenter, S. M.
(1980). The identity of the dingo. I. Morphological discriminants of dingo and dog skulls. *Australian Journal of Zoology*, 28 (4), 615–625.

Nicastro, N. & Owren, M. J.
(2003). Classification of domestic cat (*Felis catus*) vocalizations by naive and experienced human listeners. *Journal of Comparative Psychology*, 117(1), 44–52.

Nie, N. H., Hull, C. H., Jenkins, J. G., Steinbrenner, K. & Bent, D. H.
(1975). *Statistical Package for the Social Sciences* (2nd edition), New York City: McGraw Hill.

Nielsen, E. K. & Petersen, E. B.
(1993). Burials, people and dogs. In *Digging into the Past – 25 Years of Archaeology in Denmark*, ed. S. Haas, B. Størgaard & U. L. Hansen, pp. 76–81. Moesgaard, Denmark: Jutland Archaeological Society.

Nikolova, L.
(2005). Approach to anthropology everydayness – Symbols in the prehistoric enculturation process. In *Prehistoric Archaeology & Anthropological Theory and Education*, ed. L. Nikolova, J. Fritz & J. Higgins, pp. 101–106. New Bern, North Carolina (W. W. Kellogg Foundation), *RPRP* (Rural People, Rural Policy), 6–7.

Nobayashi, A.
(2006). An ethnoarchaeological study of chase hunting with gundogs by the aboriginal peoples of Taiwan. In *Dogs and People in Social, Working, Economic or Symbolic Interaction*, ed. L. M. Snyder & E. A. Moore, pp. 77–84. Oxford, England: Oxbow Books (Proceedings of the 9th ICAZ Conference, Durham, England, 2002).

Nobis, G.
(1979). Der älteste Haushund lebte vor 14 000 Jahren. *Umschau*, 79(19), 610 (in German).

Nobis, G.
   (1986). Die Wildsäugetiere in der Umwelt des Menschen von Oberkassel bei Bonn und das Domestikationsproblem von Wölfen im Jungpaläolithikum, pp. 367–376. Bonn, Germany: *Bonner Jahrbücher* (in German).

Nobis, G.
   (1996). Vom Wolf zum Schloßhund – Zur Enstehungsgeschichte unserer Haustier. *Tier und Museum*, 5(2), 35–47 (in German).

Noe-Nygaard, N.
   (1988). δ^{13}C-values of dog bones reveal the nature of changes in man's food resources at the Mesolithic-Neolithic transition, Denmark. *Chemical Geology*, 73, 87–96.

Noe-Nygaard, N.
   (1989). Man-made trace fossils on bone. *Human Evolution*, 4, 461–491.

Nolte, J.
   (1993). *The Human Brain: An Introduction to Its Functional Anatomy* (3rd edition). St. Louis, Missouri: Mosby-Year Book.

Novikov, G. A.
   (1962). *Carnivorous Mammals of the Fauna of the U.S.S.R.* Jerusalem, Israel: Israel Program for Scientific Translations.

Nowak, R. M.
   (1979). *North American Quaternary* Canis. Lawrence, Kansas: The University of Kansas, Museum of Natural History, Monograph, 6.

Nowak, R. M.
   (2002). The original status of wolves in eastern North America. *Southeastern Naturalist*, 1(2), 95–130.

Nowak, R. M.
   (2003). Wolf evolution and taxonomy. In *Wolves – Behavior, Ecology, and Conservation*, ed. L. D. Mech & L. Boitani, pp. 239–258. Chicago, Illinois: University of Chicago Press.

Nyegaard, G.
   (1985). Faunalevn fra yngre stenalder på øerne syd for Fyn. *Meddelelser fra Langelands Museum* (Rudkøbing, Denmark), 1985, 426–457 (in Danish).

Nyegaard, G.
   (1995). Qaqortoq Katersugaasivia: Julianehåb Museum. In *Archaeological Fieldwork in the Northwest Territories in 1994, and in Greenland in 1993 and 1994*, ed. M. Bertulli, J. Berglund & H. Lange. pp. 100, 103. Yellowknife, Northwest Territories: Prince of Wales Northern Heritage Centre, Archaeology Report, 16.

References

Oakleaf, J. K., Mack, C. & Murray, D. L.
  (2003). Effects of wolves on livestock calf survival and movements in central Idaho. *Journal of Wildlife Management*, 67(2), 299–306.

Oakleaf, J. K., Murray, D. L., Oakleaf, J. R., et al.
  (2006). Habitat selection by recolonizing wolves in the northern Rocky Mountains of the United States. *Journal of Wildlife Management*, 70(2), 554–563.

O'Brien, M. J. & Lyman, R. L.
  (2000). *Applying Evolutionary Archaeology – A Systematic Approach*. New York City: Kluwer Academic/Plenum.

O'Connell, A.
  (2007). The elusive Iron Age: a rare and exciting site type is uncovered at Lismullin, Co. *Meath. Seanda*, 2, 52–54.

Olsen, J. W.
  (1985). Prehistoric Dogs in Mainland East Asia. In *Origins of the Domestic Dog – The Fossil Record*, by S. J. Olsen, pp. 47–70. Tucson, Arizona: The University of Arizona Press.

Olsen, J. W.
  (1990). *Vertebrate Faunal Remains from Grasshopper Pueblo, Arizona*. Ann Arbor, Michigan: University of Michigan, Museum of Anthropology, Anthropological Papers, 83.

Olsen, S. J.
  (1968). Canid remains from Grasshopper Ruin. *The Kiva*, 34(1), 33–40.

Olsen, S. J.
  (1970). Two pre-Columbian dog burials from Georgia. *Bulletin of the Georgia Academy of Sciences*, 28, 69–72.

Olsen, S. J.
  (1974). Early domestic dogs in North America and their origins. *Journal of Field Archaeology*, 1, 343–345.

Olsen, S. J.
  (1976). The dogs of Awatovi. *American Antiquity*, 41(1), 102–106.

Olsen, S. J.
  (1979). Archaeologically, what constitutes an early domestic animal? In *Advances in Archaeological Method and Theory*, Volume 2, ed. M. B. Schiffer, pp. 175–197. New York City: Academic Press.

Olsen, S. J.
  (1985). *Origins of the Domestic Dog – The Fossil Record*. Tucson, Arizona: University of Arizona Press.

Olsen, S. L.
(2000). The secular and sacred roles of dogs at Botai, North Kazakhstan. In *Dogs through Time: An Archaeological Perspective*, ed. S. J. Crockford, pp. 71–92. Oxford, England: BAR International Series, 889.

Olsen, S. J. & Olsen, J. W.
(1977). The Chinese wolf, ancestor of New World dogs. *Science*, 197, 533–535.

Onar, V.
(2005). Estimating the body weight of dogs unearthed from the Van-Yoncatepe Necropolis in Eastern Anatolia. *Turkish Journal of Veterinary Animal Science*, 29, 495–498.

Onar, V & Belli, O.
(2005). Estimation of shoulder height from long bone measurements on dogs unearthed from the Van-Yoncatepe early Iron Age necropolis in eastern Anatolia. *Revue Medicina Veterinaria*, 156, 53–60.

Ortiz, R. & Liporace, J.
(2005). "Seizure-alert dogs": Observations from an inpatient video/EEG unit. *Epilepsy & Behavior*, 6(4), 620–622.

O'Shea, J.
(1984). *Mortuary Variability: An Archaeological Investigation*. Orlando, Florida: Academic Press.

Ostermann, H.
(1938, editor). Knud Rasmussen's Posthumous Notes on the Life and Doings of the East Greenlanders in Olden Times. *Meddelelser om Grønland* (Copenhagen, Denmark), 109(1).

Ostrander, E. A., Giger, U. & Lindblad-Toh, K.
(2006, editors). *The Dog and Its Genome*. Cold Spring Harbor, New York: Cold Spring Harbor Laboratory Press.

Packard, J. M.
(2003). Wolf behavior: Reproductive, social, and intelligent. In *Wolves – Behavior, Ecology, and Conservation*, ed. L. D. Mech & L. Boitani, pp. 35–65. Chicago, Illinois: University of Chicago Press.

Palmer, N.
(1993). Bones and joints. In *Pathology of Domestic Animals*, Volume 1 (4th edition), ed. K. V. F. Jubb, P. C. Kennedy & N. Palmer, pp. 1–181. New York City: Academic Press.

Paris, F.
(2000). African livestock remains from Saharan mortuary contexts. In *The Origins and Development of African Livestock: Archaeology, Genetics, Linguistics, and Ethnography*, ed. R. M. Blench & K. C. MacDonald, pp. 111–126. London, England: UCL Press.

References

Park, R. W.
(1987). Dog remains from Devon Island, N.W.T.: Archaeological and osteo-logical evidence for domestic dog use in the Thule Culture. *Arctic*, 40(3), 184–190.

Parker, J. D., Keightly, M. L., Smith, C. T. & Taylor, G. J.
(1999). Interhemispheric transfer deficit in alexithymia: An experimental study. *Psychosomatic Medicine*, 61, 464–468.

Parmalee, P. W.
(1959). Appendix II – Animal remains from the Modoc rock shelter site, Randolph County, Illinois. In *Summary Report of Modoc Rock Shelter 1952, 1953, 1955, 1956*, ed. M. L. Fowler, pp. 61–63. Springfield, Illinois: Illinois State Museum, Report of Investigations, 8.

Parmalee, P. W.
(1962). The faunal complex of the Fisher Site, Illinois. *American Midland Naturalist*, 68(2), 399–408.

Parmalee, P. W.
(1966). Vertebrate remains from the Bible Site (40Mi15), Franklin County, Tennessee. In *Highway Salvage in the Nickajack Reservoir*, ed. C. H. Faulkner & J. B. Graham, pp. 81–83. Knoxville, Tennessee: University of Tennessee, Department of Anthropology, Report of Investigations, 4.

Parmalee, P. W.
(1979). Inferred Arikara subsistence patterns based on a selected faunal assemblage from the Mobridge site, South Dakota. *The Kiva*, 44(2–3), 191–218.

Parmalee, P. W. & Bogan, A. E.
(1978). Cherokee and Dallas dog burials from the Little Tennessee River Valley. *Tennessee Anthropologist*, 3(1), 100–112.

Passingham, R.
(2002). The frontal cortex: does size matter? *Nature Neuroscience*, 5(3), 190–192.

Pavao-Zuckerman, B.
(2001). A dog burial excavated at the Amorel site, Arkansas (3MS23). *The Arkansas Archaeologist*, 40, 8–10.

Perdikaris, S., Amundsen, C. & McGovern, T. H.
(2002). *Report of Animal Bones from Tjarnargata 3C, Reykjavik, Iceland*. New York City: City University of New York, Northern Science and Education Center (NORSEC).

Petersen, P. V.
(1977). Vedbaek Boldbaner – Endu engang. In *Vedbaekprojektet – I Marken og i Museerne*, ed. E. B. Petersen, J. H. Jønsson, P. V. Petersen &

K. Aaris-Sørensen, pp. 131–170. Søllerød Kommune, Denmark: Søllerød-bogen, 1977 (in Danish).

Peterson, R. O.
(1977). *Wolf Ecology and Prey Relationships on Isle Royale*. Washington, D.C.: U. S. National Park Service, Scientific Monograph Series, 11.

Peterson, R. O. & Ciucci, P.
(2003). The wolf as a carnivore. In *Wolves – Behavior, Ecology, and Conservation*, ed. L. D. Mech & L. Boitani, pp. 104–130. Chicago, Illinois: University of Chicago Press.

Pierotti, R. & Wildcat, D.
(2000). Traditional ecological knowledge: The third alternative (commentary). *Ecological Applications*, 10(5), 1333–1340.

Pilot, M., Jedrzejewski, W., Branicki, W., Sidorovich, V. E., Jedrzejewska, B., Stachura, K. & Funk, S. M.
(2006). Ecological factors influence population genetic structure of European grey wolves. *Molecular Ecology*, 15(14), 4533–4553.

Pimlott, D. M.
(1975). The ecology of the wolf in North America. In *The Wild Canids*, ed. M. W. Fox, pp. 280–285. New York City: Van Nostrand Reinhold.

Pitul'ko, V. & Kasparov, A. K.
(1996). Ancient Arctic hunters: Material culture and survival strategy. *Arctic Anthropology*, 33(1), 1–36.

Polhemus, R. R.
(1969). Historical material. In *Archaeological Investigations in the Tellico Reservoir Interim Report 1969*, ed. P. Gleeson, pp. 81–99. Knoxville, Tennessee: University of Tennessee, Report of Investigations, 8.

Pongrácz, P., Molnár, C., Miklósi, Á., & Csányi, V.
(2005). Human listeners are able to classify dog (*Canis familiaris*) barks recorded in different situations. *Journal of Comparative Psychology*, 119 (2), 136–144.

Potter, M. A. & Baby, R. S.
(1964). Hopewellian dogs. *The Ohio Journal of Science*, 64(1), 36–40.

Potvin, M. J., Drummer, T. D., Vucetich, J. A., Beyer, D. E., Peterson, R. O. & Hammill, J. H.
(2005). Monitoring and habitat analysis for wolves in upper Michigan. *Journal of Wildlife Management*, 69(4), 1660–1669.

Potvin, M. J., Peterson, R. O. & Vucetich, J. A.
(2004). Wolf homesite attendance patterns. *Canadian Journal of Zoology*, 82(9), 1512–1518.

Povinelli, D. J.
 (2004). Behind the ape's appearance: Escaping anthropocentrism in the study of other minds. *Dædalus*, winter, 29–41.

Poyser, F., Caldwell, C. & Cobb, M.
 (2006). Dog paw preference shows lability and sex differences. *Behavioural Processes*, 73, 216–221.

Poznań, M. S.
 (2006). Dead animals and living society. On-line: www.jungsteinSITE.de, December 15, 2006, 1–10.

Price, E. O.
 (1984). Behavioral aspects of animal domestication. *The Quarterly Review of Biology*, 59(1), 1–32.

Prothmann, A., Bienert, M. & Ettrich, C.
 (2006). Dogs in child psychotherapy: Effects on state of mind. *Anthrozoös*, 19(3), 265–277.

Prummel, W.
 (1992). Early Medieval dog burials among the Germanic tribes. *Helinium*, 32(1–2), 132–194.

Prummell. W.
 (2006). Bronze Age dogs from graves in Borger (Netherlands) and Dimini (Greece). In *Dogs and People in Social, Working, Economic or Symbolic Interaction*, ed. L. M. Snyder & E. A. Moore, pp. 67–76. Oxford, England: Oxbow Books (Proceedings of the 9th ICAZ Conference, Durham, England, 2002).

Pulliainen, E.
 (1967). A contribution to the study of the social behavior of the wolf. *American Zoologist*, 7(2), 313–317.

Pulliainen, E.
 (1975). Wolf ecology of northern Europe. In *The Wild Canids*, ed. M. W. Fox, pp. 292–299. New York City: Van Nostrand Reinhold.

Purtill, M. P.
 (2000). A paradoxical circumstance: Recent happenings and archaeological investigations at the famed Madisonville Site. *The Ohio Archaeological Council, Newsletter*, 12(2), 8–13.

Putnam's Sons, G. P.
 (2002). *The Oxford American College Dictionary*. New York City: G. P. Putnam's Sons.

Quaranta, A., Siniscalchi, M., Frate, A. & Vallortigara, G.
 (2004). Paw preference in dogs: Relations between lateralised behaviour and immunity. *Behavioural Brain Research*, 153(2), 521–525.

Quimby, G. I.
(1966). *Indian Culture and European Trade Goods – The Archaeology of the Historic Period in the Western Great Lakes Region*. Madison, Wisconsin: University of Wisconsin Press.

Radinsky, L.
(1973). Evolution of the canid brain. *Brain, Behavior and Evolution*, 7(3), 169–202.

Radovanoviç, I.
(1999). "Neither person nor beast" – Dogs in the burial practice of the Iron Gates Mesolithic. *Documenta Praehistorica*, 26, 71–87.

Raised by Wolves, Inc.
(2006). Essential Wolves. Thoreau, New Mexico: http://raisedbywolves10 .tripod.com/essentialwolvesword.pdf (23 pages).

Raisor, M. J.
2005. *Determining the Antiquity of Dog Origins – Canine Domestication as a Model for the Consilience between Molecular Genetics and Archaeology*. Oxford, England: BAR International Series, 1367.

Randi, E.
(2008). Detecting hybridization between wild species and their domesticated relatives. *Molecular Ecology*, 17, 285–293.

Rausch, R. A.
(1969). A summary of wolf studies in south central Alaska, 1957–1968. *Transactions of the North American Wildlife and Natural Resources Conference*, 34, 117–131.

Redding, R. W.
(1971). Neurological examination. In *Canine Neurology: Diagnosis and Treatment*, ed. B. F. Hoerlein, pp. 53–66. Philadelphia, Pennsylvania: W. B. Saunders.

Reed, C. A.
(1969). The pattern of animal domestication in the prehistoric Near East. In *The Domestication and Exploitation of Plants and Animals*, ed. P. J. Ucko & G. W. Dimbleby, pp. 361–380. Chicago, Illinois: Aldine.

Reed, C. A.
(1984). The beginnings of animal domestication. In *Evolution of Domesticated Animals*, ed. I. L. Mason, pp. 1–6. London, England: Longman.

Reinold, J.
(2005). Note sur le monde animal dans le funéraire néolithique du Soudan. *Revue de Paléobiologie, Genève*, 10, 107–119 (in French).

Reisner, I. R., Mann, J. J., Stanley, M., Huang, Y. & Houpt, K. A.
(1996). Comparison of cerebrospinal fluid monamine metabolite levels in dominant-aggressive and non-aggressive dogs. *Brain Research*, 714(1–2), 57–64.

Reynolds, R. L.
(1985). Domestic dog associated with human remains at Rancho la Brea. *Bulletin of the Southern California Academy of Sciences*, 84(2), 76–85.

Reynolds, S.
(1909). The Canidae. In *Monograph on the British Mammalia of the Pleistocene Period*, ed. W. B. Dawkins, pp. 1–28. London, England: The Paleontological Society, 2(3).

Richardson, S. J.
(2007). *Review of: Rhythms of Life: Thyroid Hormone and the Origin of Species* (S. Crockford), Victoria, British Columbia: Trafford, 2006. *The Quarterly Review of Biology*, 82(2), 149.

Ridley, F.
(1954). The Frank Bay Site, Lake Nipissing, Ontario. *American Antiquity*, 20(1), 40–50.

Rightmire, G. P.
(1991). Comparative studies of late Pleistocene human remains from Klasies River Mouth, South Africa. *Journal of Human Evolution*, 20, 131–156.

Rindos, D.
(1984). *The Origins of Agriculture: An Evolutionary Perspective*. Orlando, Florida: Academic Press.

Ritchie, J. N. G.
(1981). Excavations at Machrins, Colonsay. *Proceedings of the Antiquarian Society of Scotland*, 111, 263–281.

Rolingson, M. A. & Schwartz, D. W.
(1966). *Late Paleo-Indian and Early Archaic Manifestations in Western Kentucky*. Lexington, Kentucky: University of Kentucky Press.

Rudling, D. & Butler, C.
(2004). Barcombe Villa 2003. East Sussex, England: Mid Sussex Field Archaeological Team, *Newsletter*, 52, 3–5.

Ruspoli, M.
(1987). *The Cave of Lascaux: The Final Photographs*. New York City: Harry N. Abrams.

Sablin, M. H. & Khlopachev, G. A.
(2002). The earliest ice age dogs: Evidence from Eliseevichi I. *Current Anthropology*, 43(5), 795–799.

Sacchetti, B., Scelfo, B. & Strata, P.
(2005). The cerebellum: Synaptic changes and fear conditioning. *The Neuro-scientist*, 11(3), 217–227.

Saetre, P., Lindberg, J., Leonard, J. A., Olsson, K., Pettersson, U., Ellegren, H., Bergström, T. F., Vilà, C. & Jazin, E.
(2004). From wild wolf to domestic dog: Gene expression changes in the brain. *Molecular Brain Research*, 126, 198–206.

Saladin d'Anglure, B.
(1984). Inuit of Quebec. In *Handbook of North American Indians*, Volume 5: *Arctic*, ed. D. Damas, pp. 476–507. Washington, D.C.: Smithsonian Institution.

Sand, H., Wikenros, C., Wabakken, P. & Liberg, O.
(2006). Effects of hunting group size, snow depth and age on the success of wolves hunting moose. *Animal Behaviour*, 72, 781–789.

Sandell, H. T. & Sandell, B.
(1991). Archaeology and environment in the Scoresby Sund Fjord – Ethno-archaeological investigations of the last Thule culture of northeast Green-land. *Meddelelser om Grønland, Man & Society* (Copenhagen, Denmark), 15.

Sandweiss, D. H. & Wing, E. S.
(1997). Ritual rodents: The guinea pigs of Chincha, Peru. *Journal of Field Archaeology*, 24(1), 47–58.

Santure, S. K. & Esarey, D.
(1990). Analysis of artifacts from the Oneota mortuary component. In *Archaeological Investigations at the Morton Village and Norris Farms 36 Cemetery*, ed. S. K. Santure, A. D. Harn & D. Esarey, pp. 75–110. Springfield, Illinois: Illinois State Museum, Report of Investigations, 45.

SAS Institute
(1985). *SAS User's Guide*. Cary, North Carolina: SAS Institute.

Savishinsky, J. S.
(1983). Pet ideas: The domestication of animals, human behavior, and human emotions. In *New Perspectives on Our Lives with Companion Animals*, ed. A. H. Katcher & A. M. Beck, pp. 112–131. Philadelphia, Pennsylvania: University of Pennsylvania Press.

Savolainen, P.
(2007). Domestication of dogs. In *The Behavioural Biology of Dogs*, ed. P. Jensen, pp. 21–37. Wallingford, England, and Cambridge, Massachusetts: CAB International.

Savolainen, P., Leitner, T., Wilton, A. N., Matisoo-Smith, E. & Lundeberg, J.
(2004). A detailed picture of the origin of the Australian dingo, obtained
from the study of mitochondrial DNA. *Proceedings of the National Academy
of Sciences*, 101(33), 12387–12390.

Savolainen, P., Zhang, Y., Luo, J., Lundeberg, J. & Leitner, T.
(2002). Genetic evidence for an East Asian origin of domestic dogs. *Science*,
298, 1610–1613.

Schassburger, R. M.
(1987). Wolf vocalization: An integrated model of structure, motivation, and
ontogeny. In *Man and Wolf: Advances, Issues and Problems in Captive Wolf
Research*, ed. H. Frank, pp. 313–347. Dordrecht, the Netherlands: Dr. W.
Junk.

Schledermann, P.
(1975). *Thule Eskimo Prehistory of Cumberland Sound, Baffin Island, Canada*.
Ottawa, Ontario: National Museum of Man, Mercury Series, Archaeological Survey of Canada, Paper, 38.

Schledermann, P.
(1996). *Voices in Stone – A Personal Journey into the Arctic Past*. Calgary, Alberta:
The University of Calgary, Komatik Series, 5.

Schleidt, W. M.
(1998). Is humaneness canine? *Human Ethology Bulletin*, 13(4), 1–4.

Schleidt, W. M. & Shalter, M. D.
(2003). Coevolution of humans and canids – An alternative view of dog
domestication: Homo Homini Lupus? *Evolution and Cognition*, 9(1), 57–72.

Schonberg, H. C.
(1971). Music: Soulful glissando swoops. *New York Times* (New York
City newspaper), 15 April, 1, 34 (http://select.nytimes.com/gst/abstract.
html?res=F70817F73A5F127A93C7A8178FD85F458785F9&emc=etal).

Schotté, C. S. & Ginsburg, G.
(1987). Development of social organization and mating in a captive wolf pack.
In *Man and Wolf: Advances, Issues and Problems in Captive Wolf Research*, ed.
H. Frank, pp. 349–374. Dordrecht, the Netherlands: Dr. W. Junk.

Schroedl, G. F. & Breitburg, E.
(1986). Burials. In *Overhill Cherokee Archaeology at Chota-Tanasee*, ed. G. F.
Schroedl, pp. 125–206. Knoxville, Tennessee: University of Tennessee,
Department of Anthropology, Report of Investigations, 38.

Schroedl, G. F. & Parmalee, P. W.
(1997). Dog burials from the eighteenth century Cherokee town of Chattooga,
South Carolina. *Brimleyana*, 24, 7–14.

References

Schulting, R.
(1994). The hair of the dog; The identification of a Coast Salish dog hair blanket from Yale, British Columbia. *Canadian Journal of Archaeology*, 18, 57–76.

Schulting, R. J. & Richards, M. P.
(2002). Dogs, ducks, deer and diet: New stable isotope evidence on early Mesolithic dogs from the Vale of Pickering, north-east England. *Journal of Archaeological Science*, 29, 327–333.

Schwab, C. & Huber, L.
(2006). Obey or not obey? Dogs (*Canis familiaris*) behave differently in response to attentional state of their owners. *Journal of Comparative Psychology*, 120(3), 169–175.

Schwartz, J. H. & Maresca, B.
(2006). Do molecular clocks run at all? A critique of molecular systematics. *Biological Theory*, 1(4), 357–371.

Schwartz, M.
(1997). *A History of Dogs in the Early Americas*. New Haven, Connecticut: Yale University Press.

Schwartz, M.
(2000). The form and meaning of Maya and Mississippian dog representations. In *Dogs through Time: An Archaeological Perspective*, ed. S. J. Crockford, pp. 217–226. Oxford, England: BAR International Series, 889.

Scott, J. P.
(1950). The social behavior of dogs and wolves: An illustration of sociobiological systematics. *Annals of the New York Academy of Sciences*, 51, 1009–1021.

Scott, J. P.
(1967). The evolution of social behavior in dogs and wolves. *American Zoologist*, 7, 373–381.

Scott, J. P.
(1968). Evolution and domestication of the dog. In *Evolutionary Biology*, Volume 2, ed. T. Dobzhansky, M. K. Hecht & W. C. Steere, pp. 243–275. New York City: Appleton-Century-Crofts.

Scott, J. P. & Fuller, J. L.
(1965). *Genetics and Social Behavior of the Dog*. Chicago, Illinois: The University of Chicago Press.

Sears, W. H.
(1960). *The Bayshore Homes Site, St. Petersburg, Florida*. Gainesville, Florida: Contributions of the Florida State Museum, Social Sciences, 6.

331

Seeman, M. F.
> (1994). Review of: *Middle and Late Woodland Research in Virginia: A Synthesis*, ed. T. R. Reinhart & M. E. N. Hodges, Archaeological Society of Virginia, Special Publication, 29, 1992. *American Antiquity*, 59(3), 582.

Sejersen, F.
> (2004). Horizons of sustainability in Greenland: Inuit landscapes of memory and vision. *Arctic Anthropology*, 41(1), 71–89.

Serpell, J.
> (1989). Pet-keeping and animal domestication: A reappraisal. In *The Walking Larder*, ed. J. Clutton-Brock, pp. 10–21. London, England: Unwin Hyman.

Serpell, J.
> (1995a, editor). *The Domestic Dog: Its Evolution, Behaviour and Interactions with People*. Cambridge, England: Cambridge University Press.

Serpell, J.
> (1995b). The hair of the dog. In *The Domestic Dog: Its Evolution, Behaviour and Interactions with People*, ed. J. Serpell. pp. 257–262. Cambridge, England: Cambridge University Press.

Serpell, J.
> (1995c). Introduction. In *The Domestic Dog: Its Evolution, Behaviour and Interactions with People*, ed. J. Serpell, pp. 1–4. Cambridge, England: Cambridge University Press.

Serpell, J.
> (1995d). From paragon to pariah: some reflections on human attitudes to dogs. In *The Domestic Dog: Its Evolution, Behaviour and Interactions with People*, ed. J. Serpell, pp. 245–256. Cambridge, England: Cambridge University Press.

Serpell, J. & Jagoe, J. A.
> (1995). Early experience and the development of behaviour. In *The Domestic Dog: Its Evolution, Behaviour and Interactions with People*, ed. J. Serpell, pp. 79–102. Cambridge, England: Cambridge University Press.

Shea, B. T.
> (1981). Relative growth in the limbs and trunk in the African apes. *American Journal of Physical Anthropology*, 56, 179–201.

Shea, B. T.
> (1985). Ontogenetic allometry and scaling: A discussion based on the growth and form of the skull in African apes. In *Size and Scaling in Primate Biology*, ed. W. L. Jungers, pp. 175–205. New York City: Plenum.

Shea, B. T., Hammer, R. E., Brinster, R. L. & Ravosa, M. R.
> (1990). Relative growth of the skull and postcranium in giant transgenic mice. *Genetic Research, Cambridge*, 56, 21–34.

Sherlock, H. & Pikes, P. J.
>(2002). *First Century Cemetery, Dewsall Court, Herefordshire: Interim Report*. Herfordshire, England: Archenfield Archaeology Ltd.

Shigehara, N. & Hongo, H.
>(2000). Ancient remains of Jomon dogs from Neolithic sites in Japan. In *Dogs through Time: An Archaeological Perspective*, ed. S. J. Crockford, pp. 61–67. Oxford, England: BAR International Series, 889.

Shigehara, N., Matsu'ura, S., Nakamura, T. & Kondo, M.
>(1993). First discovery of the Dingo-type dog in Polynesia (Pukapuka, Cook Islands). *International Journal of Osteoarchaeology*, 3, 315–320.

Sidorovich, V. E., Stolyarov, V. P., Vorobei, N. N., Ivanova, N. V. & Jędrzejewska, B.
>(2007). Litter size, sex ratio, and age structure of gray wolves, *Canis lupus*, in relation to population fluctuations in northern Belarus. *Canadian Journal of Zoology*, 85(2), 295–300.

Siegel, J. M.
>(1990). Stressful life events and use of physician services among the elderly: The moderating role of pet ownership. *Journal of Personality and Social Psychology*, 58(6), 1081–1086.

Silverman, R.
>(1984, editor). *The Dog Observed – Photographs, 1844–1983*. New York City: Alfred A. Knopf.

Singer, R. & Wymer, J.
>(1982). *The Middle Stone Age at Klasies River Mouth in South Africa*. Chicago, Illinois: University of Chicago Press.

Singh, M. & Kumara, H. N.
>(2006). Distribution, status and conservation of Indian gray wolf (*Canis lupus pallipes*) in Karnataka, India. *Journal of Zoology*, 270(1), 164–169.

Sinoto, Y. H.
>(1966). A tentative prehistoric cultural sequence in the northern Marquesas Islands, French Polynesia. *Journal of the Polynesian Society*, 75, 287–303.

Sisson, S. & Grossman, J. D.
>(1953). *The Anatomy of the Domestic Animals* (4th edition). Philadelphia, Pennsylvania: W. B. Saunders.

Skaggs, O.
>(1946). A study of the dog skeletons from Indian Knoll with special reference to the coyote as progenitor. In *Indian Knoll, Site Oh2, Ohio County, Kentucky*, ed. W. S. Webb, pp. 341–55. Lexington, Kentucky: University of Kentucky, Department of Anthropology, Reports in Anthropology and Archaeology, 4(3), Part 1.

Smith, B. D.
(1975). *Middle Mississippi Exploitation of Animal Populations*. Ann Arbor, Michigan: University of Michigan, Museum of Anthropology, Anthropological Papers, 57.

Smith, D. W., Drummer, T. D., Murphy, K. M., Guernsey, D. S. & Evans, S. B.
(2004). Winter prey selection and estimation of wolf kill rates in Yellowstone National Park, 1995–2000. *Journal of Wildlife Management*, 68(1), 153–166.

Smythe, R. H.
(1970). *The Dog – Structure and Function*. New York City: Arco.

Snyder, L. M.
(1991). Barking mutton: Ethnohistoric and ethnographic, archaeological, and nutritional evidence pertaining to the dog as a Native American food resource on the plains. In *Beamers, Bobwhites, and Blue-points – Tributes to the Career of Paul W. Parmalee*, ed. J. R. Purdue, W. E. Klippel & B. W. Styles, pp. 359–378. Springfield, Illinois, and Knoxville, Tennessee: Illinois State Museum, Scientific Papers, 23, and University of Tennessee, Department of Anthropology, Report of Investigations, 52.

Snyder, L. M.
(1995). Assessing the Role of the Domestic Dog as a Native American Food Resource in the Middle Missouri Subarea A.D. 1000–1840. Knoxville, Tennessee: Unpublished Ph.D. Dissertation, Department of Anthropology, University of Tennessee.

Snyder, L. M. & Moore, E. A.
(2006, editors). *Dogs and People in Social, Working, Economic or Symbolic Interaction*. Oxford, England: Oxbow Books (Proceedings of the 9th ICAZ Conference, Durham, England, 2002).

South, S.
(2005). *An Archaeological Evolution*. New York City: Springer.

Spence, L. J. & Kaiser, L.
(2002). Companion animals and adaptation in chronically ill children. *Western Journal of Nursing Research*, 24(6), 639–656.

Spencer, D. C.
(2007). Understanding seizure dogs. *Neurology*, 68(4), 308–309.

Sprott, J.
(1997). Christmas, basketball, and sled dog races: Common and uncommon themes in the new seasonal round in an Iñupiaq village. *Arctic Anthropology*, 34(1), 68–85.

Stafford, C. R., Richards, R. L. & Anslinger, C. M.
(2000). The Bluegrass fauna and changes in Middle Holocene hunter-gatherer foraging in the southern Midwest. *American Antiquity*, 65(2), 317–336.

Stager, L. E.
   (1991). Why were hundreds of dogs buried at Ashkelon? *Biblical Archaeology Review*, May/June, 26–42.

Stahl, P. W.
   (2003). The zooarchaeological record from Formative Ecuador. In *Archaeology of Formative Ecuador*, ed. J. S. Raymond & R. L. Burger, pp. 175–212. Washington, D.C.: Dumbarton Oaks Research Library and Collection.

Stearns, S. C.
   (1976). Life history tactics: A review of the ideas. *The Quarterly Review of Biology*, 51, 3–47.

Stearns, S. C.
   (1977). The evolution of life history traits: A critique of the theory and a review of the data. *Annual Review of Ecology and Systematics*, 8, 145–171.

Stephenson, R. O. & Ahgook, B.
   (1975). The Eskimo hunter's view of wolf ecology and behavior. In *The Wild Canids*, ed. M. W. Fox, pp. 292–299. New York City: Van Nostrand Reinhold.

Stewart, F. L. & Stewart. K. M.
   (1996). The Boardwalk and Grassy Bay Sites: Patterns of seasonality and subsistence on the northern northwest coast, B.C. *Journal Canadien d'Archéologie*, 20, 39–60.

Stewart, S.
   (1982). Wolf music. *Michigan Natural Resources Magazine*, 5(16), 40–49.

Stockhaus, K.
   (1965). Metrische untersuchungen an Schädeln von Wölfen und Hunden. *Zeitschrift für Zoologische Systematik un Evolutionsforschung*, 3, 157–258 (in German, with an English summary).

Street, M.
   (2002). Ein Wiedersehen mit dem Hund von Bonn-Oberkassel. *Bonner Zoologische Beitrage*, 50, 269–290 (in German).

Strong, V., Brown, S. W. & Walker, R.
   (1999). Seizure-alert dogs – Fact or fiction? *Seizure*, 8(1), 62–65.

Struever, S. & Holton, F. A.
   (1979). *Koster: Americans in Search of Their Prehistoric Past*. Garden City, New York: Anchor Press/Doubleday.

Stuckenrath, R., Adovasio, J. M., Donahue, J. & Carlisle, C.
   (1982). The stratigraphy, cultural features and chronology at Meadowcroft Rockshelter, Washington County, southwestern Pennsylvania. In *Meadowcroft: Collected Papers on the Archaeology of Meadowcroft Rockshelter and the*

Cross Creek Drainage, ed. R. C. Carlyle & J. M. Adovasio, pp. 69–90. Pittsburgh, Pennsylvania: Department of Anthropology, University of Pittsburgh.

Studzinski, C. M., Araujo, J. A. & Milgram, N. W.
(2005). The canine model of human cognitive aging and dementia: Pharmacological validity of the model for assessment of human cognitive-enhancing drugs. *Progress in Neuro-Psychopharmacology & Biological Psychiatry*, 29(3), 489–498.

Sullivan, J. O.
(1978). Variability in the wolf, a group hunter. In *Wolf and Man: Evolution in Parallel*, ed. R. L. Hall & H. S. Sharp, pp. 31–40. New York City: Academic Press.

Sutter, N. B., Bustamante, C. D., Chase, K., & 18 others
(2007). A single IGF1 allele is a major determinant of small size in dogs. *Science*, 316, 112–115.

Takabatake, É. Y.
(2007). *Musicoterapia para Cães com Depressão*. Sao Paulo, Brazil: Faculdade Paulista de Artes Curso de Musicotherapia (in Spanish, includes an English abstract).

Tan, Ü.
(1987). Paw preferences in dogs. *International Journal of Neuroscience*, 32(3–4), 825–829.

Tarcan, C., Cordy, J.-M., Bejenaru, L., & Udrescu, M.
(2000). Consommation de la viande de chien: Le vicus de braives (Belgique) et les sites geto-daces et Romains de Roumanie. In *Dogs through Time: An Archaeological Perspective*, ed. S. J. Crockford, pp. 123–128. Oxford, England: BAR International Series, 889 (text in French only, with the English title provided as well, and an English abstract only).

Tatsuoka, M. M.
(1971). *Multivariate Analysis*. New York City: John Wiley & Sons.

Tchernov, E. & Horwitz, L. K.
(1991). Body size diminution under domestication: Unconscious selection in primeval domesticates. *Journal of Anthropological Archaeology*, 10, 54–75.

Tchernov, E. & Valla, F. F.
(1997). Two new dogs, and other Natufian dogs, from the southern Levant. *Journal of Archaeological Science*, 24, 65–95.

Theall, L. A. & Povinelli, D. J.
(1999). Do chimpanzees tailor their gestural signals to fit the attentional states of others? *Animal Cognition*, 2(4), 207–214.

Theberge, J. B.
  (1971). Wolf music. *Natural History*, 80(4), 37–43.

Theberge, J. B. & Falls, J. B.
  (1967). Howling as a means of communication in timber wolves. *American Zoologist*, 7, 331–338.

Theuerkauf, J., Jędrzejewski, W., Schmidt, K. & Gula, R.
  (2003). Spatiotemporal segregation of wolves from humans in the Bialowieza Forest (Poland). *Journal of Wildlife Management*, 67(4), 706–716.

Thomas, E. M.
  (2000). *The Social Lives of Dogs – The Grace of Canine Company*. New York City: Pocket Books.

Thompson, S. M.
  (2000). *I Will Tell My War Story: A Pictorial Account of the Nez Perce War*. Seattle, Washington: University of Washington Press, with the Idaho State Historical Society, Boise, Idaho.

Thoms, J.
  (2005). Faunal and human remains from the Iron Age settlement. *Scottish Archaeological Internet Report*, 24, 86–90.

Thurston, M.
  (1996). *The Lost History of the Canine Race: Our 15,000-Year Love Affair with Dogs*. Kansas City, Missouri: Andrews and McMeel.

Tinbergen, N.
  (1951). *The Study of Instinct*. Oxford, England: Oxford University Press.

Titcomb, M.
  (1969). *Dog and Man in the Ancient Pacific, with Special Attention to Hawaii*. Honolulu, Hawaii: Bernice P. Bishop Museum, Special Publication, 59.

Tooze, Z. J., Harrington, F. H. & Fentress, J. C.
  (1990). Individually distinct vocalizations in timber wolves, *Canis lupus*. *Animal Behavior*, 40, 723–730.

Topál, J., Gácsi, M., Miklósi, Á., Virányi, Z., Kubinyi, E. & Csányi. V.
  (2005). Attachment to humans: A comparative study on hand-reared wolves and differently socialized dog puppies. *Animal Behaviour*, 70, 1367–1375.

Trantalidou, K.
  (2006). Companions from the oldest times: Dogs in ancient Greek literature iconography and osteological testimony. In *Dogs and People in Social, Working, Economic or Symbolic Interaction*, ed. L. M. Snyder & E. A. Moore, pp. 96–120. Oxford, England: Oxbow Books (Proceedings of the 9th ICAZ Conference, Durham, England, 2002).

Trut, L. N.
   (1999). Early canid domestication: The farm-fox experiment. *American Scientist*, 87, 160–169.

Trut, L. N.
   (2001). Experimental studies of early canid domestication. In *The Genetics of the Dog*, ed. A. Ruvinsky & J. Sampson, pp. 15–42. Wallingford, England and New York City: CAB International.

Trut, L. N., Pliusina, I. Z. & Os'kina, I. N.
   (2004). An experiment on fox domestication and debatable issues of evolution of the dog. *Russian Journal of Genetics*, 40(6), 644–655.

Tsuda, K., Kikkawa, Y., Yonekawa, H. & Tanabe, Y.
   (1997). Extensive interbreeding occurred among multiple matriarchal ancestors during the domestication of dogs: evidence from inter- and intraspecies polymorphisms in the D-loop region of mitochondrial DNA between dogs and wolves. *Genes & Genetic Systems*, 72(4), 229–238.

Tuck, J. A.
   (1976). *Ancient People of Port au Choix – The Excavation of an Archaic Indian Cemetery in Newfoundland*. St. John's, Newfoundland: Memorial University of Newfoundland, Institute of Social and Economic Research, Newfoundland Social and Economic Studies, 17.

Tuovinen, T.
   (2002). The Burial Cairns and the Landscape in the Archipelago of Åboland, SW Finland, in the Bronze Age and the Iron Age. Oulu, Finland: Academic Dissertation, Department of Art Studies and Anthropology, University of Oulu.

Turnbull, P. F. & Reed, C. A.
   (1974). The fauna from the terminal Pleistocene of Palegawra Cave, a Zarzian occupation site in northeastern Iraq. *Fieldiana, Anthropology*, 63(4), 81–146.

Tylor, E. B.
   (1871). *Primitive Culture*. London, England: Murray.

Udell, M. A. R., Giglio, R. F. & Wynne, C. D.
   (2008). Domestic dogs (*Canis familiaris*) use human gestures but not nonhuman tokens to find hidden food. *Journal of Comparative Psychology*, 122(1), 84–93.

Udell, M. A. R. & Wynne, C. D.
   (2008). A review of domestic dogs' (*Canis familiaris*) human-like behaviors: Or why behavior analysts should stop worrying and love their dogs. *Journal of the Experimental Analysis of Behavior*, 89, 247–261.

Valadez, R.
   (1996). The Pre-Columbian Dog. *Voices of Mexico*, July–September, 49–53.

References

Valadez, R., Peredes, B., & Rodriguez, B.
(1999). Entierros de perros descubiertos en la antigua ciudad de Tula. *Latin American Antiquity*, 10(2), 180–200 (in Spanish, with an English abstract).

Valadez, R., Rodríguez, B., Manzanilla, L. & Tejeda, S.
(2006). Dog–wolf hybrid biotype reconstruction from the archaeological city of Teotihuacan in prehispanic central Mexico. In *Dogs and People in Social, Working, Economic or Symbolic Interaction*, ed. L. M. Snyder & E. A. Moore, pp. 121–131. Oxford, England: Oxbow Books (Proceedings of the 9th ICAZ Conference, Durham, England, 2002).

Vallee, F. G., Smith, D. G., and Cooper J. D.
(1984). Contemporary Canadian Inuit. In *Handbook of North American Indians*, Volume 5: *Arctic*, ed. D. Damas, pp. 662–675. Washington, D.C.: Smithsonian Institution.

Van Ballenberghe, V. & Mech, L. D.
(1975). Weights, growth, and survival of timber wolf pups in Minnesota. *Journal of Mammalogy*, 56(1), 44–63.

Van de Noort, R.
(2007). Digging the Dutch mountains: Recent work by Leendert Louwe Kooijmans. *Journal of Wetland Archaeology*, 7, 83–88.

VanDerwarker, A. M.
(2001). An archaeological study of Late Woodland fauna in the Roanoke River Basin. *North Carolina Archaeology*, 50, 1–46.

Vasil'evskiy, R.
(1998). The Upper Paleolithic of Kamchatka and Chukotka. In *The Paleolithic of Siberia*, ed. A. P. Derev'anko, D. B. Shimkin & W. R. Powers (translated by I. P. Laricheva), pp. 290–328. Urbana and Chicago, Illinois: University of Illinois Press.

Vellanoweth, R. L., Bartelle, B. G., Ainis, A. F., Cannon, A. C. & Schwartz, S. J.
(2008). A double dog burial from San Nicolas Island, California, USA: Osteology, context, and significance. *Journal of Archaeological Science*, 35(12), 3111–3123.

Verardi, A., Lucchini, V. & Randi, E.
(2006). Detecting introgressive hybridization between free-ranging domestic dogs and wild wolves (*Canis lupus*) by admixture linkage disequilibrium analysis. *Molecular Ecology*, 15(10), 2845–2855.

Verginelli, F., Capelli, C., Cola, V., et al.
(2005). Mitochondrial DNA from prehistoric canids highlights relationships between dogs and South-East European wolves. *Molecular Biology & Evolution*, 22(12), 2541–2551.

Vigne, J.-D., Guiliane, J., Debue, K. & Gérhard, P.
(2004). Early taming of the cat in Cyprus. *Science*, 304, 259.

Vilà. C. & Leonard, J. A.
(2007). Origin of dog breed diversity. In *The Behavioural Biology of Dogs*, ed. P. Jensen, pp. 38–58. Wallingford, England, and Cambridge, Massachusetts: CAB International.

Vilà, C., Savolainen, P., Maldonado, J. E., Amorim, I. R., Rice, J. E., Honeycutt, R. L., Crandall, K. A., Lundeberg, J. & Wayne, R. K.
(1997). Multiple and ancient origins of the domestic dog. *Science*, 276, 1687–1689.

Vilà, C., Sundqvist, A.-K., Flagstad, Ø., & seven others
(2003). Rescue of a severely bottlenecked wolf (*Canis lupus*) population by a single immigrant. *Proceedings of the Royal Society of London B*, 270, 91–97.

Vilà, L., Maldonado, J. E. & Wayne, R. K.
(1999). Phylogenetic relationships, evolution, and genetic diversity of the domestic dog. *The Journal of Heredity*, 90(1), 71–77.

Vinicius, L. & Lahr, M. M.
(2003). Morphometric heterochrony and the evolution of growth. *Evolution*, 57(1), 2459–2468.

Virányi, Z., Topál, J., Gácsi, M., Miklósi, Á. & Csányi, V.
(2004). Dogs respond appropriately to cues of humans' attentional focus. *Behavioural Processes*, 66(2), 161–172.

Voigt, E. A.
(1983). *Mapungubwe: An Archaeozoological Interpretation of an Iron Age Community*. Pretoria, South Africa: Transvaal Museum, Monograph, 1.

Von den Driesch, A.
(1976). *A Guide to the Measurement of Animal Bones from Archaeological Sites*. Cambridge, Massachusetts: Harvard University, Peabody Museum of Archaeology and Ethnology, Bulletin, 1.

Vouvé, J., Brunet, J., Vidal, P. & Marsal, J.
(1982). *Lascaux en Périgord Noir – Environnement, Art Pariétal et Conservation*. Périgueux, France: Pierre Fanlac (text in French, English, German, and Spanish).

Walker, D. N. & Frison, G. C.
(1982). Studies on Amerindian dogs, 3: Prehistoric wolf/dog hybrids from the Northwestern plains. *Journal of Archaeological Science*, 9, 125–172.

Walker, R. B., Morey, D. F. & Relethford, J. H.
(2005). Early and mid-Holocene dogs in southeastern North America: Examples from Dust Cave. *Southeastern Archaeology*, 24(1), 83–92.

Walthall, J. H.
(1980). *Prehistoric Indians of the Southeast – Archaeology of Alabama and the Middle South.* Tuscaloosa, Alabama: The University of Alabama Press.

Wang, X. & Tedford, R. H.
(2008). *Dogs: Their Fossil Relatives and Evolutionary History.* New York City: Columbia University Press.

Wapnish, P. & Hesse, B.
(1993). Pampered pooches or plain pariahs? The Ashkelon dog burials. *The Biblical Archaeologist,* 56(2), 55–80.

Warren, D. M.
(2000). Paleopathology of Archaic Period dogs from the North American southeast. In *Dogs through Time: An Archaeological Perspective,* ed. S. J. Crockford, pp. 105–114. Oxford, England: BAR International Series, 889.

Warren, D. M.
(2004). Skeletal Biology and Paleopathology of Domestic Dogs from Prehistoric Alabama, Illinois, Kentucky, and Tennessee. Bloomington, Indiana: Unpublished Ph.D. Dissertation, Department of Anthropology, University of Indiana, (Ann Arbor, Michigan: University Microfilms International, Order Number 3122750).

Washburn, S. & Lancaster, C. S.
(1968). The evolution of hunting. In *Man the Hunter,* ed. R. B. Lee. & I. DeVore, pp. 293–303. Chicago, Illinois: Aldine.

Wasser, S. K., Davenport, B., Ramage, E. R., Hunt, K. E., Parker, M., Clarke, C. & Stenhouse, G.
(2004). Scat detection dogs in wildlife research and management: Application to grizzly and black bears in the Yellowhead Ecosystem, Alberta, Canada. *Canadian Journal of Zoology,* 82(3), 475–492.

Watson, P. J.
(1995). Archaeology, anthropology, and the culture concept. *American Anthropologist,* 97(4), 683–704.

Wayne, R. K.
(1986a). Cranial morphology of domestic and wild canids: The influence of development on morphological change. *Evolution,* 40(2), 243–260.

Wayne, R. K.
(1986b). Limb morphology of domestic and wild canids: The influence of development on morphologic change. *Journal of Morphology,* 187(3), 301–319.

References

Wayne, R. K.
    (1986c). Developmental constraints on limb growth in domestic and some wild canids. *Journal of Zoology, London (A)*, 210, 381–399.

Wayne, R. K.
    (1993a). Molecular evolution of the dog family. *Trends in Genetics*, 9(6), 218–224.

Wayne, R. K.
    (1993b). Phylogenetic relationships of canids to other carnivores. In *Miller's Anatomy of the Dog* (3rd edition), by H. E. Evans, pp. 15–21. Philadelphia, Pennsylvania: W. B. Saunders.

Wayne, R. K.
    (2001). Consequences of domestication: Morphological diversity of the dog. In *The Genetics of the Dog*, ed. A. Ruvinsky & J. Sampson, pp. 43–60. Wallingfor, England, and New York City: CAB International.

Wayne, R. K., Lehman, N., Allard, M. W. & Honeycutt, R. L.
    (1992). Mitochondrial DNA variability of the gray wolf: Genetic consequences of population decline and habitat fragmentation. *Conservation Biology*, 6, 559–569.

Wayne, R. K., Leonard. J. A. & Vilà, C.
    (2006). Genetic analysis of dog domestication. In *Documenting Domestication: New Genetic and Archaeological Paradigms*, ed. M. A. Zeder, D. G. Bradley, E. Emshwiller. & B. D. Smith, pp. 279–293. Berkeley, California: University of California Press.

Wayne, R. K. & O'Brien, S. J.
    (1987). Allozyme divergence within the Canidae. *Systematic Zoology*, 36(4), 339–355.

Wayne, R. K., & Ostrander, E. A.
    (2007). Lessons learned from the dog genome. *Trends in Genetics*, 23 (11), 557–567.

Wayne, R. K. & Vilà, C.
    (2001). Phylogeny and origin of the domestic dog. In *The Genetics of the Dog*, ed. A. Ruvinsky & G. Sampson, pp. 1–14. Wallingford, England, and New York City: CAB International.

Wayne, R. K. & Vilà, C.
    (2003). Molecular genetic studies of wolves. In *Wolves – Behavior, Ecology, and Conservation*, ed. L. D. Mech & L. Boitani, pp. 218–238. Chicago, Illinois: University of Chicago Press.

Webb, W. S.

(1938). *An Archaeological Survey of the Norris Basin in Eastern Tennessee.* Washington, D.C.: Smithsonian Institution, Bureau of American Ethnology, Bulletin, 118.

Webb, W. S.

(1946). *Indian Knoll, Site Oh2, Ohio, County, Kentucky.* Lexington, Kentucky: University of Kentucky, Reports in Anthropology and Archaeology, IV(3), Part 1 (republished in 1974, by the University of Tennessee Press, Knoxville, Tennessee).

Webb, W. S.

(1950a). *The Carlson Annis Mound, Site 5, Butler County, Kentucky.* Lexington, Kentucky: The University of Kentucky, Reports in Anthropology, VII(4).

Webb, W. S.

(1950b). *The Read Shell Midden, Site 10, Butler County, Kentucky.* Lexington, Kentucky: The University of Kentucky, Reports in Anthropology, VII(5).

Webb, W. S. & DeJarnette, D. L.

(1942). *An Archaeological Survey of Pickwick Basin in the Adjacent Portions of the States of Alabama, Mississippi, and Tennessee.* Washington. D.C.: Smithsonian Institution, Bureau of American Ethnology, Bulletin, 129.

Webb, W. S. & DeJarnette, D. L.

(1948a). *The Perry Site Lu25, Units 3 and 4, Lauderdale County, Alabama.* Tuscaloosa, Alabama: Geological Survey of Alabama, Alabama Museum of Natural History, Museum Paper, 25.

Webb, W. S. & DeJarnette, D. L.

(1948b). *The Flint River Site. Ma48.* Tuscaloosa, Alabama: Geological Survey of Alabama, Alabama Museum of Natural History, Museum Paper, 23.

Webb, W. S. & DeJarnette, D. L.

(1948c). *The Whitesburg Bridge Site, Ma10.* Tuscaloosa, Alabama: Geological Survey of Alabama, Alabama Museum of Natural History, Museum Paper, 24.

Webb, W. S. & DeJarnette, D. L.

(1948d). *Little Bear Creek Site Ct8, Colbert County, Alabama.* Tuscaloosa, Alabama: Geological Survey of Alabama, Alabama Museum of Natural History, Museum Paper, 26.

Webb, W. S. & Haag, W. G.

(1939). *The Chiggerville Site, Site 1, Ohio County, Kentucky.* Lexington, Kentucky: University of Kentucky, Reports in Anthropology, IV(1).

Webb, W. S. & Haag, W. G.

(1940). *Cypress Creek Villages, Sites 11 and 12, McLean County, Kentucky.* Lexington, Kentucky: University of Kentucky, Reports in Anthropology, IV(2).

Webb, W. S. & Haag, W. G.
  (1947). *Archaic Sites in McLean County, Kentucky*. Lexington, Kentucky: University of Kentucky, Reports in Anthropology, VII(1).

Wedel, W. R.
  (1941). *Archeological Investigations at Buena Vista Lake, Kern County, California*. Washington, D.C.: Smithsonian Institution, Bureau of American Ethnology, Bulletin, 130.

Wedel, W. R.
  (1986). *Central Plains Prehistory – Holocene Environments and Culture Change in the Republican River Basin*. Lincoln, Nebraska: University of Nebraska Press.

Weisman, B. R.
  (2000). Fig Springs Mission. In *The Complete Lamar Briefs*, ed. M. Williams, pp. 149–150. Savannah, Georgia: Lamar Institute (University of Georgia), Publication, 48.

Wells, D. L.
  (2003). Lateralised behaviour in the domestic dog, *Canis familiaris*. *Behavioural Processes*, 61(1–2), 27–35.

Weltfish, G.
  (1965). *The Lost Universe – Pawnee Life and Culture*. Lincoln, Nebraska: University of Nebraska Press.

Werner, E. E. & Gilliam, J. F.
  (1984). The ontogenetic niche and species interactions in size-structured populations. *Annual Review of Ecology and Systematics*, 15, 393–425.

Werth, E.
  (1944). Die primitiven Hunde und die Abstammungsfrage des Haushundes. *Zeitschrift für Tierzüchtung und Züchtungsbiologie*, 56(3), 213–260 (in German).

White, A. A., McCullough, D. & McCullough, R. G.
  (2002). *An Archaeological Evaluation of Late Prehistoric Village and Subsistence Patterns in North-Central and Northeastern Indiana*. IPFW Archaeological Survey, Indiana University and Purdue University at Fort Wayne, Report of Investigations, 216.

White, D.
  (1991). *Myths of the Dog-Man*. Chicago, Illinois: The University of Chicago Press.

Whittemore, H. & Hebard, C.
  (1996). *So That Others May Live – Caroline Hebard and Her Search-and-Rescue Dogs*. New York City: Bantam Books (mass market edition).

References

Wilford, L. A.
(1950). The prehistoric Indians of Minnesota: The Rainy River Aspect. *Minnesota History*, 31, 163–171.

Wilkens, B.
(2006). The sacrifice of dogs in ancient Italy. In *Dogs and People in Social, Working, Economic or Symbolic Interaction*, ed. L. M. Snyder & E. A. Moore, pp. 132–137. Oxford, England: Oxbow Books (Proceedings of the 9th ICAZ Conference, Durham, England, 2002).

Wilks, K.
(1999). *When Dogs Are Man's Best Friend – The Health Benefits of Companion Animals in Modern Society*. Queensland, Australia: Proceedings of the 8th Urban Animal Management Conference.

Will, U. V.
(1973). Untersuchungen zur taxonomischen Bedeutung des Kleinhirns der Gattung Canis. *Zeitschrift für Zoologische Systematik und Evolutionsforschung*, 11(1), 61–73 (in German, with an English summary).

Willis, C. M., Church, S. M., Guest, C. M., Cook, W. A., McCarthy, N., Bransbury, A. J., Church, M. R. T. & Church, J. C. T.
(2004). Olfactory detection of human bladder cancer by dogs: proof of principle study. *British Medical Journal*, 329(7468), 712–714.

Wilson, E. O.
(1975). *Sociobiology – The New Synthesis*. Cambridge, Massachusetts: Harvard University Press.

Wilson, G. L.
(1924). *The Horse and the Dog in Hidatsa Culture*. New York City: American Museum of Natural History, Anthropological Papers, 15(2).

Wilson, P. J., Grewal, S., McFadden, T., Chambers, R. C. & While, B. N.
(2003). Mitochondrial DNA extracted from eastern North American wolves killed in the 1800s is not of gray wolf origin. *Canadian Journal of Zoology*, 81(5), 936–940.

Wing, E. S.
(1978). Use of dogs for food: An adaptation to the coastal environment. In *Prehistoric Coastal Adaptation: The Economy and Ecology of Maritime Middle America*, ed. B. L. Stark & B. Voorhies, pp. 29–41. New York City: Academic Press.

Wing, E. S.
(1984). Use and abuse of dogs. In *Contributions to Quaternary Vertebrate Paleontology – A Volume in Memorial to John E. Guilday*, ed. H. H. Genoways & M. R. Dawson, pp. 228–232. Pittsburgh, Pennsylvania: Carnegie Museum of Natural History, Special Publications, 8.

Wing, E. S.
(1986). Domestication of Andean mammals. In *High Altitude Tropical Biogeography*, ed. P. Vuillemier & M. Monastario, pp. 246–263. Oxford, England: Oxford University Press.

Wing, E. S.
(1991). Dog remains from the Sorcé Site on Vieques Island, Puerto Rico. In *Beamers, Bobwhites, and Blue-points – Tributes to the Career of Paul W. Parmalee*, ed. J. R. Purdue, W. E. Klippel & B. W. Stiles, pp. 379–386. Springfield, Illinois, and Knoxville, Tennessee: Illinois State Museum, Scientific Papers, 23, and University of Tennessee, Department of Anthropology, Report of Investigations, 52.

Wing, E. S.
(2001). Native American use of animals in the Caribbean. In *Biogeography of the West Indies – Patterns and Perspectives* (2nd edition), ed. C. A. Woods & F. E. Sergile, pp. 481–518. Boca Raton, Louisiana: CRC Press.

Wing, E. S.
(2007). Pets and camp followers in the West Indies. In *Case Studies in Environmental Archaeology* (2nd edition), ed. E. Reitz, C. M. Scarry & S. J. Scudder, pp. 405–426. New York City: Plenum.

Witthoft, J., Kinsey, F. W. III & Holzinger, C. H.
(1959). A Susquehannock cemetery: The Ibaugh site. *Susquehannock Miscellany*, pp. 19–119. Harrisburg, Pennsylvania: Pennsylvania Historical and Museum Commission.

Wood-Jones, F.
(1931). The cranial characters of the Hawaiian dog. *Journal of Mammalogy*, 12(1), 39–41.

Woolpy, J. H. & Ginsburg, B. E.
(1967). Wolf socialization: A study of temperament in a wild social species. *American Zoologist*, 7(2), 357–363.

Wozencraft, W. C.
(1993). Carnivora: Canidae. In *Mammal Species of the World – A Taxonomic and Geographic Reference* (2nd edition), ed. D. E. Wilson & D. M. Reeder, pp. 279–288. Washington, D.C.: Smithsonian Institution Press.

Wright, G. J., Peterson, R. O., Smith, D. W. & Lemke, T. O.
(2006). Selection of northern Yellowstone elk by gray wolves and hunters. *Journal of Wildlife Management*, 70(4), 1070–1078.

Yates, B. C. & Koler-Matznick, J.
(2006). The evidentiary dog: A review of anthrozoological cases and archaeological studies. In *Dogs and People in Social, Working, Economic or Symbolic*

Interaction, ed. L. M. Snyder & E. A. Moore, pp. 138–46. Oxford, England: Oxbow Books (Proceedings of the 9th ICAZ Conference, Durham, England, 2002).

Yin, S.
 (2002). A new perspective on barking in dogs (*Canis familiaris*). *Journal of Comparative Psychology*, 116(2), 189–193.

Yin, S. & McCowan, B.
 (2004). Barking in domestic dogs: Context specificity and individual identification. *Animal Behaviour*, 68, 343–355.

Yohe, R. M. II & Pavesic, M. G.
 (2000). Early Archaic domestic dogs from western Idaho, USA. In *Dogs through Time: An Archaeological Perspective*, ed. S. J. Crockford, pp. 93–104. Oxford, England: BAR International Series, 889.

Young, S. P. & Goldman, E. A.
 (1944). *The Wolves of North America*. Washington, D.C.: American Wildlife Institute.

Zeder, M. A.
 (2006). Archaeological approaches to documenting animal domestication. In *Documenting Domestication: New Genetic and Archaeological Paradigms*, ed. M. A. Zeder, D. G. Bradley, E. Emshwiller & B. D. Smith, pp. 171–180. Berkeley, California: University of California Press.

Zeder, M. A., Bradley, D. G., Emshwiller, E. & Smith, B. D.
 (2006). Documenting domestication – Bringing together plants, animals, archaeology, and genetics. In *Documenting Domestication: New Genetic and Archaeological Paradigms*, ed. M. A. Zeder, D. G. Bradley, E. Emshwiller & B. D. Smith, pp. 1–12. Berkeley, California: University of California Press.

Zelditch, M. L., Sheets, H. D. & Fink, W. L.
 (2000). Spatiotemporal reorganization of growth rates in the evolution of ontogeny. *Evolution*, 54(4), 1363–1371.

Zeuner, F. E.
 (1963). *A History of Domesticated Animals*. New York City: Harper and Row.

Zimen, E.
 (1981). *The Wolf – A Species in Danger*. New York City: Delacorte Press (Originally published in German in 1978 in Vienna, Austria, and Munich, Germany, by Meyester Publishing).

INDEX

Printed in the United States
by Baker & Taylor Publisher Services